BACTERIA IN NATURE

Volume 1

Bacterial Activities in Perspective

BACTERIA IN NATURE

BACTERIA IN NATURE

Volume 1
Bacterial Activities in Perspective

Edited by

Edward R. Leadbetter
University of Connecticut
Storrs, Connecticut

and

Jeanne S. Poindexter
The Public Health Research Institute of The City of New York, Inc.
New York, New York

PLENUM PRESS • NEW YORK AND LONDON

Library of Congress Cataloging in Publication Data

Main entry under title:

Bacteria in nature.

Includes bibliographies and index.
Contents: v. 1. Bacterial activities in perspective.
1. Bacteria—Ecology—Collected works. 2. Bacteria—Collected works. I. Leadbetter, Edward R. II. Poindexter, Jeanne S. (Jeanne Stove)
QR100.B33 1985 589.9′05 85-3433
ISBN 0-306-41944-0 (v. 1)

A Division of Plenum Publishing Corporation
233 Spring Street, New York, N.Y. 10013

Printed in the United States of America

PREFACE TO THE TREATISE

The effects of bacteria on their environments were known and variously explained by human societies long before these microorganisms were recognized. Even after they had been detected microscopically, nearly two centuries elapsed before it was demonstrated that bacteria were causes, rather than effects, of fermentations, infectious diseases, and transformations of both organic and inorganic materials in soils, waters, and sediments. It was these demonstrations of the ecological roles of bacteria that gave birth to bacteriology as an experimental science. The applications of the understanding of ecological activities of bacteria have in no small part been responsible for this century's revolution in human health and longevity through changes in agricultural, medical, and sanitation practices.

However, the ecology of bacteria has only relatively recently emerged as a science in itself, having as its goal the elucidation of the interactions of bacteria and their habitats whether or not those activities appear immediately relevant to human affairs. In this ten-volume treatise, it is our intention to present this broadened view of bacterial existence, that the work may serve as a synthesis of current ideas and information that will be valuable both to basic scientists and to those directly engaged in applications of science to specific problems of human existence. Our hope is that the completed project will expose and explore the diversity of bacterial capabilities and culpabilities, limitations and sensitivities, and will imply the equally diverse ways in which they can be exploited. We hope, especially, that investigators trained in other disciplines—clinicians, oceanographers, molecular biologists, engineers—who may not expect that their disciplines are interrelated with bacterial ecology, will find this treatise both stimulating and valuable.

The introductory volume traces main points in the history of bacteriology that have led to the present state of bacterial ecology, to the awareness that bacteria constitute distinctive populations that separately and in concert affect the physiochemical conditions of the biosphere and interact, sometimes intimately, with other organisms. The second volume will review and evaluate the technical and philosophical tools presently available to the student of bacteria in nature. This volume is intended to provide evaluations of methods and not to serve as a procedural manual.

Although the initial stimulus to interest in bacteria arose from attempts to understand natural phenomena and to distinguish abiotic from biotic causes of these phenomena, most of the progress in managing and learning about bacteria has been accomplished in the laboratory, very largely through the study of pure cultures. While some devoted naturalists eschew the study of monotypic populations as artificialities, it would not be possible to understand the activities of a bacterial community in ignorance of the separate, respective potential activities of the members of the community. For this reason, two volumes will comprise information regarding structure, composition, genetics, physiology, and biochemistry of bacteria, obtained predominately from pure culture studies, that is essential to unraveling the interactions of bacteria with their environment and with each other. The goal of those volumes is not simply to review the information, but to demonstrate its importance to inferences based on studies of natural, polytypic populations.

The remaining six volumes in this series will explore bacterial habitats. Because a bacterial activity of any type—polymer solubilization, oxygen consumption, toxin production, or other—is not confined to one type of habitat, we anticipate that some groups of bacteria and some bacterial activities will be mentioned in more than one ecological context. However, since the emphasis throughout the series will be on interactions, the role of the habitat in influencing the extent and consequence of bacterial activities will vary with its own inherent stability, its resilience to the effects of bacteria, and its capacity for supporting and restricting those activities. An additional reason for organizing the treatise principally around habitats reflects the fact that human problems and advantages that arise from bacterial activities are most often met within the context of a particular kind of site. Similarly, the general ecological significance of a bacterial activity is proportional to the rate of activity allowed by a habitat, the geographical extent of the habitat, and the degree of dependence of other forms of life on the condition of the habitat affected by that activity. Accordingly, for all conceptual and practical purposes other than classification, the study of bacterial ecology is, we believe, most usefully presented by grouping the information into volumes that reflect the manner in which bacterial communities are gathered and interact in nature.

The treatise will conclude with a consideration of the frontiers and the relic habitats of the biosphere, those environments inhabited almost solely by bacteria. Modern biology recognizes bacteria as the pioneering populations—demonstrably so in the past and predictably so in the future—of this and possibly also of other worlds. It is the humble yet confident hope of the editors that the insights and experimental results of today's bacterial ecologists, compiled in these volumes, will contribute significantly to the continuing elucidation of the roles and potentials of our bacterial cohabitants—so long a major influence on this earth, yet only so recently appreciated.

This series was conceived partly as a result of our participation as instructors in the summer program in microbial ecology at the Marine Biological Laboratory, Woods Hole, Massachusetts. Our interests in bacterial ecology antedate that participation by many years, having been earlier stimulated and guided by R. A. Slepecky and J. W. Foster (E.R.L.), by W. A. Konetzka and R. Y. Stanier (J.S.P.), and by one of the greatest appreciators of microorganisms, C. B. van Niel. To our colleagues—students and faculty—in the M.B.L. course, and to our several mentors, we dedicate this series.

The Editors

CONTRIBUTORS

MARY MENNES ALLEN, *Department of Biological Sciences, Wellesley College, Wellesley, Massachusetts 02181*

DAVID R. BENSON, *Microbiology Section, University of Connecticut, Storrs, Connecticut 06268*

PATRICIA H. CLARKE, *Department of Biochemistry, University College London, London WC1E 6BT, England*

HENRY L. EHRLICH, *Department of Biology, Rensselaer Polytechnic Institute, Troy, New York 12181*

D. M. GRIFFIN, *Department of Forestry, Australian National University, Canberra 2601, Australia*

GEORGE HEGEMAN, *Microbiology Group, Biology Department, Indiana University, Bloomington, Indiana 47405*

ROBERT E. HUNGATE, *Department of Bacteriology, University of California at Davis, Davis, California 95616*

NORBERT PFENNIG, *Faculty of Biology, Konstanz University, D-7750 Konstanz, Federal Republic of Germany*

PREFACE

Any branch of biology depends for its progress on the development of new concepts and to a lesser, but sometimes crucial, extent on the elimination of erroneous notions. Understanding the roles of bacteria required first the observation that such minute creatures existed, and subsequently the experimental demonstrations that their presence was necessary for the occurrence of particular phenomena.

In this first volume, the authors review the development of scientific understanding of the role of microbes as agents of diverse natural processes. Notably absent is a separate review of the history of microbes as agents of disease, a history available in many other publications. Regrettably absent is a review of the history of microbes as agents of inorganic transformations, a serious omission that resulted from the illness of the prospective author late in the preparation of this volume. The topic will of course be treated in later volumes, although not predominantly in a historical manner. Otherwise, the emphasis in this volume is on the history of understanding interrelationships between modes of bacterial existence and the inanimate environment. These relationships were established long before multicellular, differentiated organisms appeared as potential microbial habitats, and their recognition and elucidation contributed greatly to the widened appreciation of bacterial diversity and the importance of these simpler creatures to the physiochemical conditions of the biosphere.

As bacterial ecology matures as a microbiological discipline, an appreciation of the foundation constructed during the past two centuries is indispensable to its students, and so we offer this first volume to acknowledge the historical context in which current studies proceed. Some of the concepts that arose in the nineteenth century proved important to the development of several subdisciplines, and so are reviewed by more than one author; their recurrent mention serves to emphasize their central role in the development of our present understanding.

We especially thank the authors in this volume for giving time, energy, and consideration to topics which are not the immediate concern of their present research. None is a historian by trade, but all have responded to a call to review the background of today's view of bacterial activities in nature.

The Editors

CONTENTS

CHAPTER 1

The Scientific Study of Bacteria, 1780–1980

PATRICIA H. CLARKE

CHAPTER 2

Anaerobic Biotransformations of Organic Matter

ROBERT E. HUNGATE

CHAPTER 3

The Mineralization of Organic Materials under Aerobic Conditions

GEORGE HEGEMAN

CHAPTER 4

Stages in the Recognition of Bacteria Using Light as a Source of Energy

NORBERT PFENNING

CHAPTER 5

Oxygenic Photosynthesis in Prokaryotes

MARY MENNES ALLEN

CHAPTER 6

Consumption of Atmospheric Nitrogen

DAVID R. BENSON

CHAPTER 7

The Position of Bacteria and Their Products in Food Webs

HENRY L. EHRLICH

CHAPTER 8

A Comparison of the Roles of Bacteria and Fungi

D. M. GRIFFIN

1

THE SCIENTIFIC STUDY OF BACTERIA, 1780–1980

Patricia H. Clarke

INTRODUCTION

This chapter covers a period of approximately 200 years that begins around 1780. The concept of "the bacterium" changed and developed during this time, and strong and often conflicting views were held by the leading scientists of the day. Leeuwenhoek's (1677) animalcules (described in his famous letter of 1676) are now considered to have included bacteria, and although at the beginning of the period under review there were those who thought that small living organisms might cause disease and bring about fermentations, there was no firm scientific evidence for either of these activities. The methods of the natural scientists, particularly the botanists, were used to observe and describe the teeming life that could be seen with the help of the improved microscopes. From a time when all very small organisms, including worms, protozoa, microfungi, and bacteria were regarded as similar creatures to be observed and described, we can trace an increasing awareness of the special attributes of bacteria. It took many years to settle the two major controversies over whether bacteria arose by spontaneous generation and whether fermentation was carried out by living organisms. The supporters of the spontaneous-generation origin of microorganisms had arguments of common sense to sustain them, together with weak experimental technique. Until good methods for sterilization were available, it was difficult to refute their claims, and until the heat resistance of bacterial spores had been recognized, it was difficult to ensure that sterilization methods were adequate. The most vociferous supporters of spontaneous generation were mainly biologists, and a few chemists, while the main opposition to the concept of the microbiological basis of fermentation came from the leading organic chemists. Berzelius, Liebig, and

Patricia H. Clarke • Department of Biochemistry, University College London, London WC1E 6BT, England.

Wöhler thought that fermentation and putrefaction were due to the chemical decomposition of organic matter and poured scorn on the conclusions of Schwann, and later Pasteur, that fermentation did not occur in solutions that had been sterilized by heating unless some living organisms had been added. They did not deny that yeasts, or other organisms, were present in fermentations, but considered them to be casual rather than causal.

The developments in experimental methods during the first half of the period under review were very important in the approach to the concept of bacteria as agents of disease. Pasteur, Koch, Loeffler, and other distinguished bacteriologists made it possible for most of the pathogenic bacteria to be isolated and identified by the turn of the century.

At about that time, Winogradsky and Beijerinck began studies on the biochemical activities of bacteria isolated from nature by enrichment culture and showed that they were able to carry out a variety of biochemical transformations. The biochemical investigations of bacteria drew attention to metabolic pathways, growth requirements, and the variations in enzyme activities with changes in growth conditions. Biochemistry and genetics were developing quite separately during those years with little communication between them. When at last the genetic analysis of bacteria became possible, it ushered in an era in which bacteria were regarded in a totally new light. Bacteria have become tools for the investigation of biochemical pathways and enzyme mechanisms and for studying the processes of nucleic acid and protein synthesis. In applied microbiology, they have provided the means for carrying out chemical transformations that were difficult, and sometimes impossible, by standard chemical methods.

The new concept of bacteria as hosts for genetic material was reached by intensive investigations of the structure of the genetic material and the mechanism of gene expression. The differences between eukaryotes and prokaryotes in these respects had emphasized the separateness of bacteria. However, the techniques of genetic engineering have now made it possible to make partial genetic hybrids between totally unrelated species. By making use of the appropriate gene sequences, it is possible to clone and express the genetic information of eukaryotes, as well as that of other prokaryotes, in bacterial hosts. Because of the potential of such in vitro genetic manipulation, the bacteria have now come to occupy a central position in research in molecular biology. The bacteria have assumed an importance in fields of biology that would astonish the early bacteriologists who painstakingly laid the foundation for this work.

This chapter traces the ways in which ideas about bacteria have changed during the last 200 years. The work of the giants of the nineteenth century is well documented and has been reviewed on many occasions. Discussion of the conflicting theories about the role of bacteria extended over many years, and it has been possible to draw on the writings of contemporary scientists and their immediate successors. The twentieth century has seen exponential growth in all fields of science, and spectacular advances in biology have oc curred during the last thirty years. Bacteriology has become once again part of biology, and, at this present date, some of the activities of bacteria are of interest to many different kinds of biologists.

HOW WERE THE ACTIVITIES OF BACTERIA RECOGNIZED?

Steps toward a Science of Bacteriology

The nineteenth century saw a succession of major advances in biology. The publication of *On The Origin of Species* by Charles Darwin in 1859 marked a dramatic change in patterns of thought concerning the living world. The weight of evidence from studies of the fossil record, and from observations on the immense variety of plants and animals, had made it difficult to maintain belief in the biblical account of creation. This was not abandoned without a struggle, but it was inevitable, beginning from the mid-nineteenth century, that the theory that present-day species evolved from ancestral forms should become accepted. Darwin based his theory of evolution of species by natural selection on observations made on plants and animals, but he was certainly aware of the existence of microorganisms and gave them an important role in the scheme of things. He went so far as to say that "all the organic beings which have ever lived on this earth may be descended from some one primordial form" (Darwin, 1859). In the *Origin,* he did not equate bacteria with these primordial forms, but in a letter to G. J. Romanes in 1881, he recalled a meeting at which they discussed bacteria (Freeman, 1982).

Bacteria had been seen two hundred years earlier by Anthony van Leeuwenhoek, who referred to the small creatures he observed with his microscope as "animalcules." He made no direct attempts to examine the chemical activities of the animalcules but did carry out a number of experiments in which he put samples of pepper, ginger, cloves, and nutmegs into bowls of rain water, snow water, or well water and observed that small creatures appeared after a few days or a few weeks (Leeuwenhoek, 1677). Bulloch (1938) mentioned that Christian Lange (1619–1662) had held the view that disease was caused by the entry of minute living agents into the body; however, this idea was based on mystical belief, not solid evidence, and need not be taken very seriously. Tyndall (1881) also pointed out that a germ theory of disease was favored by Linnaeus.

Although discussions about "contagion" and "infection" preceded Leeuwenhoek's observations, and continued thereafter, there was no way in which conclusions could be reached about microorganisms as causative agents of disease until adequate experimental methods had been developed. Although it is now generally accepted that some of the creatures observed by Leeuwenhoek were bacteria, the name itself did not come into use until much later. Linnaeus thought that the very small organisms found in infusions were impossible to classify and should be left to future generations to fathom.

Meanwhile, he put the "infusoria" in a group called *Chaos,* a conclusion that would be received with wry satisfaction by many bacterial taxonomists today.

As microscopes improved, the observations and descriptions of microbial forms become more recognizable. Müller (1730–1784) described and classified organisms we now designate as protozoa and bacteria, but without making any distinction between these groups. Two of his genera, *Vibrio* and *Monas,* included bacterial forms. Ehrenberg (1795–1876) produced excellent drawings and made a more detailed classification into genera and species on the basis of shape. He, also, did not distinguish between protozoa and bacteria, and classified them all as Infusoria together with rotifers, diatoms, and desmids. Among his various groups was his genus *Bacterium,* with three species of rodlike forms. Unfortunately, these could not be identified by later bacteriologists (Cohn, 1872). In the first half of the nineteenth century, there were many others who attempted to describe and classify bacteria, but the major contribution to establishing bacteriology on a sound basis was the work of Ferdinand Cohn (1828–1898).

In his very lively history of bacteriology, Bulloch (1938) gives Cohn the credit for disentangling correct observations about bacteria from the mass of incorrect and confusing statements that had been made up to that time. In his paper of 1872, Cohn raised the fundamental question of whether it was appropriate to arrange bacteria into genera and species as had been done for plants and animals. Eventually, he came down on the side of doing so, but said that future investigators would have to enquire whether the genera and species that could be delineated by morphology were natural genera and species. This question is still being asked, but in different ways, and will be discussed again later in this chapter. D. H. Howard (1982, 1983) has drawn our attention to a remarkable history of bacteriology written at a time when many of the major discoveries were still to come. Friedrich Loeffler (1852–1915) published his history of bacteriology in 1887 when he was a lecturer in the University of Berlin; he gave an account of the ideas prevailing at that time and the tortuous route that had led to them.

Loeffler (1887) was concerned to trace the development of ideas about "distinctiveness." Even a superficial glance at the earlier periods reveals that the best that could be done was to observe as carefully as possible and to describe as accurately as possible. Sometimes the observer felt impelled to draw up a new classification based mainly on botanical nomenclature or to extend the classifications of previous workers. The main contributions to bacteriological studies of classification up to that time were the observations on morphology and motility. Since the number of forms which bacteria can take are limited, there were many opportunities for arbitrary decisions. Loeffler (1887) recognized the importance of morphology but also sought evidence for distinctiveness from other areas: chemical activities as shown by fermentation, pigment production, and the association of distinguishable bacteria with specific diseases. In all these areas there had been much confusion and muddle, but by 1887 things were becoming much clearer. The rapidity of developments at that time was noted by Flügge (1847–1923) in the preface to the second edition of his textbook on microorganisms (Flügge, 1886).

Constancy of Form

Cohn (1872) was definite in his view that careful observation of the various morphological types of bacteria had established that there was a constancy of form, irrespective of the external conditions. In taking this view, he was in opposition to those who thought that the shape of bacteria was inconstant and variable. It is not surprising that there was uncertainty about the shape or form of bacteria. The earlier observations were made with more or less natural fluids, including infusions of various kinds, water from ponds, and exudates from wounds. These substances were likely to contain mixtures of many organisms, and the proportions would change over a period of time. It had been observed that the microfungi changed in shape during their life cycles; this feature of polymorphism, or pleomorphism, was thought by von Nägeli (1817–1891) to extend to other microorganisms. Von Nägeli thought that the microscopic fungi originated spontaneously and that this made it reasonable to suppose that there would be no constancy of form among this group.

Smith (1932), in an account of Koch's views on the stability of species, recorded that in 1877 von Nägeli had been still convinced of the inconstancy of form and had written that although it was impossible to observe "a micrococcus form, a bacterium form and a vibrio and spirillum form, we must not overlook the fact that the objects corresponding to these conceptions are inconstant. . . . The same organism in milk produces lactic acid and in meat putrefaction and in wine, Gummi." The idea that a single bacterium could give rise to different forms lingered on and could only be finally dispelled when techniques had been adopted for working with pure cultures. Bulloch (1938) mentioned that Lankester (1873) had not believed in the constancy of form of bacteria and had cited as evidence an organism that he had called *Bacterium rubescens,* which was peach-colored and appeared in a variety of shapes, "sphaerous, filamentous, acicular, bacillar, serpentine, spiroid and helicoid." He was taking as a criterion for distinctiveness the production of a pigment which he named "bacteriopurpurin." Others continued to maintain that bacteria were pleomorphic and changed from rods to cocci or from rods to spirals, but by the end of the century such beliefs had been abandoned. Dubos (1945) pointed out that the emphasis on constancy of form later became a hindrance to recognizing that variation of shape could occur with pure cultures.

Spontaneous Generation

In the nineteenth century, many influential figures believed that microorganisms arose by spontaneous generation. Looking back, it seems odd that at a time when a belief in creation was giving way to the theory of the evolution of higher plants and animals from more primitive forms, the smaller microorganisms should be thought to appear spontaneously in a suitable environment. Spontaneous generation was a very old belief and can be found in Aristotle and Homer and in the Old Testament. A very pleasing version was that of van Helmont (1577–1644), who believed that mice could be generated spontaneously from corn in old shirts. The size of creatures thought to originate spontaneously became smaller. Francesco Redi (1626–1697) is credited with the first clear demonstration that maggots did not arise spontaneously from meat. The idea that microscopic organisms might arise in that way was more difficult to disprove, and these were the last to be liberated. By the nineteenth century, the spontaneous generation of microorganisms had become associated with the questions of the causes of fermentation and putrefaction and of the nature of infectious disease.

Leeuwenhoek had observed that rain water freshly collected did not contain animalcules but that these appeared after some time, especially if organic substances had been added. He thought that the animalcules were probably already present as "seeds" or "germs" in the surrounding air, but, during the following years, theories of spontaneous generation of microorganisms were widely held. Spallanzani (1729–1799) followed Leeuwenhoek in finding animalcules in ordinary water, but not in distilled water, and showed that heating prevented the appearance of these creatures. Nevertheless, when Pasteur (1860), following Schwann (1810–1882) and others, put forward sound evidence against the spontaneous generation of microorganisms, his work was not immediately accepted and the dispute continued for several more years. Singer (1959) points out that the barrier to acceptance was that "common experience" showed that microorganisms appeared very readily in infusions and that those who denied that this was an example of spontaneous generation had the difficult task of attempting to prove a universal negative. Apparently definitive experiments were criticized as giving negative results because of the exceptional conditions used by the experimenters.

Pasteur's comment has elegance and conviction: "I have taken my drop of water from the immensity of creation, and I have taken it full of the elements appropriate to the development of inferior beings. And I wait, I question it, begging it to recommence for me the beautiful spectacle of the first creation. But it is dumb, dumb since these experiments were begun several years ago; it is dumb because I have kept it from the only thing man cannot produce, from the germs which float in the air, from Life, for Life is a germ and a germ is Life" (Dubos, 1945). Pasteur came to these experiments through his studies of fermentation. At that time, the view of Liebig that fermentation was a peculiarity of organic chemistry was widely held. Pasteur (who started as a chemist) thought that fermentation, and by extension putrefaction, was due to the activity of living organisms. By the time of the *Origin of Species,* it was accepted that bacteria increased in numbers by dividing into two and that a flask of broth exposed to the air would "go bad." According to Pasteur, the bacteria came from the surrounding air. It was already known that if the flask of broth was sealed and adequately heated, the broth would not become turbid. The supporters of spontaneous generation claimed that this was because the air required for bacterial growth had become altered by heating. With his famous swan-necked flask, Pasteur showed that broth could remain sterile in an open flask provided that the neck was drawn out in such a way that bacteria in the air would be trapped in the neck. Some of the flasks were left undisturbed in the Institut Pasteur and remained sterile for a century. Although the French Academy had stated that "Les expériences de M. Pasteur sont décisives," this was not the opinion of all of his contemporaries. Bulloch (1938) describes the ingenious ways in which several of his critics attempted to keep alive the doctrine of spontaneous generation. We can now see that most of these experiments that sought to demonstrate spontaneous generation of microorganisms failed because the experimenters thought they had sterilized their vessels, but had not done so. In some cases, they had used conditions that would have been suitable for smaller vessels but were inadequate for sterilizing larger volumes in larger vessels. The only major contribution to the question of spontaneous generation after Pasteur was made by the physicist John Tyndall (1820–1893). He was interested in the dust particles that were revealed by a beam of light and how these were destroyed by heating. This led him into the controversy and resulted in the publication in 1881 of his book *Essays on the Floating Matter of the Air in Relation to Putrefaction and Infection* (Tyndall, 1881). One of the stalwart supporters of spontaneous generation, Bastian (1837–1915), objected strongly to Tyndall's line of argument because he felt such matters "appertained to the biologist and physician and not to a physicist;" this was almost the last pocket of resistance, although Bastian continued to hold to his opinions (Bastian, 1905). Tyndall examined the effect of discontinuous heating and, in the same year that Cohn discovered the heat resistance of spores of bacilli, described his method in the following words: "Before the latent period of any of the germs has been completed (say a few hours after the preparation of the infusion) I subject it for a brief interval to a temperature which may be under that of boiling water. Such softened and vivified germs as are on the point of passing into active life are thereby killed; others not yet softened remain intact. I repeat this process well within the interval necessary for the most advanced of the others to finish their period of latency. . . . " He showed that discontinuous boiling for one minute on each of five successive occasions could sterilize an infusion that might not be sterilized by a single boiling. This method—Tyndallization—was used by bacteriologists for many years to sterilize solutions without recourse to excessive heating of thermolabile components.

Bulloch (1938) thought that Tyndall's contribution was particularly important in demolishing the spontaneous generation theory in that he expounded his results in newspapers and magazines, as well as in lectures and discussions.

The situation now is very different, and we can contemplate with equanimity the spontaneous origin of bacteria in the remote past while recognizing its improbability under present conditions or in the recent past.

THE PROCESS OF FERMENTATION

Pasteur and Fermentation

Senez (1968) divides the history of microbiology into three periods: that of the naturalists, before Pasteur; that of Pasteur and his contemporaries; and that after Pasteur. We have seen how important were Pasteur's experiments in settling the question of the spontaneous generation of microorganisms. His work was equally important in the related problem of the nature of fermentation. At the time he came on the scene, the opposing camps were the biologists, including Schwann (1810–1882), and a few chemists who thought that yeasts were responsible for alcoholic fermentation, versus some of the important organic chemists, including Liebig (1803–1873), who thought that all fermentations were carried out by " ferments" that consisted of decaying organic matter. Any decomposing matter was thought to "impart to other substances the state of decomposition in which it finds itself." Thimann (1963) remarks that such views were widespread among chemists at that time and that this was known as the "molecular-physiological theory of fermentation."

Alcoholic fermentations were known and practiced from prehistoric times and traditional methods gave good results. Leeuwenhoek had seen yeast "globules," but there was no advance in understanding the process until the chemists were able to identify the gas produced as carbon dioxide and to demonstrate that it was derived from the sugar. Lavoisier (1743–1794) concluded that the process of fermentation was essentially the same as an ordinary chemical reaction but recognized very clearly that fermentation did not occur in the absence of the yeast. He stated that if the sugar and water were mixed, no fermentation will occur, but that if a little yeast was added to give "le premier mouvement à la fermentation," then the fermentation would continue.

Gay-Lussac, who was also a chemist, took Lavoisier's work a step further and established the quantitive nature of fermentation in the equation which was given his name:

$$C_6H_{12}O_6 = 2CO_2 + 2C_2H_5OH$$

Bulloch (1938) stated that Gay-Lussac's interest in fermentation had arisen from experiments with grape fluid preserved by Appert's method (see below). The role of yeast was demonstrated by Cagniard-Latour (1777–1859), Schwann (1810–1882), and Kützing (1807–1893). Independently, they found that if the yeast was killed by heating or treatment with arsenate, the fermentation was stopped. Their work gave good evidence that yeast was a living organism and that it was responsible for the chemical reactions of fermentation. This conclusion was not acceptable to the supporters of spontaneous generation or to the chemists of the time who dominated the scientific scene. Berzelius (1779–1848), Wöhler (1800–1882), and Liebig (1803–1873) made consistent and savage attacks on the biological theory of fermentation, dismissing Schwann's experiments as worthless and his conclusions as frivolous (Bulloch, 1938).

Pasteur published his first paper on fermentation under the title "*Mémoire sur la Fermentation Appelée Lactique*" (Pasteur, 1857). He had already shown that the racemic mixture of tartaric acid consisted of an equimolar mixture of dextro- and levorotatory forms. Following these observations, he began to look at other fermentations. He was interested in the lactic acid fermentation since the principal product was an optically active molecule. His conclusions were that the lactic acid fermentation was due to a living organism much smaller than yeast but visible under the miscroscope as "petit globules." He referred to the "ferment lactique comme organisme vivant," thus using the terms ferment and organism interchangeably. He stated that "it seems to me from the standpoint that I have reached in my knowledge of the subject that whoever will impartially judge the results of this work. . . . will agree with me in recognising that fermentation is associated with the life and structural integrity of the cells and not with their death and decay; neither is it a contact phenomenon in which the change in the sugar takes place in the presence of a ferment without the latter giving or gaining anything" (Stephenson, 1949).

Ferments and Enzymes

The final part of Pasteur's statement that fermentation is not a contact phenomenon reads oddly today, when we can distinguish without difficulty between an organism and its enzymes and when we consider that chemical changes brought about by enzymes do take place as the result of contact phenomena. But in 1857 there was no understanding of the nature of enzymes, and the dispute was between explanations based on vague chemical activities and the microbial theory of fermentation. Stephenson (1949) states that it was not surprising that Pasteur's paper provoked a first-class scientific row all over Europe. Liebig was no more convinced by Pasteur than he had been by Schwann, and, according to Senez (1968), Liebig maintained until the end of his life that fermentation was not a biological process.

One way of describing the difference of opinion between Pasteur and Liebig is to describe Pasteur as a vitalist and Liebig as a reductionist. Pasteur's identification of a "ferment" as a living organism and of fermentation as "la vie sans air" derived from his observations that the yeast (or another microorganism) was essential to the fermentation process and increased in numbers as the fermentation proceeded. The opposition of the chemists was based on their dislike of "vital forces," and Liebig sought to interpret the process of fermentation as due to chemical instability imparted by the "decaying albuminous matter." Stephenson (1938) pointed out that the chemists had been badly confused for a long time by the idea of "vital forces" in organic molecules. Wöhler had synthesized urea in 1828, but Karrer (1938) states that even in 1847 Berzelius held to the idea of a vital force and that in 1842 Gerhard doubted the possibility of obtaining important vital products of organisms, such as sugars, by synthetic chemistry. Pasteur's firmly held conviction that fermentation required the presence of specific living organisms was far more productive than the diffuse theories of Liebig and colleagues. However, in 1869 Pasteur stated: "I am of the opinion that alcoholic fermentation never occurs without simultaneous organization, development, multiplication of cells, or the continuous life of cells already formed. The results expressed in this memoir seem to me to be completely opposed to the opinions of Liebig and Berzelius. If I am asked *in what* consists the chemical act whereby the sugar is decomposed and what is its *real cause,* I reply that I am completely ignorant of it" (Thimann, 1963). In the reference to the "real cause" (Thimann, 1963), Pasteur left open the door for more detailed investigation of the chemical transformations occurring during fermentation.

A contemporary of Pasteur and Liebig, who has been largely forgotten, was Traube (1826–1894), who studied under Liebig and later worked in the wine industry. He suggested in 1858 that ferments were substances allied to proteins and that all fermentations produced by living organisms were in reality caused by ferments within the organism. Berthelot isolated invertase from yeast in 1860 and described it as a soluble ferment that could be isolated from the mycodermic plant (sic). He stated that there were also insoluble ferments that remained attached to the tissues and could not be separated from them: "Hence the success of M. Pasteur's very important experiments on the sowing of ferments, or in my opinion, of the organised beings which secrete the actual ferments" (Stephenson, 1938). The enzymes known in Pasteur's time were all simple hydrolytic enzymes such as invertase and diastase. This knowledge was a long way from the concept of the complex metabolic pathways in which enzymes work together to carry out a series of chemical transformations. The term "enzyme" was first used in 1876 by Kühne (Gutfreund, 1976) to describe a biological catalyst. Kühne (1876) reported the isolation of trypsin and proposed that such unorganized ferments should be called enzymes (from *in yeast*). In 1895, Pasteur died, and two years later the Büchners prepared a cell-free extract that fermented sugar. Here at last was fermentation without a living cell, but this discovery did not lead immediately to a new understanding of living processes. More sophisticated methods were needed to study enzyme catalysis, and meanwhile there were new microorganisms to be discovered. Pasteur's example was very significant in stimulating these new discoveries.

THE GROWTH OF MEDICAL BACTERIOLOGY

Bacteria and Disease

The earlier, rather nebulous views on the germ theory of disease were given substance and form by the discovery of the role of microorganisms in fermentation. The association of cholera with specific water sources by John Snow (1813–1858) had directed attention to ways in which microbial infections might be spread but did not identify the organisms responsible. The use of antiseptics by Lister around 1864 was based on the germ theory of wound infection and was influenced by Pasteur's discoveries on fermentation (Bulloch, 1938). The meticulous work of Cohn had contributed to understanding the individuality of the bacteria that might be responsible for specific diseases, and this knowledge made it possible to obtain more precise data.

Davaine (1812–1882) made extensive studies on the disease known as anthrax and concluded that it was caused by a bacillus. These observations were taken further by Pasteur and by Koch (1843–1910). Koch, who was a country practitioner, reported to Cohn in 1876 that he had observed the rods of anthrax developing into long filaments and forming spores. He also observed that the spores were very resistant and observed their germination into rods. This was the first observation of the life cycle of a pathogenic bacterium. He considered that pastures became infected with anthrax spores from the excreta and blood of infected sheep and that this could account for the etiology of the disease. Koch's work laid the foundations of classical laboratory bacteriology. He improved staining methods and, even more important, he devised methods for growing bacteria on solid media. This advance made it possible to obtain "pure cultures." Although Pasteur established convincingly that fermentations, such as the lactic acid fermentation, were carried out by specific ferments, he had grown his organisms in liquid cultures by transferring very small volumes. In 1881, Koch introduced "nutrient gelatin" on sterile glass slides to be inoculated with a sterile needle or wire. In Koch's laboratory, Petri later invented the "plate" that continues to bear his name.

The influence of Koch on the now rapid developments in medical bacteriology are undisputed. The isolation and identification of pure cultures on solid medium was an essential part of establishing that specific diseases were caused by specific organisms. The use of agar, introduced by Frau Hesse in Koch's laboratory, was an improvement on gelatin; agar medium in Petri dishes remains a basic tool of microbiologists a century later.

Koch is most remembered for "Koch's postulates." These emerged from his efforts to establish unequivocally that the organisms he had isolated were the causative agents of the disease with which they appeared to be associated. His long paper "*The Investigation of Pathogenic Organisms*" (Koch, 1881) makes the points that were to be more explicit in his paper "*The Etiology of Tuberculosis*" (Koch, 1884). The form in which they have been handed down to successive generations of bacteriologists is as follows:

1. The specific organism must always be present in the case of the specific disease.
2. It must always be possible to isolate and cultivate the organism in a pure state.
3. A culture of the pure organism isolated in this way must be able to produce the disease in an experimental animal.

Bulloch (1938) points out that Henle (1809–1885), under whom Koch had studied in Göttingen, had made what were essentially these statements in his *Pathologische Untersuchungen,* published in 1840. However, at that time the propositions were academic because until Koch had devised reproducible methods for the isolation of pathogenic bacteria, they could not be tested in a practical manner. Koch devoted most of his attention to pathogenic bacteria, but in his 1881 paper he describes the bacteria he isolated from air and water and stated that "*Bacilli* occur constantly and in large numbers around dwelling houses and where gardening and agriculture is carried on."

The last decades of the nineteenth century saw the golden age of microbiology. Koch isolated the tubercle bacillus in 1882 and the cholera vibrio a year later; Loeffler isolated the diphtheria bacillus in 1884, and Shiga isolated the causative agent of dysentery in 1898. Kitasato isolated the anaerobic tetanus bacterium in 1889, and this was soon followed by the isolation of the anaerobic clostridia causing gas gangrene by Welch in 1892. Although other microbial agents of disease were to be isolated later, these twenty years saw the identification of most of the bacteria that cause diseases of man and animals. An organism that was to assume an importance out of all proportion to its medical significance or its role in nature was *Bacterium coli commune,* isolated in 1886 by Escherich. It now bears his name, although to many it is known only by his initial.

Pure Cultures and Mixed Cultures

Loeffler's three criteria for distinctiveness were morphology, chemical activities, and specific pathogenicity. Koch was most concerned with morphology and disease and, with Cohn, had firm convictions about constancy of microbial form. Earlier, reference was made to the comments of Dubos (1945) that the emphasis of Cohn (1872) and Koch on constancy of form had an inhibiting effect on later considerations of bacterial variation. Another school of thought has suggested that Koch's insistence on pure cultures diverted attention from the consideration of bacteria in nature. Bacteria living in the natural environment, rather than in laboratory medium, grow in association with other types of organisms and not in isolation. In 1976, the centenary of the year in which Koch became associated with Cohn, Penn and Dworkin (1976) drew attention to the other type of microbiology that was being developed by Winogradsky, whose work on the isolation of organisms from soil is discussed in more detail below. Penn and Dworkin (1976) emphasized that at any time in the nineteenth century there was more than one view of the role and importance of bacteria. It had been less important to Pasteur that the cultures he worked with should be completely pure than that their activities, such as fermentation, should be shown to be biological and not chemical, and that the microorganisms concerned should be shown to be derived from the air and not by spontaneous generation. Pasteur was also concerned to show that bacteria and other microorganisms were able to carry out very specific chemical transformations.

Bull and Slater (1982) have reviewed the development of Koch's methods from the earlier techniques used by the mycologists and have discussed the way in which the emphasis on pure cultures may have hindered the understanding of the interactions that may occur in a natural microbial community. The idea that microorganisms in nature existed in mixed communities is implicit in the observations of the naturalists of the earlier part of the century who saw a mixture of forms under the microscope that gave some support for theories of pleomorphism. However, during Koch's lifetime a very elegant study on the nature and activities of a specialized mixed community was carried out by Ward (1854–1906). Ward (1892) decided to investigate the "ginger-beer plant." This was a traditional cottage fermentation, which is still carried out in some rural communities. A small portion of the "plant" is added to a mixture of sugar with ginger in water and allowed to ferment. The culture becomes very turbulent and produces a lot of gas; when the mass of the "plant" sinks to the bottom, the supernatant is poured off into another bottle and more water and sugar may be added. The origin of the ginger-beer plant is lost in history, and Ward mentioned that although stories said that it might have been brought back to England from the Crimean war or from Italy, there was nothing to substantiate either claim. Ward obtained samples from various parts of England and from North America and published his conclusions in 1892. He describes the typical culture as white, solid, semitranslucent lumps. He devised a dozen or so different media, including gelatin, starch paste, and yeast-water, and set out to obtain pure cultures. He reported that all samples contained two organisms that were essential: a new yeast species that he named *Saccharomyces pyriformis* and a new bacterial species that he named *Bacterium vermiforme.* Later, he reconstituted a ginger-beer plant with these two organisms and showed that the yeast cells were entrapped in the network formed by the bacteria with their gelatinous sheaths. He also found that the samples usually contained three other yeasts and two other bacterial forms. This group of four yeasts and three bacteria was probably the normal mixed community involved in the fermentation. Ward stresses that the fermentation carried out by his reconstituted association was comparable with the original ginger-beer plant and much more effective than that carried out by either of the two principal organisms inoculated singly. Ward investigated the materials used to start a ginger-beer plant and suggested that the yeast was introduced with the brown sugar used at that time and the bacteria with the ginger. When established, the culture was maintained by occasional subculture as required. This study provides a model for the investigation of mixed communities in action, but, although it attracted attention at the time, it was not until much later that mixed cultures were again investigated in a serious manner. Kluyver (1952) mentioned that Ward's bacterium had been recognized by Perquin as responsible for some instances of clogging in the pipelines of sugar refineries.

Public Health Aspects

The Preservation of Food

Traditional methods of food preservation included heating and drying, salting, pickling in acid solutions, covering with butter, and the addition of sugar. Before the microbiological basis of putrefaction was understood, a new method of heating and sealing from the air had been devised. This is attributed to Appert who published a book in 1810 entitled *L'art de Conserver pendant plusieurs années toutes les substances animales et végétales.* Bulloch (1938) gives his first name as Nicolas and says that he has been incorrectly cited as François by other writers. According to Drummond (1957), Appert was awarded a prize of 12,000 francs for improving the provisions of the French Army. Appert's method was to place the food in glass jars, which were loosely corked, immersed in boiling water, and finally sealed as tightly as possible. This simple method made it possible to preserve fruit and vegetables at harvest time for use later in the year. Until the advent of deep freezers, this remained the predominant method in domestic use for protecting foodstuffs from bacterial deterioration. When this type of preservation was introduced, it was generally thought that the important feature was the exclusion of air.

The use of cans for food preservation was developed in England in 1810 and they were used to preserve meat for the British navy. The advantage of using higher temperatures was soon realized, and Appert used a type of autoclave for this. Fifty years later, one of his family fitted them with pressure gauges for more exact control (Derry and Williams, 1960). An alternative method, used in Britain, was to stand the can in a solution of boiling calcium chloride. The success of canning resulted in an expansion of the market, and the manufacturers started to use much larger cans. The result of this was that the contents were not completely sterilized and went bad. Drummond (1957) mentions that in 1850 one supplier had 111,108 pounds of tinned meat condemned by the Admiralty. There were two problems, neither of which was properly understood at that time. Some of the meat provided may have contained bacterial spores, and thus prolonged heating would have been required to kill them. Cohn's identification of the heat-resistant bacterial spore and Tyndall's observations on the heat resistance of some bacteria did not take place until 1877. The other problem that needed to be solved was the relationship of the time of heating to the size of the container in achieving complete sterilization. The immediate problem was overcome by reducing the size of the container, but it was not until the end of the century that studies at the Massachusetts Institute of Technology established that it was essential to take into account the size of the container and the nature of the commodity. Extended heat sterilization was essential with products of low acidity, while fruits with high acidity could be satisfactorily sterilized in boiling water by the method originally devised by Appert. Bastian, who had remained a consistent supporter of spontaneous generation, reported that growth often occurred in urine that had been made alkaline before boiling. This observation caused

Pasteur to reexamine the requirements for sterilization, and he established the practice of heating fluids under pressure, thus introducing autoclaving into laboratory procedure (Topley and Wilson, 1936).

Milk and Its Distribution

The dissemination of scientific knowledge about bacteria and disease in a way that affected daily life was slow. Milk in the cities during the first part of the nineteenth century was obtained from town dairies, where cows were kept under poor conditions. With the growth of the railways, milk was brought into the cities by trains, and by 1866 over two million gallons were transported annually to London from the country. By 1860 Pasteur had shown that many organisms were killed by heating to about 60°C, but the process of pasteurization was not used by the dairy industry in England until 1890. Its purpose was to prolong the life of the milk and to prevent its going bad before it reached the consumer. A few years later it began to be realized that mild heat treatment could also protect against disease from pathogenic bacteria carried in milk.

Water Supplies

At the beginning of the nineteenth century, sewerage systems were nonexistent and water was frequently contaminated. Concern about the effects of insanitary and unpleasant conditions on the health of the inhabitants of large towns began to lead to improvements. The privy bucket was replaced by enclosed sewers; raw sewage contamination of rivers used for water supplies was reduced; piped water replaced pumps. In London an outbreak of cholera had been traced to a particular pump by Snow in 1849; by this time the idea that infections could be caused by bacteria had taken root, although it was not until 1884 that *Vibrio cholerae* was isolated by Koch. Most credit for the improvements in the water and sewerage systems of cities goes to the pioneering water engineers, but the increasing confidence of the bacteriologists enabled the bacteriological purity of the water systems to be checked. In England, the chemical and bacteriological analysis of water contributed to the improvement in the standard of water supplies.

BACTERIA AND THE NATURAL CYCLES

Isolates from Nature

In the last decades of the nineteenth century, the study of disease-causing microorganisms became concentrated on bacteria and largely institutionalized. In earlier years, the scientists interested in bacteria had originally been chemists like Pasteur, physicists like Tyndall, or botanists like Cohn and Ward. The science of medical bacteriology was now becoming so well established as to become a separate discipline. Within a short period it was to extend into immunization and virology. The demands on medical bacteriology laboratories for identification of clinical isolates led to the development of new culture techniques and diagnostic methods. Penn and Dworkin (1976) saw a sharp dichotomy between the direct influence of Koch on medical microbiology, with its emphasis on pure cultures and constancy of form, and the studies of microorganisms in their environment, which they associated with the work of Winogradsky. They suggested that emerging alongside the Koch tradition was "another competing, distinctive and less conspicuous tradition associated with the name of Sergei Winogradsky." Another interpretation is that the investigations of Winogradsky carried on the tradition established by Pasteur and naturalists of the earlier part of the century, whereas Koch represented the specialization into bacteriology applied to medicine. The benefits of investigations into medical bacteriology were obvious, and such work received support and public acclaim. Other types of bacteriology did not appear at that time to have much practical importance. Also, Koch's legacy was mainly to the scientific study of bacteria, thus making a separation between bacteria and other microorganisms. In this respect it is interesting to note that the term "microbe" was first used by Sédillot in 1878 at a meeting of the Académie des Sciences in Paris, and that Pasteur in 1882 suggested that the science of microbial life should be known as "microbie" or "microbiologie" (Gale, 1970). The pure culture technique was originally developed by de Bary and Brefeld and others for use with fungi (Bull and Slater, 1982; Bulloch, 1938). They succeeded in isolating single cells and also cultivated fungi on solid medium, but the techniques were unsuitable for use with bacteria. Many methods were tried, but Koch and his colleagues deserve most of the credit for the methods used today. We have seen that Ward (1892) had no difficulty in isolating and identifying both yeasts and bacteria and establishing their role in the mixed culture of the ginger-beer plant. The progress in understanding the role of microorganisms in the natural cycles of the elements also depended on identifying and growing the organisms concerned in these activities in pure cultures.

Nitrifiers

The importance of maintaining soil fertility was recognized long before the presence of microorganisms in the soil was known. Drummond (1957) states that at the end of the eighteenth century the value of the contents of privies as manure was well known, even though microbial degradation and recycling were not suspected. It was also known that ammonia was oxidized to nitrate in the soil, and in 1862 Pasteur thought that this might be a biological activity similar to the oxidation of alcohol to acetic acid. In 1877, Schloesing and Muntz carried out an experiment in which a tube of sand and chalk was fed with sewage water. After some days, nitrate appeared, the ammonia content fell, and the process was maintained for 20 days. When they added chloroform to the tube, the nitrification ceased (Stephenson, 1949). All attempts to obtain cultures of the nitrifying organisms were unsuccessful until Winogradsky thought it might be helpful to omit an organic source of nitrogen. The first isolates were made in liquid medium, but Winogradsky (1890) succeeded in obtaining colonies on a solid medium which contained "silica jelly" instead of gelatin and with ammonium chloride as the nitrogen source. This was the first isolation of *Nitrosomonas* and *Nitrobacter*.

Nitrogen Fixation

Pasteur's observations on the oxidative and fermentative activities of microorganisms stimulated research on soil problems. The first report on biological nitrogen fixation was by Jodin in 1862, although it is unclear whether he had been influenced by Pasteur (Stephenson, 1949). Bertholot observed nitrogen fixation in soil left uncultivated in pots and found that it did not occur if the soil had been sterilized. The first isolation of a nitrogen-fixing organism in pure culture was reported by Winogradsky (1894), and this anaerobic bacterium was given the name *Clostridium pasteurianum*. A decade later, the Dutch microbiologist Beijerinck (1901) isolated the aerobic *Azotobacter* and showed that nitrogen fixation was not confined to anaerobic organisms. Earlier, symbiotic nitrogen fixation had been demonstrated with leguminous plants. Boussingault had shown in 1838 that clover, cultivated in sand, gained in both carbon and nitrogen, but could not do so if the sand had been "very completely ignited." Although nodules were observed on the roots of leguminous plants, their significance was not recognized until many years later (see Benson, this volume). Hellriegel and Wilfarth reported experimental evidence for the view that nodulation was essential for nitrogen fixation to take place and deduced that a "soil ferment" was involved in this process. Soon after, Beijerinck (1888) isolated *Rhizobium* and succeeded in growing it as a laboratory culture (Stephenson, 1938; Thimann, 1963).

Autotrophic Sulfur Bacteria

Winogradsky is especially remembered for his work on autotrophic bacteria. The sulfur-oxidizing bacterium *Beggiotoa* had been observed in hot springs, and Cohn (1875) showed that the refractile granules contained elementary sulfur. (Since Koch encountered Cohn in the following year, this observation was probably known to him.) The main outlines of the physiology of *Beggiotoa* were set out by Winogradsky (1889), although it was not isolated in pure culture until some years later. The discovery of bacteria that obtained their energy by the oxidation of inorganic compounds (ammonia, nitrite, ferrous iron, and the sulfur compounds) introduced a new type of metabolic activity. Pasteur had shown that some microorganisms could live and grow in the absence of oxygen with organic compounds providing the carbon and energy sources. Winogradsky now identified microorganisms that existed by oxidizing inorganic compounds and assimilating carbon dioxide as their carbon source. Studies on autotrophic bacteria were extended and developed in several laboratories, but particularly by Beijerinck who, with Winogradsky, contributed to a new approach to the study of bacteria. Winogradsky, in a paper on the nitrifying organisms, set out very clearly the nature of the autotrophic mode of life:

> two groups of organisms. . . . I have designated them by the names sulfur bacteria and iron bacteria. The first group live in natural waters which contain hydrogen sulfide and do not grow in media lacking this substance. This gas is absorbed extensively and oxidized by their cells and is converted into sulfur granules. These latter are in turn degraded and sulfuric acid is excreted. The second group are able to oxidize iron salts and their life is also closely connected with the presence of these compounds in their nutrient medium.

He proceeded to describe the nitrifiers and how it was essential to exclude organic matter from the culture medium. Since the only sources of carbon in the medium were carbon dioxide and carbonates, it was possible for him to conclude that the cell carbon was obtained from carbon dioxide (Brock, 1961).

Photosynthetic Bacteria

While Winogradsky was demonstrating that some bacteria assimilated carbon dioxide without the intervention of chlorophyll, attention was being drawn by Engelmann (1883) to bacterial photosynthesis. He described a culture of pigmented organisms (purple bacteria) that had a well-defined absorption spectrum and observed that on a microscope slide illuminated through a prism the organisms collected in the region of their own absorption bands. The nature of anaerobic bacterial photosynthesis was not understood until the work of van Niel (1941). Engelmann made other interesting observations on phototaxis and chemotaxis and was the first to show that motile aerobic bacteria accumulated, along the long spinal chloroplast of *Spirogyra* (Stanier et al., 1957; see Chapters 4 and 5, this volume).

CHEMICAL TRANSFORMATIONS CARRIED OUT BY BACTERIA

Oxidations

Apart from the traditional alcoholic fermentations, the oldest chemical transformation for which microorganisms were employed is the manufacture of vinegar. Pasteur (1862) studied the process in response to a request to investigate the diseases of wine and identified an organism later designated as *Acetobacter aceti.* For vinegar production, the wines were stored in vats; the oxidation occurred at the surface, while the vinegar was withdrawn from the bottom of the vat. In the quick vinegar process, the wine or beer was trickled continuously through vessels packed with twigs or shavings and the bacteria attached themselves to the twigs forming a thin film. Thus, this process made use of both continuous culture and immobilized cells.

Boutroux (1880) showed that a similar type of bacterium could oxidize glucose to gluconic acid, and Brown (1886) reported the bacterial oxidation of mannitol to fructose and *n*-propanol to propionic acid. Pelouze (1852) had attempted to obtain succinic acid "par l'action de l'air sur la jus de sorbier." This experiment in early biotechnology was not successful, but he did obtain a new compound which he named sorbine. Almost fifty years later, it was identified as sorbose. Bertrand (1896) showed that sorbose was produced by the oxidation of sorbitol by an *Acetobacter* strain, although this organism was subsequently lost. Beijerinck, Kluyver, and their colleagues carried out extensive investigations of the biochemical activities of the *Acetobacter* and studied the growth conditions under which these bacteria carry out incomplete oxidations, such as the oxidation of glucose to gluconic acid and sorbitor to sorbose.

Stodola (1957) commented that although observations on microbial reactions continued to be reported, it was not until the 1930s that they became of more than academic interest. He traces the upsurge of interest to the commercial use of *Acetobacter suboxydans* for the oxidation of sorbitol to L-sorbose for the manufacture of ascorbic acid. This comment by Stodola is supported by the view of Topley and Wilson (1936) that although the first few decades of the century had seen few new and dramatic discoveries in medical bacteriology, the indications were that with the rise of biochemistry "bacteriology would be on the march again."

However, much had been accomplished in applied microbiology earlier than this. Beijerinck and his colleagues had continued from 1880 onward to expand their collection of bacteria and other organisms from the soil and ditches of Holland. The laboratory at Delft was the center for the isolation of many different types of bacteria with novel metabolic activities. From this type of work came the aphorism "everything is everywhere, the environment selects." It is related that in his lecture on Beijerinck's elective culture methods, Kluyver said that many requests for cultures were for bacteria that could be found on the shoes of the petitioner (Kamp et al., 1957).

Anaerobic Fermentations

Kluyver succeeded Beijerinck as director of the Microbiology Laboratory of the Technical University of Delft in 1922. In his inaugural lecture, he discussed the part that microbiology might play in the education of the chemical engineer (Kluyver, 1957). Among the useful compounds produced by microorganisms he listed alcohol, lactic acid, butanol, butyric acid, and acetone. He cited the export of 1500 tons of lactic acid from Germany in 1909 and the large-scale production of acetic acid. He pointed out that in 1918 the

British Government appointed an "Alcohol Motor Fuel Committee" that was charged with investigating how to achieve a large increase in industrial alcohol production and at the same time to investigate how to use alcohol as a fuel for the internal combustion engine.

The fermentations of the anaerobic clostridia were recognized by Pasteur (1861), but the potential of such activities became apparent only when the chemists had devised a method for the manufacture of synthetic rubber from butanol. This did not get under way at that time, but the production of acetone by *Clostridium butylicum* became extremely important for the manufacture of cordite in the First World War. The fermentation was developed into an industrial process by Weizmann in 1915. Six distilleries in England were adapted for the production of acetone, and later, acetone plants were constructed in India and in Canada. The scale of operations was such that in the Canadian plant in 1918, there were 22 functioning fermenters, each with a capacity of 30,000 gallons (Kluyver, 1957). This fermentation produced twice as much butanol as acetone, but it was not until the Second World War that the butanol was used industrially for the manufacture of synthetic rubber in the United States (this had been Weizmann's original interest).

Bacterial Metabolism and the Unity of Biochemistry

In his account of Kluyver's contribution to microbiology and biochemistry, van Niel (1957) remarked that when Kluyver had been appointed to the laboratory at Delft in 1922, the knowledge of the chemical activities of microorganisms had been virtually restricted to the awareness that a large number of chemical transformations could occur while biochemistry as such was mainly devoted to the analysis of blood and urine. While Kluyver was actively engaged in investigating microbial transformations in more detail, another important development was taking place in England. Marjory Stephenson turned from the study of mammalian biochemistry to embark on her remarkable research on bacterial metabolism. In the first instance, she looked upon bacteria as tools for the biochemist and like Pasteur before her had little interest in taxonomy. She was very interested in the way in which the biochemical activities of a culture changed at different stages of growth, and, in investigating these properties, she made use of washed suspensions. Bacteria were harvested from culture media at different stages of growth to measure enzyme activities. Critics of this type of work complained that this approach treated bacteria as if they were merely bags of enzymes. However, it turned out to be a powerful method and had the advantages that enzyme activities could be measured in intact cells and that the activities could be related to growth conditions. In most cases, the experiments were designed to study single enzymes rather than metabolic pathways.

The result of the research on the metabolism of bacteria strengthened the concept of the unity of biochemistry that had been implicit in the work of the Cambridge microbial biochemists from the early 1920s. Kluyver's research, which started from the multiplicity and variety of microbial activities, led to similar conclusions. In 1924 he gave a lecture to the Netherlands Chemical Society under the title of "*Unity and Diversity in the Metabolism of Microorganisms*" (Kluyver, 1957), and a few years later he gave a series of lectures in England on "*The Chemical Activities of Microorganisms,*" in which he extended those ideas (Kluyver, 1931).

By 1930 a number of enzymes had been purified, and substrate specificity was recognized as a characteristic of these biological catalysts. Kluyver was very concerned to reconcile the concept of limited substrate specificity with the wide variety of metabolic activities that had been reported for microorganisms. He had been particularly impressed with the enormous range of compounds that could be used as growth substrates by a single species of *Pseudomonas putida* (Den Dooren De Jong, 1926). He commented that the substrate specificity observed for hydrolases had led biochemists to consider the oxidoreductases would be equally specific, but that "it can scarcely be conceived that the cells of the bacterium in question [*P. putida*] contain as many dehydrases [dehydrogenases] as there are suitable oxidation substrates for these cells" (Kluyver, 1931). He went on to say that "there is no clear reason why one should not go further and accept the supposition that in *Pseudomonas putida* there is only one single oxido-reduction promoting agent which acts on all the substrates. . . ." This pleiocatalytic view of enzymes has some resemblance to the pleiomorphic theories of the bacteriologists of an earlier data. In fact, Kluyver was wary of ascribing metabolic reactions in terms of specific enzymes at a time when so little was known of enzyme structure and function. He had also carried out some experiments that led him to distrust the notion of the coenzyme, and it was not until many years later that metabolic enzyme studies were carried out in his laboratory (van Niel, 1957).

Biochemical Variation

Variation in morphology had been suspected by the early bacteriologists, and variation in chemical activities had been observed by Beijerinck and Kluyver. In the second edition of *Bacterial Metabolism* (Stephenson, 1938), bacterial adaptation to differences in the chemical environment was described in terms of adaptive and constitutive enzymes. These terms had been introduced by Karström (1930) to describe the fermentative properties of bacteria grown on glucose and xylose, respectively. In Cambridge, it had been noticed that washed suspensions of *E. coli* grown in tryptic broth did not liberate hydrogen from formate until they had been in contact with formate for about 24 hr. Following Karstrom's idea that some enzymes were adaptive, they were able to show that the enzyme formic hydrogenlyase appeared in washed suspensions of resting cells without any appreciable increase in cell numbers. They found that a source of nitrogen (in this case they added some tryptic broth) was essential for the appearance of the new enzyme activity (Stephenson and

Strickland, 1933). The emphasis in the early experiments on the absence of growth during the production of adaptive enzymes now seems somewhat excessive, but, at the time, it was critical to establish this point. These observations represented another example of variability, and it was essential to define the phenomenon clearly. In experiments of this sort, which continued with other enzymes, the objective was to distinguish reversible and temporary changes in the enzyme activities of bacterial cultures from those permanent changes that were due to the appearance of new variants. The implications of these findings will be discussed in more detail below.

Genetic Variation

Biochemistry was not the only science that was converging on bacteriology. Genetic experiments in the 1930s were being carried out with *Drosophila*, as well as with mice and plants. In the first years of the century, Garrod (1909) had suggested that certain congenital diseases, "the inborn errors of metabolism," were due to metabolic defects, and the geneticist Bateson pointed out that the familial incidence of alkaptonuria was consistent with "a rare recessive character in the Mendelian sense" (Clarke, 1976). Mutants of *Drosophila* became amenable to biochemical analysis, and Beadle came to the conclusion that there were two genes in control of the biochemical reactions leading to the synthesis of the eye pigments. Beadle (1974) recalled that he first had thought of changing over from working with a fly to working with a mould when he had sat in on some lectures by Tatum on comparative biochemistry and had realized that it would be easier to examine the relationship between genes and enzyme activities if he looked for mutants in a particular biochemical pathway. Within a few years, Beadle and Tatum (1941) had shown that the genes for the separate steps in a biosynthetic pathway could be identified by mutations. They concluded that one gene, and one only, was concerned with each step of the pathway. In many cases, the details of the biosynthetic pathways with which they were concerned had not been fully established, and an indirect result of their work was to improve this position. The direct result was the one gene–one enzyme hypothesis that was to survive (with minor modifications) as one of the fundamental concepts of molecular biology.

The bacteria had been considered to be too small and too insignificant to have a complicated replication system, and reproduction by binary fission was generally considered quite adequate for them. Dubos (1945) discussed the origin of variants in bacterial populations, such as loss of pigment, loss of virulence, loss of metabolic activities, and so on, and pointed out that it was very difficult to explain any of these inherited changes in terms of classical genetics.

He discussed the observations of Griffith (1928) that avirulent pneumococci could be transformed to virulence by extracts from heat-killed virulent strains and mentioned other cases where bacteria grown in the presence of extracts of related virulent forms were said to have acquired the virulence properties of the latter. He was also able to discuss the results of Avery et al. (1944), published in the previous year, on transformation of pneumococci by DNA. He commented that "assuming that the substance which induces transformation is really a deoxyribonucleic acid, as the evidence strongly suggests, then nucleic acids of this type must be regarded not merely as structurally important but as functionally active in determining the biochemical activities and specific characteristics of the pneumococcal cells; they possess a biological iological charactersspecificity the chemical basis of which is as yet undetermined." Within a few years, Tatum and Lederberg (1947) had demonstrated that *Escherichia coli* could exchange genetic material by conjugation, and thus began the revolution in genetics that has made *E. coli* the most widely cultivated organism in the world. The genetics of bacteria turned out not to be classical in the Mendelian sense but to have an elegant simplicity that facilitated fundamental discoveries in biochemistry and molecular biology. Nucleic acid had been discovered by Miescher in 1869, at about the time that Mendel had arrived at his laws of inheritance, but it was not until the techniques for studying the genetics of bacteria and then viruses became available that the chemical basis of inheritance could be firmly established. Curiously, although most of Meischer's education was undertaken at Basel Medical School, Portugal and Cohen (1977) mention that he spent a semester at Göttingen in 1865. At that time, Koch was completing his medical studies at Göttingen under Wöhler and Henle and graduated as a doctor of medicine in 1866 (Bulloch, 1938).

LEVELS OF UNDERSTANDING

From Nature to the Laboratory

By 1945 the study of microorganisms was beginning to be accepted as a separate discipline. This was marked in Britain by the establishment of the Society for General Microbiology. At the inaugural meeting, the first president, Marjory Stephenson, made an analysis of progress so far. She distinguished five levels of investigation and understanding of the nature of bacteria and their activities. At the first level was the initial exploration of the activities of mixed cultures in their natural environment. The second level could be considered to have been initiated by Pasteur and Koch and led to the examination of the activities and properties of organisms in pure culture. At that time, the special study of medical bacteriology was initiated, and many new species were isolated. The third level was identified as the use of washed suspensions of nongrowing bacteria to study enzyme reactions. This type of investigation led directly to the fourth level, at which nutritional studies were carried out with cultures grown on defined media, as in the work of Fildes, Knight, and colleagues. At the fifth level the investigations began to involve more detailed biochemistry using cell-free extracts. While these five levels could be seen as a historical development, they could also be taken as guidelines for starting any new investigation. She considered that research must occur at all levels to obtain an understanding of bacteria as they are found in nature (Elsden and Pirie, 1949).

In a brief review of the history of microbiology, Gale (1970) commented that the list could be continued with a sixth level, starting with the genetic studies of Tatum and Beadle and leading to bacterial genetics.

What about Bacterial Species?

Cohn (1872) had asked whether it was appropriate to give species names to bacteria according to the Linnaean system adopted for plants and animals. The medical bacteriologists had given generic and species names to pathogenic bacteria, and taxonomy and identification had gone hand in hand. The criteria used for taxonomic purposes began to include simple biochemical reactions, such as sugar fermentation; as time went on these biochemical tests became more elaborate, and the creation of new species increased in proportion to the number of bacteriologists. Critical voices were raised at various times; Pasteur is reported as saying in response to a comment about the correct classification of a bacterium, "How little that means to me!" and these sentiments were echoed almost a century later by Marjory Stephenson. The Dutch microbiologists, Beijerinck and Kluyver, devised their own taxonomy for the bacteria isolated by enrichment culture and used biochemical activities as their main criteria for allocating names of genera and species. The genera *Nitrosomonas, Nitrobacter, Acetobacter, Thiobacillus,* and *Photobacterium* were given names that accorded with their most striking characteristics. This reliance on physiological characters was found to present problems at a later date but was convenient and workable. Beijerinck himself pointed out that strains of luminous bacteria might lose the ability to emit light, which was the main feature that had led to their being given their names.

Attempts were made at various times to reconcile taxonomy and phylogeny, and among these one could mention those of Kluyver and van Niel (1936) and Stanier and van Niel (1941). Cowan (1962) made a general attack on the species concept. He quoted John Locke (1689) (who was not of course referring to bacteria but to ideas): "genera and species. . . . depend on such collections of ideas as men have made and not on the real nature of things"; and Darwin (1859), who said, "Nor shall I here discuss the various definitions which have been given of the term species" and "From these remarks it will be seen that I look at the term species as one arbitrarily given, for the sake of convenience, to a set of individuals closely resembling each other. . . ." If the author of the *On the Origin of Species* had difficulty at arriving at a definition, it should not be too worrying to bacteriologists that the definition of a bacterial species still presents problems. In practice, two empirical solutions appear to have been adopted. Bacteriologists, particularly those engaged with pathogenic organisms, have adopted, as far as they could, the classifications of International Commissions on Nomenclature, such as appear in the contemporary edition of *Bergey's Manual of Determinative Bacteriology*. The microbial biochemists who isolated new strains from nature frequently obtained cultures that could not be fitted into known slots and contented themselves with generic identification, e.g., *Pseudomonas sp.*, or, if the identification was even more problematical, the new strain would be called Strain X. Sneath (1962), and others, reintroduced the Adansonian principle of the analysis of a large number of characters, all given equal weight, for analyzing the natural relationships between bacteria, and de Ley (1962) extended this list of characters to include metabolic pathways. From such analysis came the idea of related "clusters" of bacterial strains.

Cowan (1962) reached the conclusion that the bacterial species is a myth: that it is impossible to define except in terms of a nomenclatural type. He also said that the characterization and separation of bacteria (presumably establishing distinctiveness in Leoffler's sense) depended on progress in biochemistry and genetics. The concept of the species did not become much stronger in the next 20 years. On the other hand, relatedness in terms of comparisons of details of molecular structure could be examined in much greater detail. It became possible to make comparisons between sequences of homologous proteins and between sequences of nucleic acids. The discovery of bacterial plasmids and the identification of transposons and insertion elements (Bukhari et al., 1977) added to the complexity of the bacterial genome. The study of the inconstancy of bacteria has now become more interesting than the study of constancy (Clarke, 1982).

The Impact of Genetics

Kaplan (1952) reviewed the genetics of microorganisms and discussed methods of obtaining mutants of *E. coli* and the nature of mutagenesis. It is clear that the mechanism of gene transfer in *E. coli* was far from understood at that time, since he stated, "Evidence is continually accumulating to show that bacteria possess normal chromosome-like arrangements and a Mendelian mechanism for exchange of their genetic factors." He suggested: "The mechanism of bacterial gene recombination is analogous to a bimolecular reaction. Long times and high cell concentrations reveal a saturation effect; whereas short times and low concentrations shown an induction effect perhaps caused by production of a copulation hormone." However, he did say that "no special unknown basic mechanisms e.g. inheritance of acquired characters, are responsible for this but the same causes known in the genetics of higher organisms, mainly mutation and selection."

The elucidation of the structure of DNA, giving a molecular basis for the faithful replication of the genetic material, made it possible to think of genetic mechanisms in a new way (Watson and Crick, 1953). The existence of bacterial genetic material became less nebulous with improved staining methods for DNA. Dubos (1945) had envisaged that the bacterial chromosome could consist of "a single gene string existing as a small granule or rod-like body rather than as a definite vesicle separating it from the cytoplasm," and events were to prove that he was on the right track. Hayes (1960) said that "so far as *Escherichia coli* K12 is concerned we are now in a position to say definitely that it has only one chromosome" and that "it is probable that the chromosome is normally circular." By that time, sex factors had been identified and transduction by bacteriophages was also known.

Genetic studies with bacteria were smoldering away in 1947 and were not to burst into flames for another few years. Meanwhile, so rewarding had been the studies on bacterial nutrition and the assay of enzyme activity under different growth conditions that bacterial experiments became very important in understanding biochemical reactions. Woods (1947) remarked in the first volume of the *Annual Review of Microbiology* that the trend towards detailed studies on bacterial growth factors and their role in metabolism had continued and that there had been much to report. The list of microbial growth factors that had been, or were about to be, identified as coenzymes was now considerable and included thiamin, nicotinic acid, riboflavin, pyridoxal phosphate, pantothenic acid, and biotin; and the role of *p*-aminobenzoic acid and folic acid was about to be elucidated. There were some references to the biosynthesis of amino acids based on studies on the nutrition of an exacting strain of bacteria, but this was limited to observations such as that *Eberthella typhosa (Salmonella typhi)* strains that required tryptophan could grow if provided with indole. Fildes (1945) had concluded from these experiments that indole was a precursor of tryptophan and that these bacteria were defective in the synthesis of indole. Gale (1947) pointed out that the studies of Tatum and Beadle and colleagues with *Neurospora* mutants had indicated that indole was the precursor of tryptophan and that they had also shown that L-serine was required (Tatum and Bonner, 1944). Mutants of *Neurospora* were proving to be very useful for investigations of amino acid biosynthetic pathways, and Gale (1947) thought that with the isolation of exacting mutants of *E. coli* by Tatum (1945), similar studies could be expected with bacterial mutants. Within a few years, much of the research on biosynthetic pathways was carried out with bacterial mutants. An examination of a textbook of general biochemistry, such as that by Fruton and Simmonds (1958), shows that ten years later the biosynthetic pathways of almost all the amino acids had been estblished and that, with one or two variations, the pathways for *Neurospora* were the same as those for *E. coli;* further, in so far as they could be compared with other organisms, there was a universal pattern. This demonstration of the unity of biochemistry was very satisfying to those who had maintained that investigations of bacterial metabolism would make fundamental contributions to biochemistry in general.

Exchange of genetic information by conjugation and transduction that was originally observed for *E. coli* and related enteric bacteria has now been reported in many other species. The map of the *E. coli* chromosome is still far more detailed than that of any other species, but genetic studies are now being pursued with many other groups of bacteria (Glover and Hopwood, 1981).

The discovery of the *E. coli* sex factor was the first indication of extra-chromosomal inheritance in bacteria (Hayes, 1960). The drug resistance factors were found to be able to transfer multiple resistances through a bacterial population and clearly provide an advantage to bacteria growing in the presence of various drugs and antibiotics. The existence of catabolic plasmids, first recognized in the pseudomonads, answered some of the questions that had been raised about the instability of biochemical characters. The complexity of the bacterial genome became extremely interesting with respect to the evolution of strains with new characters (Clark, 1982). The properties of the plasmids themselves assumed a new importance with the possibility of constructing recombinants by in vitro techniques. The role of transposons and insertion elements in bringing about genetic rearrangements within plasmids and within the chromosome added to this complexity (Bukhari et al., 1977).

Adaptation—Genotype or Phenotype?

Adaptation of bacteria to new growth conditions could be taken as an example of variability. Winogradsky and Beijerinck had noticed that bacteria could lose some of their properties when maintained in laboratory culture, while other investigators observed that some enzyme reactions could be related to the medium on which the bacteria had been grown. The first type of change was inherited and permanent, and the second was transient and reversible. We would now describe these two different types of adaptation as genotypic or phenotypic events, respectively. The experiments with formic hydrogenlyase (Stephenson and Strickland, 1933) had shown that de novo appearance of formic hydrogenlyase activity did not require growth of the cells. Studies on enzyme adaptation were continued by Gale (1943), who made the distinction between the potential enzymic constitution of a culture and the actual enzymic constitution produced in response to the particular growth conditions. This was a description of adaptive enzyme synthesis, but unfortunately adaptation was, and still is, used by biologists to describe both the evolutionary changes that allow organisms to occupy new ecological niches as well as the short-term response to new environment. Stanier (1951) suggested that the term "enzymatic adaptation" should be applied to "substrate-activated biochemical variation not involving changes in genotype." This definition made the problem much clearer and paved the way for the proposal by Monod and Cohn (1952) that the term "induced enzyme" should replace "adaptive enzyme," and so it did. Monod and colleagues (1952) had studied the kinetics of synthesis of the inducible enzyme, β-galactosidase, in growing cultures, and this technique enabled them to identify non-substrate inducers and to show that inducer specificity was not necessarily the same as substrate specificity.

This allowed the gratuitous induction of enzymes under conditions in which they were not required for growth.

One of the most important features of these experiments on the kinetics of enzyme induction was their precision. This made it possible to demonstrate that the addition of the inducer gave rise to de novo protein synthesis rather than the modification of existing proteins. This finding supported the template theory of protein synthesis, which was then gaining ground.

Another line of research was concerned with the "training" of bacteria. Fildes et al. (1933) had found that a bacterial strain requiring the amino acid tryptophan for growth could be "trained" to dispense with it by serial subculture in media containing decreasing amounts of that compound. Stephenson (1938) comments that "the modus operandi is at present obscure," although later this phenomenon was clearly recognizable as mutation and selection. Experiments of this sort can be considered as studies in population genetics, but the interpretation placed on them by Hinshelwood and colleagues was that changes occurred within the cells in response to the chemical composition of the medium and after a time these became more permanent. Dean and Hinshelwood (1966) said: "Long ago Penfold, speaking of acquired fermentative properties of bacteria, observed that the longer a character is impressed the longer it is retained, and that the more easily the cells take on a new character, the longer they retain it. The same, as we shall see is generally true of drug resistance." In a series of rather simple experiments using serial subculture in different media, the authors and their colleagues claimed to have observed adaptation and deadaptation of bacterial cultures. The conclusion from this type of work was that the inheritance of acquired characters could be observed in bacteria. Just as bacteria were the last group to be liberated from spontaneous generation, they became the last to be liberated from Lamarckian theories. Yet, within a few more years, a very different kind of acquired inheritance was identified with the discovery of the transmissible drug resistance plasmids, and, later, population studies on bacteria became very rewarding.

With the development of bacterial genetics, it became possible to examine the mechanism of enzyme induction in a new way. The Institut Pasteur was the main center for this activity, and this research led in 1961 to the publication of the well-known Jacob–Monod operon hypothesis. This paper was given an enthusiastic welcome by all those interested in the regulation of enzyme synthesis, and the clarity and elegance of this paper have made it a classic (Jacob and Monod, 1961). The new generation of microbiologists, followed by the biochemists, learned to discuss phenotypic adaptation and the synthesis of enzymes in term of "transcription" of DNA into RNA and the "translation" of RNA sequences into amino acid sequences of proteins.

It turned out later that not all inducible enzymes were regulated in the same way as the β-galactosidase of *E. coli,* but because of the general acceptance of the negative control in this system by the regulator gene product, it proved difficult for a while to get acceptance for positive control systems, such as those for the arabinose operon (Beckwith and Rossow, 1974). Studies extended later to other metabolic pathways and to other bacterial species led to a more thorough understanding of the many ways in which bacterial enzyme synthesis may be controlled.

Molecular Biology

There have always been at least two reasons for scientists to carry out scientific investigations with bacteria. One is to find out more about fundamental problems in biology, and the other is to find out more about the bacteria themselves. Kluyver saw this dichotomy as "unity and diversity in microbial metabolism." Pasteur saw the general case of fermentation as a biological activity balanced against the properties of particular microorganisms. At most times throughout the period under review, it has been possible to see that research carried out with bacteria has had more general implications for biology. This is particularly so for the last two decades from 1960. As the result of intensive studies, mainly carried out with *E. coli* and coliphages, it became possible to define genes in terms of DNA and to discuss gene regulation in terms of interactions between proteins and specific DNA sequences. The details of the genetic code, relating the sequences of amino acids in a protein to the triplets of nucleotides in the corresponding DNA sequences, were worked out on the basis of experiments with microorganisms. As a result of advances in these areas during the last few years, we can take the levels of investigations as formulated by Marjory Stephenson in 1945 one stage further. Genetic analysis of bacteria that developed so rapidly in the 1960s has led to the analysis of bacteria at the molecular level. The implications of experiments with bacteria for other forms of life were very clear. At one point the unity of biochemistry was so self-evident that it was widely claimed that "if it is true for *E. coli,* it is true for an elephant."

Studies of structure and biochemical activities had revealed fundamental differences between prokaryotes and eukaryotes (Stanier, 1964), but it looked for a time as if there might be universality of the genetic code and of genetic structure and function. Because of the body of knowledge about the bacterial genetic systems, it was possible to test this hypothesis. The discovery of introns in the genes of higher organisms and some other molecular differences brought back the concept of the separateness of bacteria. Also, it appears that we still have much to learn about differences at the molecular level among the bacteria themselves. The identification of the class of Archaebacteria (Woese et al., 1978) and the demonstration of significant differences between them and the eubacteria underline that there is still much to learn about the molecular basis of microbial diversity.

Bacterial Growth

Studies on bacterial growth rates were at first confined to devising complex media on which reasonable yields of cells could be obtained. Later, attempts were made to identify the stages of growth. Mean generation times were measured at different growth stages and related to the previous history of the inoculum used. Buchanan (1918) summarized previous work on these lines and distinguished seven distinct growth phases of which only the lag phase, logarithmic phase, and stationary phase became widely used by subsequent experimenters. Investigation of growth rates was not included by Stephenson among the levels of study that she discussed. Her views on such work were set out very clearly (Stephenson, 1938): "In the problem of bacterial growth advances have been made along new lines. Happily this subject now attracts mathematicians and statisticians less than formerly but has passed into the hands of biochemists interested in problems of nutrition; this has led to results of both theoretical and practical importance." In taking this view she was thinking of the values of chemically defined media for identifying growth factor requirements and for providing precise information about nutritional characteristics. In 1947, she referred with approval to Monod (1942) whose publication *Recherches sur la Croissance des Cultures Bactériennes* gave precise and careful analyses of growth in defined media on different carbon sources. He examined the effects of differences in concentration of growth substrates, rate of aeration, and temperature on growth rates and total cell yields. This study was an important landmark in ideas about bacterial growth and had considerable influence on the work of later investigators. Monod himself went on, from asking about the factors affecting the rate of increase of numbers of bacteria, to asking questions about the rate of increase of a single bacterial enzyme. The value of this line of approach became evident with the unraveling of the mechanism of induction of β-galactosidase in *E. coli* and the formulation of the operon theory (Jacob and Monod, 1961), which led to so many advances in molecular biology.

The biochemists working with bacteria were content to follow the line of working with pure cultures in defined media (if possible) and to examine enzyme activities either of bacteria at the stationary phase or of samples removed during logarithmic growth. This approach yielded data that could be used to establish both catabolic and biosynthetic pathways and could also be used to analyze the processes of induction and repression of enzyme synthesis. As genetic techniques developed, these methods were used to compare mutant and wild-type organisms, with a corresponding increase in the information about the systems being studied.

Monod (1949) in the *Annual Review of Microbiology* said: "The study of the growth of bacterial cultures does not constitute a specialized subject or branch of research: it is the basic method of Microbiology. It would be a foolish enterprise, and doomed to failure, to attempt reviewing briefly a 'subject' which covers actually our whole discipline. Unless, of course, we consider the formal laws of growth for their own sake, an approach that has frequently proved sterile." He went on to say that he would consider the quantitative aspects of growth with respect to the study of bacterial physiology and biochemistry. He extended the mathematical analysis of data derived from growth experiments but rejected oversimplification. For example, "The kinetic speculations of Hinshelwood, although interesting as empirical formulations of the problem, do not throw any light on the nature of the basic mechanisms involved in the regulation of enzyme formation by the cells."

A new development was the appearance of two papers on continuous culture (Monod, 1950; Novick and Szilard, 1950). The most widely adopted system was the chemostat, in which the culture is fed continuously with a medium that is growth-limiting for one of the essential nutrients. Chemostat culture has been reviewed on many occasions and its advantages compared with batch culture, and in a few laboratories it has become the major research tool. Tempest (1970) discussed the research developments over twenty years, pointing out that with chemostat culture it is possible to vary the growth rate between wide limits and to maintain a constant chemical environment so that bacteria in a steady state of growth can be observed over long periods of time. Herbert (1961) said that at least three, and preferably five, generations of exponential growth were needed to obtain steady-state conditions, and that while this could be achieved with care in batch culture, the steady state could be maintained in chemostat culture for as long as required. Comparisons were made between the chemical composition of cells from chemostat cultures at different growth rates, and cultures that were carbon-limited for growth were compared with cultures that were nitrogen-limited or phosphate-limited. These methods were used to examine the variations in cell-wall polymers, ribosomes, and lipids of various bacteria. Tempest (1970) asked, "Why has this technique been largely ignored up to the present time? Why are so many microbiologists so inexperienced in its usage?" and came to the conclusion that the main reason was the lack of suitable chemostat vessels.

In addition to the chemostat investigations on chemical composition, a few studies had been made on the regulation of enzyme synthesis. Tempest (1970) reported an investigation on the isocitrate lyase of *Pseudomonas ovalis* and Hamlin et al. (1967) compared results obtained for the glucose enzymes of *Pseudomonas aeruginosa* in batch and chemostat culture. Clarke and Lilly (1969) discussed the regulation of amidase synthesis during growth in batch and chemostat culture and described the variations in enzyme levels at different growth rates for wild type and regulatory mutants of *Pseudomonas aeruginosa*. In the next ten years increasing use was made of chemostat culture for studies on enzyme regulation, but the technique supplemented, rather than replaced, the traditional batch growth method. Monod et al. (1952) introduced the concept of the differential rate of enzyme synthesis, plotting enzyme activity as a function of the increase in bacterial growth, and this was rapidly adopted by those studying enzyme adaptation. If nonsubstrate inducers were available, it was possible to obtain good kinetic data on enzyme synthesis with batch cultures in the exponential growth phase, and this remained the more widely used method for investigating the kinetics and regulation of enzyme synthesis.

A doubt that was occasionally raised about chemostat culture was the validity of the steady-state concept. The bacteria in a chemostat are maintained at submaximal growth rates, and this condition might be expected to exert strong selection pressure on the bacterial population. Tempest (1970) suggested that mutant selection rarely occurred in experiments of short duration (i.e., less than one month). Other reports indicated that mutants could appear in a much shorter time. Several investigators used chemostat culture as a tool for studying microbial evolution.

Novick and Szilard (1950), in one of the classical papers on this subject, discussed the use of the chemostat for selecting spontaneous mutants. Later, Horiuchi et al. (1962) isolated mutants of *E. coli* that were hyperconstitutive for the production of β-galactosidase, and continuous culture became one of the general methods for the isolation of constitutive mutants. Inderlied and Mortlock (1977) selected mutants of *Klebsiella aerogenes* that produced ribitol dehydrogenase activities tenfold higher than the wild type after about 48 generations in the chemostat. Like the mutants isolated by Rigby et al. (1974), these mutants appeared to have duplication of the ribitol genes. These experiments in population genetics, using chemostat growth, offer opportunities for mutant selection and for studying population shifts that could not be achieved by other methods. The classical approach to chemostat culture was based on steady-state growth of pure cultures, but the selection of mutants that are able to grow faster than the parent strain offers a very useful method for studies in experimental evolution.

Several investigators have examined plasmid stability in chemostat culture under selective and nonselective conditions and with different types of growth limitation. Plasmids may be entirely lost or part of a plasmid may be lost, leaving a variety of smaller plasmids, although this does not always appear to be directly related to growth rates (Godwin and Slater, 1979; Helling et al., 1981).

The first level of study of bacteria involved a mixed population, and, recently, ideas about the value of such work have turned full circle. A few laboratory studies were carried out with chemostat cultures containing two or more bacteria, and the factors affecting the interactions that might occur were discussed by Slater and Bull (1978). More interesting, and more relevant to interactions in the natural environment, have been the studies on mixed microbial communities. One of these was the seven-membered microbial community growing on the herbicide 2,2′-dichloropropionate obtained by Senior et al. (1976). Such communities are of very great value for detailed investigations of microbial interactions and may shed light on the way in which chemical compounds are degraded in nature.

PRESENT AND FUTURE

Many of the ways in which bacteria are used at the present time are likely to continue into the future. They will continue to be used as tools of molecular biologists, and the cloning of eukaryotic genes in bacteria may enable many problems of gene expression to be tackled that would be too complex in the original host. Biotechnology based on recombinant DNA is moving from the research stage to industrial exploitation, and this is bringing problems to be solved by biochemical engineers as well as microbiologists. New strains, obtained by mutant selection or constructed by genetic transfer, will continue to be in demand for all other applications of biotechnology. Immobilized enzymes are almost all obtained from microbial sources and will continue to be needed, although there is also interest in using immobilized cells for some purposes. Suggestions have been made for using bacteria for dealing with chemical pollution. It has been proposed that a strain constructed by the transfer of several catabolic plasmids into a single strain should be used to clean up oil spills, and some firms already offer bacterial kits for dealing with effluent problems. Mixed cultures are assuming increasing importance for biodegradation.

Ideas about bacteria have changed so greatly that now they have almost acquired the status of microscopic slaves of the lamp that will always do their masters' bidding. But those scientists who work with bacteria, as distinct from those who merely exploit them, will probably continue to echo the words of Robert Hooke in a letter to Leeuwenhoek in 1678: "The prospect of those small animals have given satisfaction to all Persons that have viewed them" (Dobell, 1932).

ACKNOWLEDGMENTS. I have attempted to convey an impression of the shifting ideas about bacteria by quoting from historical accounts written at various times, as well as from original sources. The translations of the New Sydenham Society of many of the papers of Koch and his contemporaries were particularly helpful. Another useful source was *Milestones in Microbiology,* which includes translations, with comments by Brock (1961), of many important papers on general microbiology. Bulloch's *History of Bacteriology,* published in 1938, is an invaluble account of all the major advances of the nineteenth century and has the advantage of being written before biochemistry and genetics had affected ways of thinking about bacteria. Several textbooks of microbiology have brief but interesting introductions; the second editions of Topley and Wilson (1936), Stephenson (1938, 1949), Thimann (1963), Senez (1968), and Stanier et al. (1957) contained the most interesting comments of those consulted. Recent and current research in microbiology has been given very cursory treatment in this chapter. The prevailing concepts about bacteria will be familar to the reader, even though there may not be universal agreement about what importance to assign to them. Detailed analysis of current research is given in other chapters. I am also grateful to many of my microbiological friends who very kindly helped me with advice and criticism.

REFERENCES

Avery, T., MacCleod, C. M., and McCarty, M., 1944, Studies on the chemical nature of the substances inducing transformation of pneumococcal types, *J. Exp. Med.* **79:**137–158.

Bastian, H. C., 1905, *Nature and Origin of Living Matter,* Fisher Unwin, London.
Beadle, G. W., 1974, Recollections, *Annu. Rev. Biochem.* **43:**1–13.
Beadle, G. W., and Tatum, E. L., 1941, Genetic control of biochemical reactions in *Neurospora, Proc. Natl. Acad. Sci. USA* **27:**499–506.
Beckwith, J., and Rossow, P., 1974, Analysis of genetic regulatory mechanisms, *Annu. Rev. Genetics* **8:**1–13.
Beijerinck, M. W., 1888, Die Bakterien der Papilionaceen-Knollchen, *Bot. Zeit.* **46:**725–804.
Beijerinck, M. W., 1901, Über oligonitrophile Mikroben, *Zentralbl. Bakteriol. (II)* **7:**561–582.
Bertrand, G. H., 1896, Preparation biochimique du sorbose, *C. R. Acad. Sci.* **122:**900–903.
Boutroux, L., 1880, Sur une fermentation nouvelle du glucose, *C. R. Acad. Sci.* **91:**236–238.
Brock, T. D., 1961, *Milestones in Microbiology,* Prentice-Hall, Englewood Cliffs, New Jersey.
Brown, A. J., 1886, The Chemical action of pure cultivations of *Bacterium Aceti, J. Chem. Soc. Trans.* **46:**172–187.
Buchanan, R. E., 1918, Life phases in a bacterial cell, *J. Inf. Dis.* **23:**109–125.
Bukhari, A. I., Shapiro, J. A., and Adhya, S. L. (eds.), 1977, *DNA Insertion Elements, Plasmids, and Episomes,* Cold Spring Harbor Laboratory, Cold Spring Harbor, New York.
Bull, A. T., and Slater, J. H., 1982, Historical perspectives on mixed cultures and microbial communities, in: *Microbial Interactions and Communities,* (A. T. Bull and J. H. Slater, eds), Academic Press, London.
Bulloch, W., 1938, *The History of Bacteriology,* Oxford University Press, London.
Clarke, P. H., 1976, Genes and enzymes, *FEBS Lett.* (Suppl.)**62:**E37–E46.
Clarke, P. H., 1982, The metabolic versatility of pseudomonads, *Antonie van Leeuwenhoek,* **48:**105–130.
Clarke, P. H., and Lilly, M. D., 1969, Enzyme synthesis during growth, in: *Microbial Growth, Symposium 19, Society for General Microbiology* (P. M. Meadow and S. J. Pirt, eds.), Cambridge University Press, London, pp. 113–159.
Cohn, F., 1872, Untersuchen über Bakterien, *Beitr. Biol. Pflanzen, Bd.* I, *Heft* 2, 126–224.
Cohn, F., 1875, Untersuchen über Backterien II, *Beitr. Biol. Pflanzen, Bd.* I, *Heft* 3, 141–207.
Cowan, S. T., 1962, The microbial species—a macromyth? in: *Microbial Classification,* Symposium 12, Society for General Microbiology (G. C. Ainsworth and P. H. A. Sneath, eds.), Cambridge University Press, London, pp. 433–455.
Darwin, C., 1859, *On The Origin of Species,* John Murray, London.
Dean, A. C. R., and Hinshelwood, C., 1966, *Growth, Function and Regulation in Bacterial Cells,* Clarendon Press, Oxford.
de Ley, J., 1962, Comparative biochemistry and enzymology in bacterial classification, in: *Microbial Classification,* Symposium 12, Society for General Microbiology (G. C. Ainsworth and P. H. A. Sneath, eds.), University Press, Cambridge, England, pp. 164–195.
Den Dooren de Jong, L. E., 1926, *Bijdrage tot de kennis van het mineralisatie process,* Nijgh and Van Ditmar, Rotterdam.
Dobell, C., 1932, *Antony van Leeuwenhoek and his 'Little Animals,'* Dover, New York.
Derry, T. K., and Williams, T. I., 1960, *A Short History of Technology,* Oxford University Press, London.
Drummond, J. C., 1957, *The Englishman's Food,* Jonathan Cape, London.
Dubos, R. J., 1945, *The Bacterial Cell,* Harvard University Press, Cambridge, Mass.
Elsden, S. R., and Pirie, N. W., 1949, Obituary notice—Marjory Stephenson, 1885–1948, *J. Gen. Microbiol.* **3:**329–339.
Engelmann, T. W., 1883, *Bacterium photometricum.* Ein Beitrag zur Verlichenden Physiologie des Licht-und Farbensinnes, *Pflügers Arch.* **30:**95.
Fildes, P., 1945, The biosynthesis of tryptophan by *Bact. typhosum, Br. J. Exp. Pathol.* **26:**416–428.
Fildes, P., Gladstone, G. P., and Knight, B. C. J. G., 1933, The nitrogen and vitamin requirements of *B. typhosus, Br. J. Exp. Pathol.* **14:**189–196.
Flügge, C., 1886, *Micro-organisms with special Reference to the Etiology of the Infective Diseases* (Translated from *Fermente and Mikroparasiten* by W. Watson Cheyne), Volume 132, The New Sydenham Society, London.
Freeman, R. B., 1982, *Darwin and Gower Street. Exhibition, 19 April 1982, Catalogue item 14,* University College London, London.

Fruton, J. S., and Simmonds, S., 1958, *General Biochemistry,* John Wiley, New York.
Gale, E. F., 1943, Factors influencing the enzymic activities of bacteria, *Bact. Rev.* **7:**139–173.
Gale, E. F., 1947, Nitrogen metabolism, *Annu. Rev. Microbiol.* **1:**141–158.
Gale, E. F., 1970, The development of microbiology, in: *The Chemistry of Life* (J. Needham, ed.), Cambridge University Press, London.
Garrod, A. E., 1909, *Inborn Errors of Metabolism,* Hodder and Stoughton, London.
Godwin, D., and Slater, J. H., 1979, The influence of the growth environment on the stability of a drug resistance plasmid in *Escherichia coli* K12, *J. Gen. Microbiol.* **111:**201–210.
Glover, S. W., and Hopwood, D. A. (eds.), 1981, *Genetics as a Tool in Microbiology,* Symposium 31, Society for General Microbiology, Cambridge University Press, London.
Griffith, F., 1928, The significance of pneumococcal types, *J. Hyg.* **27:**8–159.
Gutfreund, H., 1976, Wilhelm Friedrich Kühne; An appreciation, *FEBS Lett.* (Suppl.) **62:**E1–E2.
Hamlin, B. T., Ng, F. M.-W., and Dawes, E. A., 1967, Regulation of enzymes of glucose metabolism in *Pseudomonas aeruginosa* by citrate in: *Microbial Physiology and Continuous Culture.* (E. O. Powell, C. G. T. Evans, R. E. Strange, and D. W. Tempest, eds.), H. M. S. O., London, pp. 217–231.
Hayes, W., 1960, The bacterial chromosome, in: *Microbial Genetics* (W. Hayes and R. C. Clowes, eds.), Symposium 10, Society for General Microbiology, Cambridge University Press, London, pp. 12–38.
Helling, R. B., Kinney, T., and Adams, J., 1981, The maintenance of plasmid-containing organisms in populations of *Escherichia coli, J. Gen. Microbiol.* **123:**129–141.
Herbert, D., 1961, The chemical composition of microorganisms as a function of their environment, in: *Microbial Reaction to Environment,* Symposium 11, Society for General Microbiology (G. G. Meynell and H. Gooder, eds.), Cambridge University Press, London, pp. 391–416.
Horiuchi, T., Tomizawa, J. I., and Novick, A., 1962, Isolation and properties of bacteria capable of high rates of β-galactosidase synthesis, *Biochim. Biophys. Acta* **55:**152–163.
Howard, D. H., 1982, Friedlich Loeffler and his history of bacteriology, *A. S. M. News* **48:**297–302.
Howard, D. H. (Trans.), 1983, *Loeffler's History of Bacteriology, 1887,* Coronado Press, Lawrence, Kansas.
Inderlied, C. B., and Mortlock, R. P., 1977, Growth of *Klebsiella aerogenes on xylitol: Implications for bacterial enzyme evolution, J. Mol. Evol.* **9:**181–190.
Jacob, F., and Monod, J., 1961, Genetic regulatory mechanisms in the synthesis of proteins, *J. Mol. Biol.* **3:**318–356.
Kamp, A. F., La Rivière, J. W. M., and Verhoeven, W. (eds.), 1957, *A. J. Kluyver: His Life and Work,* North-Holland, Amsterdam.
Kaplan, R. W., 1952, *Genetics of Microorganisms, Annu. Rev. Microbiol.* **6:**49–76.
Karrer, P., 1938, *Organic Chemistry,* Elsevier, Amsterdam.
Karstrom, H., 1930, Über die Enzymbildung in Bakterien, Thesis, Helsingfors, Finland.
Kluyver, A. J., 1957, Microbiology and industry, in: *A. J. Kluyver: His Life and Work,* (A. F. Kamp, J. W. M., La Rivière, and W. Verhoeven, eds.). North-Holland, Amsterdam, pp. 165–185.
Kluyver, A. J., 1957, Unity and diversity in the metabolism of micro-organisms, in: *A. J. Kluyver: His Life and Work,* (A. F. Kamp, J. W. M. La Rivière, and W. Verhoeven, eds.), North-Holland, Amsterdam, pp. 186–210.
Kluyver, A. J., 1931, *The Chemical Activities of Micro-organisms,* University of London, London.
Kluyver, A. J., 1952, Microbial metabolism and its industrial implications, *Chemistry and Industry,* 136–145.
Kluyver, A. J., and van Niel, C. B., 1936, Prospects for a natural system of classification of bacteria, *Zentr. Bakt. Parasitenk Abt.* II, **94:**369–403.
Koch, R., 1881, On the investigation of pathogenic organisms (Translation by V. Horsley), in: *Bacteria in Relation to Disease* (W. Watson Cheyne, ed.), Vol. 151, New Sydenham Society, London, pp. 1–64.
Koch, R., 1884, The etiology of tuberculosis (Translation by S. Boyd), in: *Bacteria in Relation to Disease* (W. Watson Cheyne, ed.), Vol. 151, New Sydenham Society, London, Vol. 151, pp. 67–201.

Kühne, W., 1876, Über das Verhalten verschiedener organisirter und sog. ungeformter Fermente, Separat-Abdruck aus den Verhandlungen des Heidelb. Naturhist.-Med. Vereins. N. S.13. [Reprinted 1976, *FEBS Lett.* (Suppl.) **62:**E3–E12].

Lankester, E. R., 1873, On a peach-coloured bacterium—*Bacterium rubescens, N. S. Quart. J. Microscop. Sci.* **13:**27–40.

Leeuwenhoek, A. van, 1677, Observations, communicated to the publisher by Mr. Antony van Leeuwenhoek, in a Dutch letter of the 9th of October, 1676, here English'd: Concerning little Animals by him observed in Rain—Well—Sea—and Snow water; as also in water wherein pepper had lain infused, *Phil. Trans. Roy. Soc. London* **12:**821–831.

Loeffler, F., 1887, *Vorlesungen über die geschichtliche Entwickelung der Lehre von den Bacterium,* Vogel, Leipzig.

Locke, J., 1689, *An Essay Concerning Human Understanding,* Book III, Ch. VI.

Monod, J., 1942, *Recherches sur la Croissances des Cultures Bactériennes,* Hermann et Cie, Paris.

Monod, J., 1949, The growth of bacterial cultures, *Annu. Rev. Microbiol.* **3:**371–394.

Monod, J., 1950, La technique de culture continue, théorie et applications, *Annal. Inst. Pasteur* **79:**390–410.

Monod, J., and Cohn, M., 1952, La biosynthèse induite des enzymes (adaptation enzymatique), *Adv. Enzymol.* **13:**67–119.

Monod, J., Pappenheimer, A. M., and Cohen-Bazire, G., 1952, La cinétique de la biosynthese de la β-galactosidase chez *E. coli* considérée comme fonction de la croissance, *Biochem. Biophys. Acta* **9:**648–660.

Novick, A., and Szilard, L., 1950, Experiments with the chemostat on spontaneous mutations of bacteria, *Proc. Natl. Acad. Sci. USA* **36:**708–719.

Pasteur, L., 1857, Mémoire sur la fermentation appelée lactique, *C. R. Acad. Sci.* **45:**913–916.

Pasteur, L., 1860, Experiences relatives aux générations spontanees, *Compt. Rend. Acad. Sci.* **50:**303–675.

Pasteur, L., 1861, Animalcules infusoires vivant sans gaz oxygenè libre et déterminant des fermentations, *C. R. Acad. Sci.* **52:**344–347.

Pasteur, L., 1862, Études sur les mycodermes. Role de ces plantes dans la fermentation acétique, *Compt. Rend. Acad. Sci.* **54:**265–272.

Pelouze, J., 1852, Sur une nouvelle matière sucrée extraite des baies de sorbier, *Ann. Chim. Phys.* **35:**222–235.

Penn, M., and Dworkin, M., 1976, Robert Koch and two visions of microbiology, *Bact. Rev.* **40:**276–283.

Portugal, F. H., and Cohen, J. S., 1977, *A Century of DNA,* MIT Press, Cambridge, Mass.

Rigby, P. W. J., Burleigh, B. D., and Hartley, B. S., 1974, Gene duplication in experimental enzyme evolution, *Nature (London)* **251:**200–204.

Sédillot, C., 1878, De l'infuence des découv̂ertes de M. Pasteur sur les progres de la chirurgie, *C. R. Acad. Sci.* **86:**634–640.

Senez, J. C., 1968, *Microbiologie Générale, Deren, Paris.*

Senior, E., Bull, A. T., and Slater, J. H., 1976, Enzyme evolution in a microbial community growing on the herbicide Dalapon, *Nature (London)* **263:**476–479.

Singer, C., 1959, *A Short History of Scientific Ideas to 1900,* Oxford University Press, London.

Slater, J. H., and Bull, A. T., 1978, Interactions between microbial populations, in: *Companion to Microbiology* (A. T. Bull and P. M. Meadow, eds.), Longman, London, pp. 181–206.

Smith, T., 1932, Koch's views on the stability of species among bacteria, *Ann. Med. Hist.* **4:**524–530.

Sneath, P. H. A., 1962, The construction of taxonomic groups, in: *Microbial Classification,* Symposium 12, Society for General Microbiology (G. G. Ainsworth and P. H. A. Sneath, eds.), Cambridge University Press, London, pp. 289–332.

Stanier, R. Y., 1951, Enzymatic adaptation in bacteria, *Annu. Rev. Microbiol.* **5:**35–56.

Stanier, R. Y., 1964, Towards a definition of the bacteria, in: *The Bacteria,* Volume 5 (I. C. Gunsalus and R. Y. Stanier, eds.), Academic Press, New York, pp. 445–464.

Stanier, R. Y., and van Niel, C. B., 1941, The main outlines of bacterial classification, *J. Bacteriol.* **42:**437–466.

Stanier, R. Y., Doudoroff, M., and Adelberg, E. A., 1957, *General Microbiology,* 3rd ed., Prentice-Hall, Englewood Cliffs, New Jersey.
Stephenson, M., 1938, *Bacterial Metabolism,* 2nd ed., Longmans Green, London.
Stephenson, M., 1949, *Bacterial Metabolism, 3rd ed., Longmans Green, London.*
Stephenson, M., and Strickland, L. H., 1933, Hydrogenlyases. Further experiments on the formation of formic hydrogenlyase by *Bact. coli, Biochem. J.* **27:**1528–1532.
Stodola, F. H., 1957, *Chemical Transformations by Microorganisms,* John Wiley, New York.
Tatum, E. L., 1945, X-Ray induced mutant strains of *Escherichia coli, Proc. Natl. Acad. Sci. USA* **31:**215–219.
Tatum, E. L., and Bonner, D., 1944, Indole and serine in the biosynthesis and breakdown of tryptophane, *Proc. Natl. Acad. Sci. USA* **30:**30–37.
Tatum, E. L., and Lederberg, J., 1947, Gene recombination in the bacterium *Escherichia coli, J. Bacteriol.* **53:**673–684.
Tempest, D. W., 1970, Continuous culture in microbiological research, *Adv. Microb. Physiol.* **4:**223–250.
Thimann, K. V., 1963, *The Life of Bacteria,* 2nd ed., MacMillan, New York.
Topley, W. W. C., and Wilson, G. S., 1936, *The Principles of Bacteriology and Immunity,* 2nd ed., Edward Arnold, London.
Tyndall, J., 1881, *Floating Matter of the Air in Relation to Putrefaction and Infection,* 2nd ed., Longmans Green, London.
van Niel, C. B., 1941, The bacterial photosyntheses and their importance for the general problem of photosynthesis, *Adv. Enzymol.* **1:**263–328.
van Niel, C. B., 1957, Kluyver's contributions to microbiology and biochemistry, in: *A. J. Kluyver, His Life and Work* (A. F. Kamp, J. W. M. La Rivière, and W. Verhoeven, eds.), North-Holland, Amsterdam, pp. 68–155.
Ward, H. M., 1892, The ginger-beer plant and the organisms composing it: A contribution to the study of fermentation-yeasts and bacteria. *Phil. Trans. Roy. Soc. London B.* **183:**125–197.
Watson, J. D., and Crick, F. H. C., 1953, A structure of deoxyribosenucleic acid, *Nature (London)* **171:**737–738.
Winogradsky, S., 1889, Recherches physiologiques sur les sulfobactéries. *Ann. Inst. Past.* **3:**49–60.
Winogradsky, S., 1890, Sur les organismes de la nitrification, *Compt. Rend. Acad. Sci.* **110:**1013–1016.
Winogradsky, S., 1894, Sur l'assimilation de l'azote gazeux de l'atmosphere par les microbes, *Compt. Rend. Acad. Sci.* **118:**353–355.
Woese, C. R., Magrum, L. J., and Fox, G. E., 1978, Archaebacteria, *J. Mol. Evol.* **11:**245–252.
Woods, D. D., 1947, Bacterial metabolism, *Annu. Rev. Microbiol.* **1:**115–140.

2

ANAEROBIC BIOTRANSFORMATIONS OF ORGANIC MATTER

Robert E. Hungate

INTRODUCTION

Transformations involve both matter and energy, energy being the capacity to do work, i.e., to exert force to cause motion. The kind and magnitude of the energy transformation depend on the materials moved. The six different categories of energy—nuclear, chemical, mechanical, radiant, electrical, and heat—are interconvertible, as shown in Figure 1, but whereas the first five forms of energy ("high" forms) can be converted completely to heat (random motion), the latter can be only partially converted to the others, the more completely the higher the temperature. Work can be defined also as the extent to which energy is converted to a high form. To the extent that high forms of energy are converted to heat, there is a degradation of energy; also, heat energy degrades as the difference diminishes.

ABIOTIC TRANSFORMATIONS

To obtain a perspective on biotransformations, it is helpful to review Earth's abiotic transformations. Cosmologists postulate that the matter of the universe originated within a small space as a product of an enormously powerful explosion that raised the temperature to millions of degrees centigrade. Immediate expansion of the created matter was accompanied by cooling during which protons, electrons, and neutrons formed spontaneously within a few minutes; the atoms within an hour, and molecules during further ex-

Robert E. Hungate • Department of Bacteriology, University of California at Davis, Davis, California 95616.

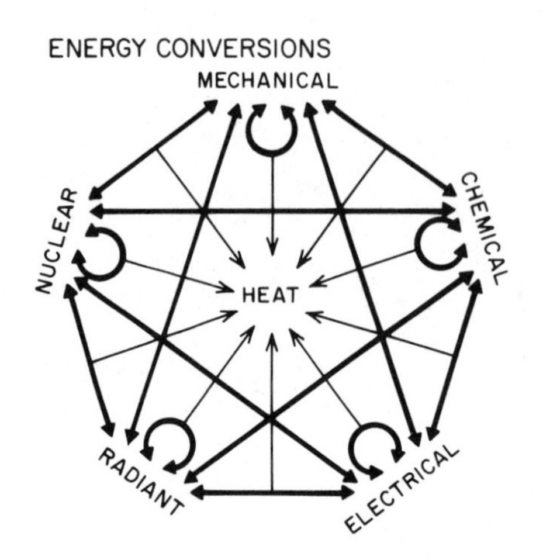

FIGURE 1. Energy diagram to show the energy conversions in which work is done (wide lines) and that degraded (narrow lines).

pansion and cooling. Heterogeneities appeared, with greater concentrations of matter in some regions, and in these a random profusion of all sorts of molecules appeared spontaneously. The magnitude of the energy transformations decreased as the temperature dropped (Figure 2).

Synthesis

The important point is that in the process of partial degradation of energy from the initial state at very high temperature, there was *synthesis* of matter, ranging from very small elements through atoms to small molecules. According to these views, synthesis of material is inherent in spontaneous reactions involving an overall degradation of energy. Some products are at a lower

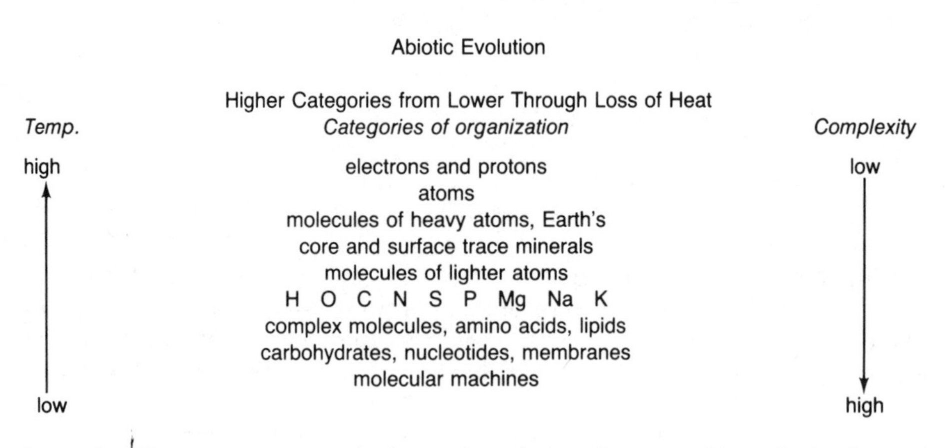

FIGURE 2. Diagram to represent the inorganic evolution of matter, with synthesis of more complex forms of matter with falling temperature.

energy level; others at a higher energy level than the individual units of the starting material (Kluyver and Donker, 1926).

Synthesis continued with formation of larger molecules. Organic compounds were among those formed, including innumerable kinds of amino acids, nucleotides, lipids, carbohydrates, and organic phosphates, from which even more complex aggregates spontaneously arose, again with an overall degradation of energy but with an ever increasing complexity of order in part of the material. Continuation of transformations and syntheses was aided by a continuous inflex of high energy in the form of incident light. Molecular systems for inheritance, protein (enzyme) formation, and growth appeared, accompanied by ever increasing complexity in at least part of the material. The configuration of DNA must have been selected very early, together with a template for preserving through replication valuable configurations once they had appeared. Selection was for both DNA self-replication and for transmission of information for protein synthesis to RNA.

Evolution of Metabolism

Growth

Initially, growth may have involved assembly of simple, already formed molecules in processes akin to crystallization, but as these molecules became less abundant, a premium was placed on ability to synthesize the essential components of protoplasm from ever simpler molecules. For these syntheses energy was needed. It was available in chemical form and as light, but light energy, though appropriate for rapid transmission, must be stored chemically. Chemical reactions are the form of energy appropriate for the synthesis of protoplasm. The reactions of metabolism have been selected to do the chemical work of synthesis of living material.

Carbon Compounds and Chemical Work: Why Carbohydrates?

Compounds of carbon, hydrogens, and oxygen were relatively abundant near Earth's surface, but no O_2 had formed. The supply of oxygen atoms was insufficient to convert all elements into their oxidized form, and compounds of carbon at intermediate states of oxidation must have been abundant. These are peculiarly fitted to react in the absence of O_2 (Hungate, 1955). A mixture of carbon, hydrogen, and oxygen atoms is at its lowest energy state anaerobically when all the carbon atoms are in the form either of CO_2 or of CH_4. Energy is available in the transformation of incompletely oxidized carbon into methane and carbon dioxide, i.e., in a simultaneous complete oxidation of some carbon atoms and complete reduction of others.

The exact intermediate state of carbon oxidation is uniquely fitted for performance of anaerobic chemical work (Table I): first, by the energy available in reactions leading toward the lowest energy state (CO_2 and CH_4), and,

TABLE I
Why Carbohydrates?

$4\ HCOOH \rightarrow 3\ CO_2 + CH_4 + 2\ H_2O$	$\Delta G^{\circ\prime} = -120$ kJ
184 D	0.65 kJ/D
$4\ CH_3OH \rightarrow 3\ CH_4 + CO_2 + 2\ H_2O$	$\Delta G^{\circ\prime} = -311$ kJ
128 D	2.43 kJ/D
$2\ CH_2O \rightarrow CH_4 + CO_2$	$\Delta G^{\circ\prime} = -176$ kJ
60 D	2.93 kJ/D
$n\ CH_2O \rightarrow (CH_2O)_n$	
soluble → insoluble, no concentration effect	

second, by its capacity to do work in those same reactions. Those reactions include synthesizing ATP, forming larger and more complex carbon compounds through C–C bonding, and maintaining reactivity through retention of intermediate states of oxidation in many of the carbon atoms while linking some to atoms of nitrogen, sulfur, phosphorus, and other elements, including many trace elements, each peculiarly suited as a component in one or more kinds of molecular reactants synthesizing the living material.

This suitability of carbohydrate for the performance of chemical work under anaerobic conditions and for providing the carbon skeletons for amino acids was in part the basis for its selection. Additionally, carbohydrates were well suited for energy storage; insoluble aggregates, such as starch and glycogen, could be accumulated without affecting osmotic pressure. Further, cellulose was strong enough to serve as skeletal material. It could have been selected initially as a wall material surrounding the aquatic cell, of sufficient strength to withstand the osmotic pressure created by an external semipermeable membrane protecting the integrity of the contained protoplasm.

The advent of photosynthesis increased the energy available, but carbohydrates remained important as material for cell synthesis and energy storage. Initially, photosynthesis may have used reduced sulfur compounds and various types of organic matter to supply the hydrogen for the reduction of carbon (van Niel, 1931). Later, when water supplied the hydrogen for photosynthesis, carbohydrate became as cheap as light, air, and water, and could be manufactured in large quantities at little expense to the cell. Cellulose, evolved as the supporting wall of aquatic cells, later became an important component of land plants, supporting the leaves to obtain light and carbon dioxide, linking them to the roots supplying water and other nutrients, and strengthening them to resist collapse under the high negative pressures of water created by transpiration. Plant cell walls are today the quantitatively most important renewable chemical energy source on Earth.

THE ADVENT OF DIOXYGEN

Photosynthetic accumulation of O_2 greatly increased the chemical work obtainable from carbohydrate and other foods. Larger organisms with greater energy needs could be supported, and new evolutionary innovations could arise. However, almost all synthetic reactions were anaerobic; the vast majority of the kinds of chemical reactions in aerobes do not involve O_2. This poses the very interesting question, How is the organism able to carry on simultaneously both aerobic and anaerobic reactions? It was relatively easy for many microbes to develop metabolic machinery allowing them to grow either in the complete absence of O_2, or in its presence. Others did not evolve this capacity. Fortunately for them, O_2 is relatively insoluble in water, and many anaerobic habitats have existed throughout organic evolution. In these and in some produced in human technology, organisms incapable of using O_2 survive.

HUMAN APPRECIATION OF ANAEROBIOSIS

Human appreciation of anaerobic life is relatively recent. Primitive humans practiced anaerobic techniques in wine and cheese production and in the leavening of bread. More recently it was found that combustible gas could arise from within the earth, as was demonstrated in 1767 when Volta ignited marsh gas. Spontaneous fires started by phosphine combustion of methane, if observed, nourished the idea of supernatural powers, as might also the spontaneous combustion of hay stored when slightly moist, becoming anaerobic, and later, when dry, exposed to the air. But identification of oxygen was really prerequisite to appreciation of its lack, and anaerobiosis was not understood until almost the nineteenth century.

Oxygen was discovered by Lavoisier and Priestley in 1790. Oxygen is the name of the element, but it has been commonly used also for the name of the molecule. The same applies to nitrogen and hydrogen, but for clarity dinitrogen and dihydrogen have been used to designate the molecule, and dioxygen is similarly useful in instances in which it could be confused with the atom. Similarly, in the interest of precision it is sometimes advantageous to employ the terms dioxic and adioxic instead of aerobic and anaerobic.

Carbon dioxide was soon shown to be a product of respiration, and dioxygen to be essential for many forms of life, including humans. But chemists were discovering also that carbon dioxide arose in the absence of air. Gay-Lussac (1810) showed that the carbon dioxide produced in the vinous fermentation was 120-fold greater than the total amount of dioxygen in the system, and that the dioxygen was completely metabolized prior to the most active fermentation. This work was the basis for the later formulation of the equation, $C_6H_{12}O_6 \rightarrow 2\ CO_2 + 2\ CH_2OH$.

Production of CO_2 and acids anaerobically in the rumen was described by Sprengel (1832) in his *Chemie für Landwirte, Forstamänner und Cameralisten.* He named the process "Selbstentmischung" (self-unraveling), recognizing the lack of participation of dioxygen in the process. Acetic and butyric acids were identified as products of the rumen fermentation even at that early time. In 1843, Gruby and Delafond reported the myriads of protozoa growing anaerobically in the rumen.

PASTEUR'S GREAT CONTRIBUTIONS

It was chiefly Pasteur (1858) who alerted the world to the actions of microorganisms in the absence of dioxygen. His phrase, "fermentation, c'est la vie sans air," widely publicized anaerobic microbial metabolism. Pasteur described the formation of lactic acid, ethanol, and butyric acid, and identified their specific causal microbial agents. Fermentation has become the almost universal term to describe these and similar processes occurring in the absence of O_2. Attention initially centered on the wastes of the fermentation, but Pasteur emphasized that the process was essential to microbial growth.

It was appropriate that Pasteur should have been the one to recognize the great difference in metabolism of yeast in the presence and in the absence of O_2, a difference called for many years the "Pasteur effect." In the absence of O_2, the yeast used sugar more rapidly and produced ethanol and CO_2. Subsequently it has been found that in O_2, the sugar is used only about two-thirds as rapidly and little ethanol appears, though complete suppression of its formation required extreme aeration. A greater quantity of yeast is formed per unit of sugar used aerobically.

Pasteur recognized also that putrefaction (fermentation of protein) occurred in the alimentary tract, and "resolved to devote his attention to its study no matter how disgusting" (English translation), a resolve never implemented. Different terms were applied to the fermentation of protein and carbohydrates because of the marked differences in odor, carbohydrates typically forming products acceptable to the olfactory sense, but putrefaction being most disagreeable in odor though not necessarily in taste (limburger cheese).

DEVELOPMENT OF BIOCHEMISTRY IN RELATION TO ANAEROBIC TRANSFORMATIONS

Buchner (1897) showed that a limited alcoholic fermentation could occur in cell-free extracts of yeast, and a short time later Harden and Young (1905) implicated phosphates in the process. Rapidly accelerating research followed, chiefly on the mechanism of ethanol formation by yeast and lactic acid formation by muscle (glycolysis). As these two fermentations were analyzed, steps common to the two processes were identified, and the Embden–Meyerhof–Parnas or EMP scheme for breakdown of carbohydrate was formulated. It is the most common pathway for conversion of sugars under both aerobic and anaerobic conditions, and pyruvic acid is the key intermediate. Less common are the Entner–Doudoroff and the phosphoketolase pathways, together with minor variations in various species.

Waste products of sugar fermentation may be CO_2, H_2, CH_4, ethanol, and formic, acetic, propionic, butyric, lactic, and succinic acids, together with many other less common products such as glycerol, 2,3-butylene glycol, butanol, acetone, and many other compounds, depending on the nature of the substrate and of the microbes present.

Hydration and Dehydrogenation

Concomitant with the sorting out of the EMP scheme, many other reactions were under study. There were two hypotheses to explain the oxidation of sugars to the waste products shown in the equation,*

$$C_6H_{12}O_6 + 6\ O_2 \rightarrow 6\ CO_2 + 6\ H_2O,\ \Delta G_o' = -2880\ kJ \qquad (1)$$

In one, espoused by Warburg (1924), O_2 combined with the substrate carbon to form CO_2. In the other, developed by Wieland (1913), oxidation was accomplished by hydration and dehydrogenation, i.e., the oxygen in the CO_2 came from water. As is so often the case in vigorous disagreements between scientists, both were right, Warburg in his claim that dioxygen combines with the hydrogen, and Wieland in his thesis that the actual mechanism for addition of oxygen to the carbon is by hydration.

Wieland's scheme for the oxidation of glucose can be written in two steps,

$$C_6H_{12}O_6 + 6\ H_2O \rightarrow 6\ CO_2 + 24\ H,\ \Delta G_o' = 24\ kJ \qquad (2)$$

$$24\ H + 6\ O_2 \rightarrow 12\ H_2O,\ \Delta G_o' = -2904\ kJ \qquad (3)$$

involving hydration, dehydrogenation and decarboxylation, and reduction of O_2 with the H, the two reactions adding up to Eq. 1. This distinction is important for our consideration of adioxic transformations because it shows that oxidation is in part an adioxic process even in aerobes. Note that dehydrogenation takes only a small amount of energy.

TYPES OF FERMENTATION OF CARBOHYDRATES

One of the early problems in the metabolism of anaerobes was to determine the types of fermentation products formed. Many of the enzymes concerned were identified by means of Thunberg (1930) tubes in which cells or cell extracts were held in a closed tube with a side arm containing a test substrate that would be added together with methylene blue or other dye that changed color when hydrogen was transferred to it from the test substrate. This was followed by the Warburg respirometer in which reactions were arranged in a fashion such that a gas was liberated or consumed, and the pressure change could be read on an attached manometer. Acid production could be measured by including bicarbonate in the medium, and calibrating to determine the amount of gas released by a given amount of each acid.

During the Warburg respirometer epoch, it was generally believed that

* The free energies are obtained from Thauer et al. (1977).

the energy available anaerobically was so small that growth was negligible, and it was assumed that the quantities of waste products should agree within narrow limits with the quantity of substrate fermented. Many a student seeking to obtain a satisfactory grade in a fermentation laboratory course must have tested his conscience severely in obtaining satisfactory balances. But through all this interest, the pathways of many of the more important types of anaerobic carbohydrate conversion patterns were firmly established (Wood, 1961).

Lactacidigenesis

The homolactic souring of milk was one of the fermentations studied by Pasteur. It is characteristic of many streptococci and lactobacilli and of a number of other organisms when the medium lacks sufficient iron. The homolactic and the homoacetic fermentations are unique in having the proportions of carbon, hydrogen, and oxygen in the single product the same as in carbohydrate. In the internal oxidation-reduction to form lactate, carbons 1 and 6 of the hexose are reduced to the methyl group and the 3 and 4 carbons become the carboxyl. This was the fermentation used by Bauchop and Elsden (1960) to show that the two ATP formed from glucose corresponded to the production of approximately 10.5 mg of cells (dry weight) produced per mole of substrate fermented, a value approximated in many experiments with other types of fermentations in which knowledge of biochemistry is sufficient to permit reasonably accurate assessment of the substrate reactions yielding ATP. However, the actual cell yield varies also with growth rate, the extent of chemiosmotic work, and nutritional factors.

The milk habitat provides abundant protein and other organic nutrients. The lactacidigenic bacteria usually require multiple amino acids and other organic nutrients in the culture medium, and, when these are supplied, the bacteria develop rapidly. They withstand acidity better than do many anaerobes, and the pH drops to around 4.5 before fermentation is inhibited. Although the streptococci cannot grow at that acidity, the lactobacilli can, and often do if sugar and protein are still available. About three fourths of the lactose in milk is unfermented when streptococcal growth ceases.

Ethanolic Fermentation

This fermentation is well known as the basis for the alcoholic beverage industry, and resembles the lactacidigenic fermentation in its occurrence at rather acid pH's. This reduces the competition from other microbes, and through manipulation of environmental factors an ethanolic fermentation relatively free of other types can be obtained without preliminary sterilization. A high percentage of sugar, almost 40% (w/v), can be fermented before carbon dioxide becomes inhibitory. This is possible because the ethanol is a neutral compound and the CO_2 is not only a very weak acid but also can escape from the fermenting mixture. Two ATP are formed also in this fermentation.

Whereas in the lactacidigenic fermentation, the two hydrogen atoms removed from 3-phosphoglyceraldehyde to NAD are returned to carbon 2 of the pyruvate, forming lactate, in the ethanolic fermentation the pyruvate is decarboxylated and the hydrogen then reduces the acetaldehyde of acetyl-CoA to ethanol (Figure 3).

Ethanol is a minor product of many fermentations by pure cultures of bacteria, but it is usually not predominant except in a few bacterial species. One of these is *Zymomonas mobilis,* another is *Thermoanaerobacter ethanolicus* (Wiegel and Ljungdahl, 1981); both can ferment sufficiently high concentrations of sugar to make their use feasible in industrial ethanol production. The 78°C maximum temperature of the latter makes tantalizing the possibility of recovering the ethanol, at least in part, by stripping it during fermentation.

It seems likely that an acidic environment selects for fermentations producing neutral products. Undissociated acids often penetrate cells more readily than do salts and could be toxic. *Sarcina ventriculi* has an almost completely ethanolic type of fermentation at very acid pH's, but produces less ethanol near neutrality (Bauchop, 1960).

Acetigenic Fermentation

In this fermentation, pyruvate is split to form acetyl-CoA, carbon dioxide and 2 H that may remain with the carbon as formate or can be released as H_2. In the rumen fermentation, 82% of the pyruvate converted to acetate was estimated to yield H_2 and CO_2, with the remaining 18% producing formate which in turn was rapidly split into H_2 and CO_2 (Hungate et al., 1970).

The acetyl-CoA can form an ATP via a phosphotransacetylase reaction,

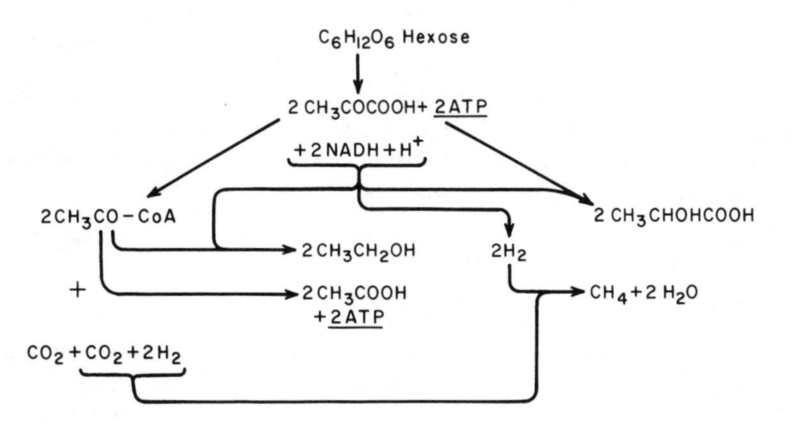

FIGURE 3. Eclectic diagram to show the ATP formed in the lactacidigenic and acetigenic fermentations.

and if each pyruvate from hexose produces acetate, a total of 4 ATP per hexose can be obtained (Figure 3), enabling the cell to accomplish more growth than is possible in the lactacidigenic and ethanolic fermentations. A great many anaerobes form at least some acetate in pure culture, and an anaerobe, strain N_2C_3, isolated from the rumen of a starving Zebu steer in Kenya (Margherita and Hungate, 1963), produced almost solely acetate, H_2, and CO_2 according to the equation,

$$C_6H_{12}O_6 + 2\ H_2O \rightarrow 2\ CH_3COOH + 2\ CO_2 + 4\ H_2 \quad (4)$$

A relatively small concentration of cellulose was fermented by this strain, possibly because the H_2 inhibited the fermentation. This is not a problem in the rumen because, as will be discussed later, the H_2 is used in producing methane from CO_2 or reducing many other compounds.

Also, in other natural, uncontrolled fermentations, in contrast to controlled fermentations in which the population is known, hydrogenotrophic bacteria maintain a low concentration of H_2, or diffusion removes it, and acetate constitutes the principal acid formed.

A fermentation in which acetic acid is the only product of carbohydrate breakdown was found in *Clostridium thermoaceticum* (Fontaine et al., 1942), with hexose yielding three acetate and very little of any other products. This fermentation involves formation from triose of two acetates (via pyruvate), two CO_2, and four H_2, with the CO_2 and H_2 used to form a third acetate. A similar reaction was found by Barker et al. (1945) in *Eubacterium limosum* (*Butyribacterium rettgeri*). Recently, evidence has been obtained (Odelson and Breznak, 1983) that a similar fermentation may characterize the agnotobiotic population in certain wood-eating termites dependent on their gut microbiota to digest and ferment cellulose.

Several species of bacteria are able to grow solely on carbon dioxide and H_2, converting them to acetate according to the equation,

$$2\ HCO_3^- + 2\ H^+ + 4\ H_2 \rightarrow CH_3COO^- + H^+ + 4\ H_2O,\ \Delta G_o' = -104.6\ kJ \quad (5)$$

These include *Clostridium aceticum* (Wieringa, 1940), *Clostridium thermoautotrophicum* (Wiegel et al., 1981), *Acetobacterium woodii* (Balch et al., 1977), and *Eubacterium limosum* (Genthner and Bryant, 1982). Acetic acid is probably the most common organic product in natural fermentations involving plant cell wall structures.

Succinigenic and Propionigenic Fermentations

Succinigenic fermentation is a common feature of the metabolism of many rumen microbes, e.g., *Bacteroides, Succinimonas, Succinivibrio, Ruminococcus flavefaciens, Anaerovibrio, Selenomonas,* and *Treponema bryantii.* Propionic acid is an important fermentation product of species of *Propionibacterium, Veillonella,* and some *Selenomonas* strains growing on lactate (Bryant et al., 1956). In all these species, the pathway to product formation includes a step in which carbon dioxide is combined with pyruvate to form a 4-carbon dicarboxylic acid, which is then reduced to the succinate stage and either released as such or decarboxylated to form propionate (Figure 4). The reduction of fumarate to succinate with NADH is sufficiently exergonic for the production of an ATP, accomplished by membrane phosphorylation (Macy et al., 1975).

In *Propionibacterium,* the carboxyl group that condenses with pyruvate to form oxaloacetate is transferred directly from succinyl-CoA to the pyruvate in a closed reaction in which the 1-C moiety does not equilibrate with CO_2. The net effect is conservation of the carboxyl without danger of loss by diffusion as CO_2. This is of adaptive value in a habitat containing little carbon dioxide, as in cheese. In the rumen, which normally is in contact with an atmosphere of about 70% carbon dioxide, there is no necessity for its conservation, and ambient CO_2 can be used to carboxylate pyruvate, presumably in a reductive carboxylation to form malate, a reaction not requiring ATP. Rumen strains producing succinate show a significantly large uptake of carbon dioxide, because more succinate than acetate is produced.

In fermentations involving the 4-C dicarboxylic acids, 4 ATP can be generated per hexose fermented, a yield equal to that generated when acetate is formed. These fermentations are not quite equivalent to the acetigenic fermentation, because, in the latter, the H_2 and CO_2 generated along with the acetate can be further converted to acetate or methane, with genesis of additional ATP. Unless H_2 is available in the environment of propionigenic cells they produce acetate as a means of obtaining the hydrogen needed to render propionic acid two hydrogens more reduced than the carbohydrate being fermented. The balance between propionate and acetate production can be influenced by the availability of reducing power in the milieu. When meth-

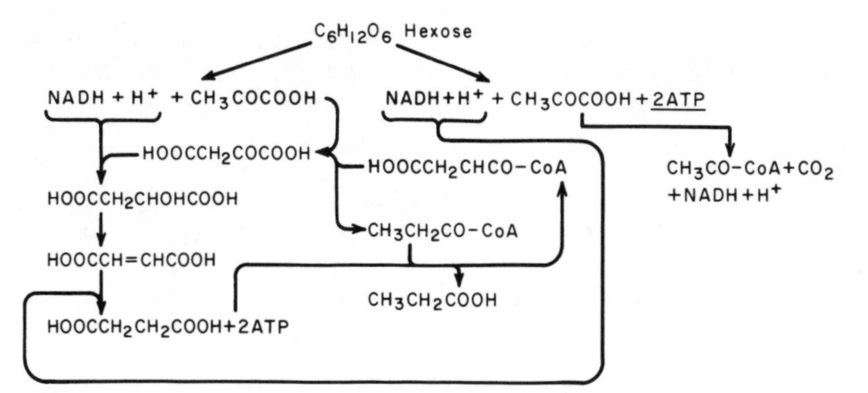

FIGURE 4. Eclectic diagram to show the ATP formed in the propionigenic fermentation.

anogenesis in the rumen is inhibited by methane analogs (Bauchop, 1967), an increased proportion of the H_2 is used in forming propionate, and the ratio of propionate to acetate increases. This can be of economic importance because propionate can generate glucose in the animal but acetate cannot.

The proportion of propionate formed in the rumen fermentation often increases when the rumen turnover is more rapid; i.e., the dilution rate is faster and the retention time shorter. A possible explanation is that the proportions of amino acids and other complex growth factors is slightly greater soon after feeding and could therefore be more available at faster turnover rates.

In *Clostridium propionicum* and *Megasphaera elsdenii*, propionic acid is formed by direct reduction of lactyl-CoA via acrylyl-CoA without a 4-C intermediate. This yields only one ATP (two per hexose) and would not be competitive with the pathway yielding four ATP. However, Counotte (1981) has found that *Megasphaera* competes effectively when the lactate concentration is high.

Butyrigenic Fermentation

This is encountered in a number of species of *Clostridium;* I once observed copious production of butyrate (identified by the smell) in some beet pulp left in a tall cylinder of water in the laboratory to allow bacteria to ferment the soluble nutrients. Butyrate is formed also by *Butyrivibrio fibrisolvens.* The butyrivibrio fermentation differs from that of most clostridia in that little acetate is formed; in fact, if the medium contains acetate, as in rumen fluid media, acetate may be taken up instead of produced. Experiments with isotopes added to rumen contents show that label is readily exchanged between acetate and butyrate.

Fermentations leading to butyrate generate only 3 ATP per hexose because the 4-C compound is produced from two acetyl-CoA, of which only one Co-A is conserved as butyryl-Co-A and can generate ATP. This appears to make butyrigenesis less valuable than aceti- and propionigenesis insofar as accomplishment of cell synthesis is concerned, and in open ecosystems it is a less common product than acetic and propionic acids. It may be a means for managing excess acidity by converting two acetate carboxyls to one butyrate.

Mixed Product Fermentations

Pure cultures of *Escherichia coli, Ruminococcus albus, Lachnospira multiparus,* and *Clostridium lochheadii,* as well as numerous other species produce a mixture of products, including formate, acetate, ethanol, lactate, H_2, CO_2, and often succinate. Except for the CO_2 and formate, these products are more reduced than acetate and pyruvate and probably represent a means for disposal of hydrogen.

Megasphaera elsdenii exhibits a mixed fermentation of a different sort. It disposes of hydrogen in the form of H_2 but forms no ethanol or lactate, common in other mixed fermentations. Instead, it forms propionic, butyric and small amounts of higher acids. It is able to ferment lactate, and according to Counotte (1981) it is the chief rumen microbe fermenting DL-lactate at acidities of pH 5.2 or less. It forms propionate and higher acids from lactate but differs from the propionibacteria and other rumen microbes producing propionate in that it does not form succinate. Rather, lactate is dehydrated to acrylate and then reduced to propionate with the hydrogen gained through oxidation of other lactate molecules to pyruvate, acetyl-CoA, and acetate. The abundance of *Megasphaera* occasionally seen in the rumen of some cattle may be due to a slight lactic acidosis from consumption of too much readily fermented carbohydrate.

DEVELOPMENT OF ANAEROBIC METHODS

Let us now turn to the problems involved in identifying and describing adioxic microbes. Pasteur's demonstration of life without air soon led to development of many methods for growing such organisms. As a substitute for Koch's pour plate method for isolating and counting bacteria, Liborius (1886) used shake cultures in tubes with two-hole rubber stoppers to allow air displacement with H_2, or used an even more rigorous method with all-glass containers having inlet and outlet capillaries that could be sealed off after all air had been displaced and the culture inoculated. He used a needle or capillary pipette for inoculating from colonies that he dispersed by mashing them between two sterile glass slides and suspended in a salt solution prior to inoculation. To recover colonies, he blew through a capillary pipette inserted to the bottom of the tube, forcing the agar into a sterile plate and slicing it for microscopic examination and picking of colonies. He tested the Koch plating method, using H_2 to displace air, but did not obtain significant growth unless the organism used O_2. Liborius distinguished between obligate anaerobes, obligate aerobes, and facultatives.

Fränkel (1888) combined the Liborius procedure with von Esmarch's (1886) roll tube technique. Roux (1887) drove gases out of tubes drawn to a capillary near the neck, and after inoculation sealed the capillary in a flame. He picked colonies with a curved capillary pipette; Veillon (1893) used capillary pipettes on a rubber tube for this purpose.

In 1900, Wright cultivated bacteria in a fluid medium by inserting a heavy plug of cotton soaked in KOH and pyrogallol and stoppering the tube; alkaline pyrogallol combines with O_2. Barker (1936a,b) substituted bicarbonate for the KOH to permit an appreciable tension of CO_2 in the tube. In the Burri (1902) method, a culture tube open at both ends was used, a rubber stopper closing the lower end, with an absorbent plug with alkaline pyrogallol inside the other end, also closed with a rubber stopper.

Ecological Considerations

With all these methods, many of them quite adequate for excluding O_2 from even the most stringently anaerobic organisms, it is surprising that until about 1940 the known obligately adioxic bacteria were chiefly the spore-formers and non-spore-formers of clinical importance (Weiss and Rettger, 1937, Prévot, 1966). Many of the anaerobes known from their products to exist in nature had not been cultured. This was probably due in part to the great popularity of the Petri dish. The ease with which aerobic bacteria could be isolated on plates led to many procedures for adapting plates to the cultivation of anaerobes, but attempts to inoculate and incubate the plates under adioxic conditions were insufficiently rigorous until the anaerobic glove box technique was perfected (Aranki et al., 1969).

A second factor concerns the culture medium. In culturing microbes of medical importance, it was correctly assumed that they lived on complex nutrients contained in the animal body. Extracts of animal tissues were included in culture media. Yeasts were grown in natural juices of fruits or with supplements of yeast extract. The lactic acid bacteria were provided sugar and proteins as in milk. In contrast, it was assumed that these materials were absent from soil and other habitats of free-living microbes, and they were not recognized as essential until much later (Lochhead, 1952).

The Liquid Enrichment Method

In soil and aquatic habitats in which the bulk of organic matter undergoing decomposition was plant bodies, it was assumed that, like the plants, the microbes would require only inorganic food plus an energy source. In many cases, preliminary enrichments using mineral medium and a single selective substrate elicited the development of desired microbes. The success of Beijerinck in growing *Azotobacter* and of Winogradsky in culturing N_2-fixing *Clostridia*, nitrifiers and cellulolytic bacteria, all in media with solely inorganic nutrients or supplemented with carbohydrate or carbohydrate derivatives, fostered the belief that given a particular organic food the free-living organisms could develop. It is understandable that complex proteinaceous nutrients were avoided. They served as substrates for many rapidly developing organisms that overgrew the desired ones.

The results of further applications of the liquid enrichment culture techniques of Beijerinck and Winogradsky seemingly reinforced this view, for inoculation of soil into enrichments containing sulfur compounds, cellulose, pectin and chitin, or other relatively purified substrate, with repeated subculture, led to rapid decomposition and in many cases to isolation of pure cultures. But in other instances, all attempts to isolate pure cultures using the enrichment method were fruitless. The primary biota attacked the selective substrate and leaked sufficient nutrients to support development of a subsidiary microflora, some of which in turn gave off metabolites required by the primary agents. When these organisms were eliminated in attempts to obtain pure cultures, no growth could occur.

In few cases was the inoculum from the habitat serially diluted and inoculated into medium containing an extract of the habitat with its essential nutrients. These were not supplied by the inorganic media, and use of rich media in liquid cultures led to overgrowth by unwanted species.

The Solid Medium Enrichment Method

The difficulties of the liquid enrichments were surmounted by three modifications in the isolation procedure. First, nutrients in addition to the energy-yielding substrate were included in the medium, thereby potentially supporting the growth of a much wider variety of microbes. Second, agar was included in the medium in order to separate the colonies and prevent overgrowth of the desired organism by the competition possible in a nutritious liquid environment. Third, the inoculum was diluted prior to inoculation, thereby providing some indication of which organisms were most abundant, yet allowing growth of others in lower dilutions.

Experience in recognition of colony characteristics in solid medium is needed if the desired organism is in low numbers. For example, methanogenic cells and colonies can be detected by their fluorescence in ultraviolet light due to their content of F_{420} and F_{342} (Doddema and Vogels, 1978; Edwards and McBride, 1975). Antibiotics for inhibition of eubacteria can also be useful in isolation (Godsy, 1980). The agar medium can elicit the development of stenotrophic organisms, for example, cellulolytic types, in a niche-stimulating medium (Hungate, 1962), and of eurytrophs when a habitat-stimulating medium is used (Bryant and Burkey, 1953).

Modern successes in growing a wide variety of adioxic microbes have resulted from judicious combinations of adioxic methods, simulation of nutrients and other factors in the natural environment, and rigorous tests of purity. The last consist of successive quantitative dilutions into clear solid media in which all colonies can be seen and counted. The colony count should be proportional to the dilution. Upon subculture of a colony, all resulting colonies should show colony and cell morphology similar to the colony used for inoculation, and further similar subculture should give identical results. In some cases, where the cells of the colony do not disperse well in the diluting fluid, the above criteria are not applicable; other methods and criteria must be developed.

Gradients in Anaerobiosis—the Redox Potential

We have used the terms anaerobic and aerobic, adioxic and dioxic, as if they described two entirely separate conditions applicable to both the organism and the environment. Originally there was sufficient reason to adopt this concept of contrasts; the organisms grew in air or they didn't. But after the discovery of the importance of dioxygen for life of many sorts, the concept that all organisms might require a little O_2 was common. Kursteiner (1907) showed this to be false when he demonstrated that at least some organisms grew without any demonstrable O_2. However, the influence of various concentrations of O_2 was not precisely expressed until the electron (oxidation-reduction) potential of pure biochemicals was established and used to predict the direction of electron flow.

The redox potential of O_2 at atmospheric pressure is approximately 0.8 V. This potential decreases with decreasing oxygen concentration according to the equation,

$$E'_{O_2} = E'_{oO_2} + \frac{0.06}{n} \text{ V} \times \log \frac{(\text{conc. atm. } O_2)}{\text{conc. } H_2O} \quad (6)$$

where n is the number of electrons involved. On the assumption that a 2-electron transfer is concerned in the reduction of one of the O_2 atoms to water the equation becomes,

$$E' = E'_o + 0.03 \text{ V} \times \log \frac{(\text{conc. atm. } O_2)}{\text{conc. } H_2O} \quad (7)$$

Since the concentration of water remains constant, a 10-fold reduction in O_2 concentration causes only a 0.03 V decrease in the oxygen potential. Some adioxic organisms will not grow at redox potentials above -0.3 V. According to the equation, the log of the O_2 concentration at -0.3 V would need to be $-0.3 - (0.8)/0.03 = -36$. The concentration of O_2 in a culture medium would need to be 10^{-36} of its concentration in air in order to have a potential of -0.3. If only one electron were involved, the concentration would still need to be 10^{-18} of that in air.

This emphasizes the difficulty in obtaining very low potentials in an environment simply by removing O_2 from a culture medium. A reduced substance having a low potential must be added. Recent techniques for measuring concentrations of dissolved O_2 (Scott et al., 1983) and H_2 (Scranton et al., 1983) have extended the lower limits of measurement for these important gases, but the O_2 method is still not sufficiently sensitive to measure the O_2 concentration (below 30 mM) at which methanogenesis is inhibited (Scott et al., 1983).

Factors Influencing Redox Potential

Some organisms liberate reducing substances, e.g. H_2 and formate, into the culture medium during growth, and measurement with a platinum electrode can show a very low redox potential. It was early (Cahen, 1887) shown that methylene blue, indigocarmine, and litmus could be reduced in bacterial cultures. Later, methylene blue (which lacks oxygen atoms) was used in many biochemical experiments to determine whether an enzyme preparation was effective in transferring electrons from a test substrate to form reduced (colorless) methylene blue.

Reduced indicators are not used to reduce media; the required concentration would be expensive and in most cases toxic. Litmus was added earlier to milk to test for its reduction due to growth of bacteria or presence of cells of bovine origin. N. S. Golding at Washington State University at Pullman made an exhaustive study (unpublished) of the redox indicators that might be used for testing milk, and found that resazurin was least inhibitory to a wide range of bacteria. It reduces irreversibly to resorufin, a red dye, and this reduces reversibly to a colorless form.

The potential of a mixture of equal parts of resorufin and the colorless form is −0.043 V. A concentration of 0.0001% in media is sufficient to detect the appearance of color on oxidation. Other redox indicators of potential value in certain instances are phenosafranin at −0.252 V, and benzylviologen at −0.357 V. The reduced form of the latter is colored and the color is not so easily detected as is that of resorufin. Phenosafranin is somewhat more toxic than resazurin. Titanium IV citrate is an excellent reducing agent and is also an indicator, the reduced form being blue. It has a very low potential of −0.48 V but tends to precipitate unless well chelated. It may also be toxic.

The potential of the medium is the best means for defining the degree of adioxia achieved in setting up a culture, but it is difficult to interpret insofar as the exact mechanism of growth inhibition is concerned. It has been asked whether it is the potential that is important or the concentration of O_2. It is impossible to separate O_2 from its potential; the potential is inherent in the structure of the O_2 molecule. If, due to lack of catalysts, this molecule cannot react to oxidize materials that are essential in the reduced state, then dioxygen will have little effect. Such a situation is improbable in most anaerobes. The closest approach may occur in the lactic acid bacteria. They grow in air but do not use dioxygen. Some form peroxides, but these are relatively nontoxic and may accumulate to measurable concentrations without impeding growth. Some lactic acid bacteria possess a peroxidase with which the peroxides can be used as electron acceptors for limited pyruvate oxidation. The lactic acid bacteria are unusual in carrying on a fermentation in air.

Oxidants, e.g. Fe^{3+}, for which the organism possesses no catalysts linking them to its metabolism can give a culture medium a high electrode potential, but the anaerobe can still grow.

Redox Potentials in the Organism

The great majority of aerobic bacteria produce a catalase and also one or more superoxide dismutases whose functions are to remove peroxides and the oxygen radical from the sites of susceptible vital anaerobic reactions. Presumably the location of oxidases at the exterior of the cell aids in protecting internal anaerobic reactions by diminishing the dioxygen tension in the interior. The extremely high O_2 consumption of *Azotobacter* may be a factor in obtaining the low potentials in the cell interior at which dinitrogen can be reduced.

In species such as the homofermentative streptococci, e.g., *Streptococcus bovis,* the culture medium does not become significantly reduced when O_2 is excluded, whereas *Escherichia coli* cultures show a drop in potential unless vigorously aerated. *E. coli* differs from the streptococcus in producing H_2 and formate as waste fermentation products and this can occur only at a very low potential. By incorporating a platinum catalyst into a dihydrogen-containing inorganic medium, Mylroie and Hungate (1954) obtained conditions sufficiently adioxic to permit growth of *Methanobacterium formicicum.* In methanogenic cultures the redox potential does not decrease further during growth (Baresi et al., 1978). The methanogen in nature cannot alone create the potential it needs in order to grow.

If H_2 is being used by the system, it is difficult to maintain equilibrium between gaseous and dissolved H_2, and the concentration of H_2 in the medium cannot be ascertained from its concentration in the gas (Hungate, 1967) when H_2 is being metabolized. This assumes that the transfer barrier between gaseous and dissolved H_2 is relatively as effective at low as at high H_2 concentrations.

CATEGORIES OF ORGANISMIC RELATIONSHIPS TO DIOXYGEN

Organisms can be categorized in their relationships to O_2 as follows:

- Anaerobes—cannot use dioxygen.
  - obligate—cannot grow in the presence of O_2.
  - oxyduric—not killed by O_2.
  - oxylabile—killed by O_2.
  - aerotolerant—can grow in the presence of O_2 but do not use it.
- Aerobes—can use O_2 as a source of energy for metabolism.
  - obligate—cannot grow without O_2.
  - euryoxic—can grow without O_2.
- Microaerophiles—grow best at a concentration of O_2 significantly lower than atmospheric.

The microaerophilic category is in the author's opinion somewhat unsatisfactory in that the concentration of O_2 in the vicinity of microaerophilic cells exhibiting maximal growth rate is not ordinarily reported. It is difficult to determine, and species are often reported as anaerobic or microaerophilic without provision of the evidence on which the classification is based. In a shake tube containing reduced medium but with the surface of the agar exposed to air, an anaerobe will show more abundant growth in a zone beneath the surface of the agar than in the depths of the tube, with no colonies appearing above the zone. The unpopulated zone provides additional substrate and removes more waste for adjacent colonies as compared to the colonies at greater depths in the agar.

Many obligate anaerobes exhibit an ability to grow at low O_2 tensions in the environment, absorbing the O_2 (Yarlett et al., 1982). The question of whether such absorbed O_2 increases growth has been studied in *Selenomonas ruminantium,* one of the less stringent anaerobes in the rumen (Wimpenny and Samah, 1978). A continuous culture was fed O_2 at partial pressures ranging from 0 to 35 mm Hg. Growth and fermentation were diminished at the lowest rates of O_2 addition even though no dissolved O_2 could be detected with a galvanic O_2 probe. The cells washed out at 36.7 mm pressure of O_2. Similar experiments with microaerophiles over a range of O_2 tensions would be useful.

The selenomonads absorbed increasing amounts of O_2 as its concentration increased, but were unable to maintain a low redox potential; it increased from -180 to $+94$ mV, at which point the culture washed out. Increased quantities of O_2 caused increases in NADH oxidase activity, with production of superoxide anion which was believed to induce production of superoxide dismutase. The dismutase production increase up to 1 mm Hg of O_2 pressure and then diminished. Apparently the selenomonad possessed a limited capacity to reduce O_2, but had no machinery for using it to generate ATP. The authors postulate that the O_2 shunted electrons from biosynthetic pathways.

Many dioxylabile anaerobes are killed extremely rapidly upon exposure to O_2. This has been particularly evident among the methanogenic bacteria, though one strain is oxyduric (Zehnder et al., 1980). The enzymes concerned with methane production are extremely sensitive to O_2. This is apparently irreversible and their inactivation can explain at least in part the toxicity of O_2. Certain cellulases of *Clostridium thermocellum* and *Ruminococcus albus* are irreversibly inactivated by O_2, presumably by oxidation of disulfide bonds.

Endospores resistant to heat and desiccation are presumably dioxytolerant. They are produced primarily by soil organisms, a habitat with frequent exposure to O_2. The author is not aware of any studies demonstrating O_2 sensitivity of endospores of obligately anaerobic bacteria.

METHANOGENIC FERMENTATIONS

The methanogenic fermentation proved more difficult to understand and analyze, both biologically and biochemically, than were the fermentations of carbohydrates. Methane was produced in anaerobic enrichment cultures that contained cellulose and a methanogenic organism attacking cellulose was postulated (Omelianski, 1902), but culture procedures and media were inadequate for isolation either of the methanogens or of the cellulolytic bacteria. Subsequently, the demonstration of bacterial photosynthetic CO_2 reduction with hydrogen from organic molecules (van Niel, 1931) led to the postulate by van Niel that reduction of CO_2 could explain the formation of methane.

This was tested by Barker (1936a,b), who obtained methane from an enrichment culture on ethanol and CO_2, and showed by analysis that the ethanol was dehydrogenated to acetate, with CO_2 reduced to methane. The later availability of isotopic carbon enabled Barker et al., (1940) to prove conclusively that the methane was derived from CO_2, and the acetate from ethanol.

Availability of isotopes permitted also the studies of Buswell and Sollo (1948) on municipal sludge. They found that methyl-labeled acetate was converted to labeled methane and unlabeled CO_2, demonstrating a second pathway for methanogenesis.

The microbiology of methanogenesis was more difficult to elucidate. Barker's (1940) *Methanobacillus omelianskii,* converting ethanol and CO_2 to acetate and methane, used for biochemical studies as a pure culture, was not subjected to critical tests for purity until 1967 (Bryant et al.) when it was found to consist of two species, one dehydrogenating ethanol to acetate and H_2, and the other, *Methanobacterium bryantii,* using the H_2 to reduce CO_2 to methane.

Schnellen (1947) was the first to obtain indisputably pure cultures. He isolated *Methanobacterium formicicum,* which used formate, and *Methanosarcina barkeri,* which could use acetate, methanol, carbon monoxide, and H_2/CO_2. The list of methanogenic substrates was extended (Hippe et al., 1979) to include methyl-, dimethyl-, and trimethylamines, the methyl groups being converted to methane. Probably the methyl group of methionine can be converted to methane; such a conversion was demonstrated in sediments by Zinder and Brock (1978).

Although the substrates are few, the habitats of methanogens are extremely varied: sediments (Barker, 1936a,b), the rumen (Smith and Hungate, 1958), fresh and salt water (Romesser et al., 1979), thermal hot springs (Stetter et al., 1981), tree trunks (Zeikus and Henning, 1975), salt evaporation ponds (Paterek and Smith, 1983), and geothermal vents (Leigh and Jones, 1983) in deep marine fissures. Magendie (1816) found methane in the gas within the large intestine of criminals examined immediately after execution.

MODERN ANAEROBIC HABITATS: THE ALIMENTARY TRACTS OF ANIMALS

The chief habitats in which obligately anaerobic microbes have survived and evolved are the alimentary tracts of animals, aquatic sediments, soil, and other habitats in which poorly aerated moist organic matter collects and temperature and moisture conditions are favorable. Aquatic habitats in which geogenic or biogenic H_2S is abundant are also anaerobic, since H_2S is spontaneously oxidized by O_2 to sulfur. Many municipal and industrial digestors are anaerobic. Conditions in some of these habitats need not initially be anaerobic but become so due to the growth of euryoxic microbes.

The alimentary tracts of animals are the easiest habitat to subject to a quantitative ecological analysis because of their relative constancy as compared to other anaerobic habitats. The temperature in homiothermic animals is relatively high and constant. The animal maintains in its gut a supply of water and other foods, dioxygen is almost totally absent, as is light, and the fermentations proceed rapidly, making the rates of various processes easier to measure.

Study of the relationships of the microbes to the host is aided by the availability of animals freed of all microbes. These may be obtained by cesarian section and subsequent germ-free handling and feeding, or by heavy doses of antibiotics coupled with subsequent aseptic handling. The gut of humans can be obtained essentially free of microbes by the latter procedure, as evidenced by the voiding of feces with olfactory characteristics resembling those of food rather than conventional feces. Obligate anaerobes could not be cultured from the feces (unpublished experiments). Axenic mice and rats are routinely raised, and sheep have been reared almost to maturity before succumbing due to unidentified causes (Lysons et al., 1977).

The animal alimentary tract evolved as a means for sequestering food from consumption by other animals, and allowing retention and digestion of digesta and absorption of digestion products as the animal moved about. But the food of animals is food also for microorganisms, and because of the potential rapidity of microbial growth under the favorable conditions in the gut, the microbes could be serious competitors for the animals' food. This microbial challenge has modified the course of evolution of animals, selecting for different types of host–microbe relationships (Hungate, 1976).

In all examples, microorganisms have become quite abundant ($>10^{10}$/g wet weight) by the time the digesta reach the caudal portion of the alimentary tract. A considerable part of the fermentation in the large intestine involves protein, as judged by the malodorous products in the feces. Methane and H_2 are among the gases produced in the fermentation in some humans (Calloway, 1968), the amount depending on the fermentable materials in the food that escape or are not susceptible to digestion or absorption by host enzymes (Salyers, 1979). Individuals tend to have either CH_4 or H_2 in the large intestine. Presumably, H_2 and CO_2 are the substrates for methanogens in individuals showing methane (Levitt and Ingelfinger, 1968).

The cephalad portion of the animal alimentary tract has evolved according to two different evolutionary strategies. In one the food is primarily of animal origin and can be digested by the enzymes of the animal consuming it. In the other, the food is primarily of plant origin, much of it not susceptible to digestion by the consumer. The evolutionary strategy in the first instance has been to compete with the microbes, in the second to cooperate with them.

The Competition Model

In the competition model, the animal secretes hydrochloric acid in microbicidal concentrations in the pyloric stomach just prior to entrance of the food into the succeeding portions of the alimentary tract where most of the animal's enzymes act. The pepsin of the stomach has apparently evolved concomitantly with the acid secretion. Most of the microorganisms are killed or their numbers are so reduced that even though the acid is neutralized in the duodenum and conditions become extremely favorable for microbial growth on the food or the products of host enzymes, the microbial numbers are insufficient to ferment appreciable quantities before the host absorbs the products of its enzymes. Some of the few microbes that are in the upper small intestine grow rapidly (Macy et al., 1978). No methane is formed in the small intestine (Levitt and Ingelfinger, 1968) and only a small amount of H_2. This portion of the alimentary tract is uniformly of small diameter, and digesta pass directly through it relatively rapidly with little mixing; this type of passage is designated plug flow (see below). The bacteria from the small intestine presumably compete in the large intestine with the microbes that develop there.

The microbes in the large intestine decompose foods not digested by host enzymes, and in addition act on the host secretions and on cells sloughed into the lumen of the gut. Vitamins are synthesized, and the hind-gut fermentation of plant fiber may be important in the health of the hind-gut wall (Hungate, 1976, 1984). The numbers of microbes in the large intestine and rectum of carnivores may be very high.

The competition model is exemplified in the vertebrate groups: Carnivora; Pinnepedia (seals, etc.); Insectivora (shrews, moles); Chiroptera (bats); Tubulidentata (anteaters); Edentata (pangolin); most primates; a number of marsupial species; and most birds, reptiles, amphibians, and fish. Most invertebrates are presumably of this type, with notable exceptions.

The Cooperation Model

Another important adaptation may be termed the cooperation model. This occurs in many animals consuming chiefly plant materials. Much of the plant consists of cell walls. Very few animals have evolved enzymes capable of digesting the plant cell wall, but they are produced by many microbes. In some of the animals consuming plant cell walls, their food is subjected to microbial fermentation prior to entrance into the acid stomach. An expanded part of the fore-gut retains the digesta long enough to permit active digestion and fermentation.

The cooperation model with an offset fermentation chamber is characteristic of ruminants (Hungate, 1966), the hippopotamus (Thurston and Noirot-Timothée, 1973), and possibly the gray whale (Herwig, et al., 1984). The cooperative model with modified plug flow occurs in the colobus and langur leaf-eating monkeys (Bauchop and Martucci, 1968), the quokka (Moir and Hungate, unpublished experiments) and other herbivorous marsupials, the 3-toed sloth, and probably the panda, which consumes chiefly bamboo leaves.

Among invertebrates the sea urchins exhibit digestion comparable to the cooperation model (Prim and Lawrence, 1975).

The Combined Competition–Cooperation Model

In other animals, the digesta are subjected first to the host acid and digestive enzymes, with absorption of the products, and then to the digestive activities of the microbes (McBee, 1977). In mammals, a large fermentation chamber, the caecum and large intestine, is provided caudad to the small intestine, and in this the digesta are retained long enough to permit microbial attack. The host obtains not only the nutrients digested by its own enzymes but also fermentation products from materials its enzymes cannot digest.

The combined model is found in the horse and its relatives, the elephant and hyrax, rodents and lagomorphs, dugong and manatee, and possibly the sloth. Members of the grouse family of birds, living at times on chiefly plant material, have well-developed caecae, long fingerlike extensions from the cloaca, in which an active fermentation occurs; however, the fermentation is not a major supplier of the bird's energy, and apparently fiber digestion is minor (McBee, 1977). Termites are a famous example of this model; they are active consumers and digesters of wood in tropical and temperate climates throughout the world (Hungate, 1946). This is probably the model for most invertebrates feeding on plant cell wall materials, e.g., the millipede (Taylor, 1982), but in some cellulose-digesting species, such as the snail, a microbial fermentation has not been demonstrated.

Occurrence of cellulase in the digestive fluid of the snail (Karrer, 1930) is an apparent exception to the generalization that animals depend on microbes for fiber digestion. Recently, Waterbury (Waterbury et al., 1983) has demonstrated that the cellulase in other molluscs, the wood-boring teredos, is elaborated by cellulolytic, nitrogen-fixing bacteria in the gland of Deshayes. This gland has been postulated as the source of cellulase and of amino acids in the teredos, and the isolation and characterization of the symbiote indicates microbial participation in these processes. The bacterium is euryoxic, but fixes N_2 only when growing anaerobically.

The question whether the cellulase produced by the protozoa of termites and cattle is elaborated by intracellular symbiotic bacteria has not been exhaustively studied, but in thin sections prepared from pure cultures of *Trichonympha sphaerica* growing on cellulose, Yamin (1981) was unable to detect any bacterialike structures.

A disadvantage with the occurrence of the fermentation after host enzymes have acted is that the microbes produced in the posterior fermentation cannot be digested there by the host, and, although the host absorbs the waste fermentation products, the microbial cells themselves cannot be used. Many examples of the competition–cooperation model have overcome this deficiency by consuming the feces containing the microbes. The termites void two kinds of feces: relatively dry pellets that accumulate as waste in old burrows and moist fecal materials that are solicited by other members of the colony and are consumed as they are voided. Rabbits also produce dry feces containing undigested feed residues and moist feces containing material from the caecum. The latter are consumed by the individual as they are voided; such coprophagy is common also among rodents.

Continuous Fermentation

The rumen microbial ecosystem has been the most studied, particularly from a quantitative kinetic standpoint, and a number of relationships of rumen microbial species to their environment have been elucidated. The system is more accessible than most alimentary microbial habitats, and because of its economic importance both the microbial and the ruminant activities have received extensive study.

In the adult ruminant, the rumen never empties, even in starved individuals. Passage of digesta through it depends on muscular contractions of the rumen wall that force rumen liquid to bathe the ingested solids and that gradually circulate the solids, mixing them. The solids particles in suspension are regurgitated and further chewed (rumination) to comminute them. Eight or ten hours after feeding, the rumen contents become more uniform, with a solids contents of 12–16% (dry wt/vol). Copious saliva is secreted during feeding and rumination, and its high content of bicarbonate plus a lesser concentration of phosphate assists in buffering the fermentation acids, maintaining usually a pH somewhat above 6.0. With the favorable body temperature of ca. 39°C, sufficient moisture, absence of O_2, absorption of acid products, and suitable feed, the microbes of the alimentary tract constitute perhaps the most active anaerobic microbial population in nature. An amount of substrate equal to 12–16% (dry wt/rumen vol) is fermented daily. This high rate of substrate conversion is possible because the ruminant maintains a continuous fermentation, supporting the concentrated microbial population in a postexponential stage of growth. The food supply is limiting. This can be readily confirmed by observing that addition of fresh food to the fermenting mass increases the fermentation rate. The limitation in food availability applies also to all the animal models and indeed to most natural anaerobic microbial ecosystems.

This limitation of food supply is prerequisite for stability in a continuous fermentation in which feed rate governs the rate of growth. The rate of growth is equal to the rate at which the microbial population leaves the rumen, usually termed the dilution rate, in terms of the fraction of population leaving (and being produced) per unit time. In animals fed once or twice a day, the rumen fermentation only approximates a continuous system, but when feed is supplied by frequent intervals throughout the 24 hr of each day, the rumen closely simulates a continuous fermentation. The dense population that is maintained is limited by the food supply, and newly ingested materials are attacked at a rapid rate. This explains its effectiveness in transforming large amounts of substrate.

Many animal-microbial relationships are based on continuous fermentation with a greater or lesser degree of mixing, combined with plug flow.

Few alimentary habitats simulate a continuous culture as closely as does the rumen. The leaf-eating colobus and langur monkeys accommodate the fermentation in an enlarged cardiac portion of the stomach, separated from the acid pyloric region by a constriction in the wall; methane is one of the fermentation products in the langur, whereas hydrogen is the final product in the colobus. The quokka (*Setonix brachyura*), a grazing marsupial, has a stomach similar to that of the monkeys, with a fermentation in which both H_2 and CH_4 are produced along with acetic, propionic, and butyric acids (unpublished experiments with R. J. Moir).

The fermentation in termites (Hungate, 1939; Breznak, 1982) also produces both H_2 (Hungate, 1939) and CH_4 (Breznak, 1982), the relative amounts of the two products varying with the species. It is surprising that the gut of the small termites such as *Reticulitermes* would be sufficiently anaerobic to permit methane production, particularly in view of their network of respiratory tracheae.

The plug-flow model for continuous fermentation envisions an initial continuous mixing of inoculum with substrate, together with passage of the mixed material into a tube along which flow is linear. It assumes little mixing after the initial inoculation, and the fermentation proceeds as in a batch culture. Passage along alimentary tracts is seldom strictly plug flow because peristaltic contractions of the gut wall cause some mixing. The long alimentary tract of the dugong without large fermentation chambers simulates a plug-flow model with acid or alkali added at certain somewhat enlarged segments.

The Rumen

For many years the analysis of microbial activities in the rumen was considered difficult because of failure to culture the myriads of bacteria and protozoa that inhabit it. But through careful application of the methods reviewed in the introduction, and by simulating the ecological conditions within the rumen, such as CO_2-bicarbonate buffer and rumen fluid as a nutrient, cultivation was improved (Hungate, 1942, 1947), though the protozoa have still not been grown axenically. In the rumen is a microcosm populated by 10^{10} or more bacteria and 10^6 protozoa/ml (Hungate, 1966; Gruby and Delafond, 1843). Most of the bacteria are non-spore-formers and at least 99% are obligately anaerobic, some extremely so. Some of the relationships of the rumen microbial ecosystem are shown in Figure 5.

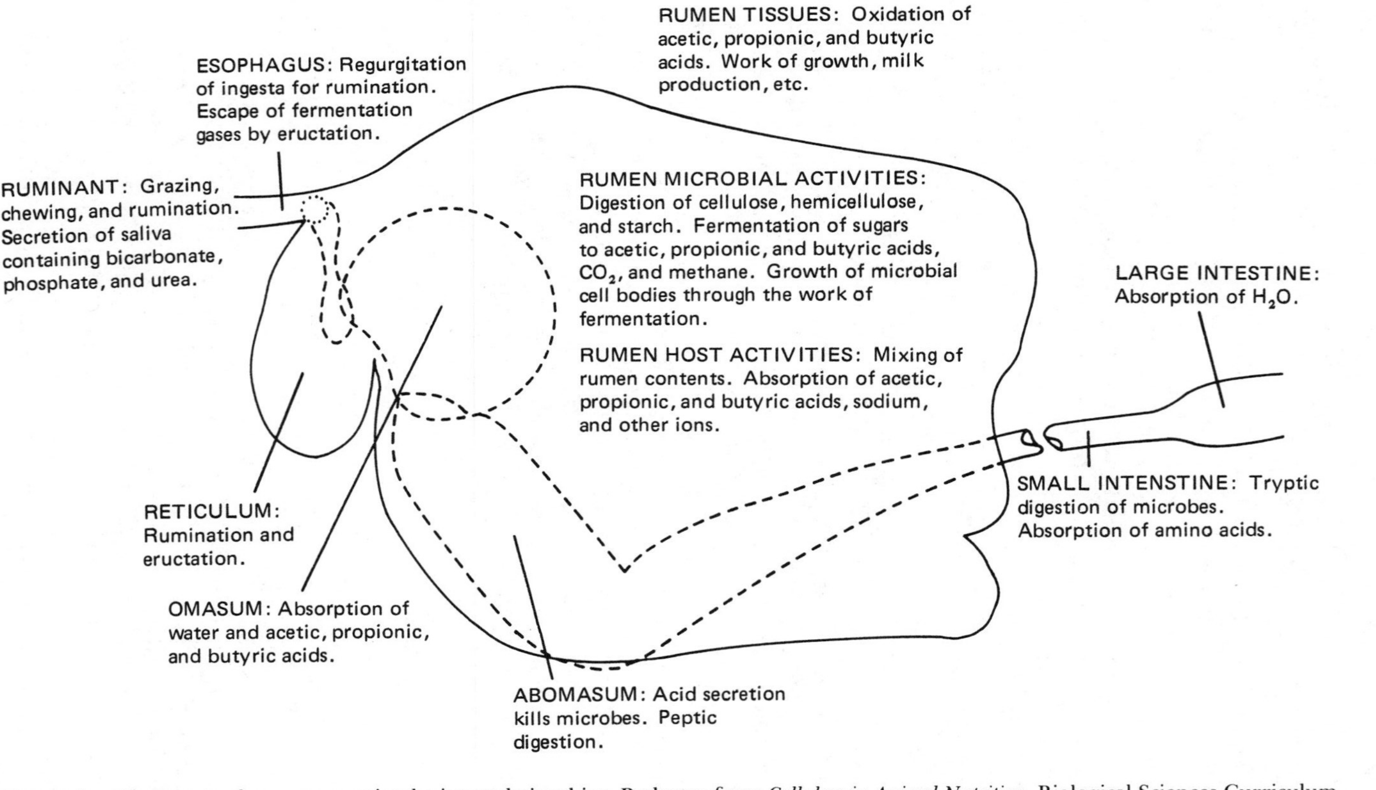

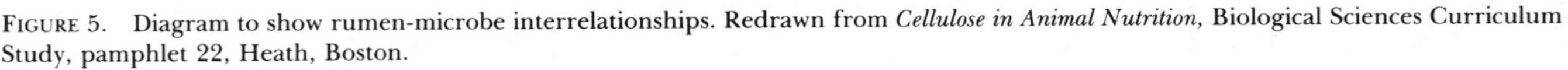
FIGURE 5. Diagram to show rumen-microbe interrelationships. Redrawn from *Cellulose in Animal Nutrition,* Biological Sciences Curriculum Study, pamphlet 22, Heath, Boston.

Substrates and Bacteria

The known substrates used by pure cultures of bacteria from the rumen include cellulose, starch, sugars, pectin, hemicelluloses, carbon dioxide, H_2, methanol, the bodies of other microbes, and a host of minor substances. Some of the protozoa consume cellulose and starch, others only starch, some live off soluble sugars, and most of them use bacteria as their proteinaceous food.

Some of the bacteria, such as *Bacteroides succinogens, Ruminococcus albus,* and *R. flavefaciens,* are quite stenotrophic, fermenting chiefly cellulose or its hydrolytic products. Most ruminococci do not even use glucose. Many of these cellulolytic species also digest pectin and xylan. *Bacteroides amylophilus* similarly utilizes starch, dextrins and maltose, but not glucose. In contrast, eurytrophic species such as *Butyrivibrio fibrisolvens* utilize a wide variety of carbohydrate substrates, including cellulose, hemicelluloses, starch, and many sugars. *Bacteroides ruminicola* is similarly eurytrophic.

With the stenotrophic species it seems relatively easy to identify the food they use in the natural habitat, but with the eurytrophic species this is difficult. By determining the soluble foods usable by each species, and the affinity of its enzymes for each food, i.e. the K_s, it should be possible to estimate the rate of utilization of each food by each species (Russell et al., 1981; Counotte and Prins, 1979; Counotte et al., 1981). The characteristics of some rumen species are summarized in Table II.

The bacteria listed in Table II do not include all of the rumen species. For some of the most abundant bacteria growing on a rumen fluid medium containing an alkaline extract of alfalfa, no specific substrates used could be identified (Chung and Hungate, 1976). Colonies of these bacteria never reached a large size.

Culture and direct counts of rumen bacteria (Bryant and Burkey, 1953; Hungate, 1966) can provide relative comparisons of the numbers of some

TABLE II
Types of Rumen Microbes

Celluloclastic: *Ruminococcus albus, R. flavefaciens, Bacteroides succinogenes, Butyrivibrio fibrisolvens, Eubacterium cellulosolvens,* and some protozoa
Amyloclastic: *Strep. bovis, Bacteroides amylophilus, Succinimonas amylolytica, Butyrivibrio* spp., and entodiniomorph protozoa
Saccharoclastic: *Lachnospira multiparus, Butyrivibrio,* holotrich protozoa, and *Bacteroides ruminicola*
Lactacidigenic: *Strep. bovis* and *Eubacterium cellulosolvens*
Lactacidiclastic: *Selenomonas ruminantium, Megasphaera elsdenii,* and *Veillonella* spp.
Succiniclastic: *Selenomonas ruminantium* and *Veillonella alkalescens*
Lipoclastic: *Anaerovibrio lipolytica*
Bactoclastic: *Anaeroplasma bactoclastica* and protozoa
Methanogenic: *Methanobacterium ruminantium* and *M. mobilis*
CO_2 utilizers: All *Bacteroides* and *Methanobacterium* spp.

species, but these estimates are complicated by the tendency for many species to attach to particles of digesta (Akin et al., 1973, 1974). The preponderance of "attached" to unattached species can be demonstrated by collecting a representative sample of rumen contents, i.e., one with the normal proportions of solids and liquid, and comparing the fermentative activity of the total with the activity of the liquid which drains from the total sample. The activity is determined by adding to the samples a finely divided excess of the ration of the ruminant and measuring the fermentation rates (El-Shazly and Hungate, 1965). Some balanced salt solution including 0.5% sodium bicarbonate buffer should also be added to accommodate the additional fermentation products from the excess substrate.

This type of experiment usually shows the fermentation rate of the whole contents/unit volume to be three to four times the rate of the liquid alone. On the assumption that the fermentation rate is the same function of population size for both solids and liquids, it is concluded that the population on the solids is three to four times as large as that in the liquid. This method of measuring population size has been used successfully also to measure the net growth of the rumen population (El-Shazly and Hungate, 1965).

The attachment measured by this method does not necessarily indicate that the microbial cells are fixed to the particulate substrate. Direct microscopic examination of rumen contents shows a very large number of unattached protozoa, many of which would be retained on the solids during gentle filtration. There are also masses of bacteria seemingly unattached and also perhaps retained during filtration. Although accurate measurement of the ratio of attached to unattached cells in the rumen is difficult, studies on pure cultures can reveal differences in ability of isolates to attach. Latham et al. (1978a,b) have shown that both *Bacteroides succinogenes* and *Ruminococcus flavefaciens* adhere to plant cell walls and digest holes (Frassbetten) in the plant cell walls corresponding to the contour of the bacterial cell. *Ruminococcus albus* does not appear to be so closely attached, but strain 8 has been shown to require a factor in rumen fluid which has been identified as phenylpropionic acid (Hungate and Stack, 1982). It is required for formation of capsular material of a high molecular weight which is actively cellulolytic (Stack and Hungate, 1984), and may be concerned with attachment to the substrate.

I have observed (1944a) that in unshaken broth cultures of *Clostridium cellobioparum* growing on finely divided cotton cellulose, the supernatant fluid remained very clear during digestion of the cellulose, and contained very few bacterial cells, but when the cellulose neared complete digestion the entire tube became turbid. Microscopic examination of a liquid culture showed that during early stages the motile cells of this species tended to remain in the vicinity of the cellulose rather than randomly distributing in the liquid, but upon complete digestion of the cellulose they distributed throughout the tube. The early behavior suggested a chemotactic reaction. The moving bacteria were relatively few compared with those within the mass of cellulose.

Rates within the Rumen Ecosystem

A goal of ecologists is to measure the rates of transformation in an ecosystem, to analyze it into its component biotic parts (for the microbiologist this means obtaining pure cultures) and determine what and how much each part accomplishes in the ecosystem, to test the validity of the analysis by adding algebraically the rates of activity of all parts and to compare the sum with the rate of the total ecosystem. Disagreement is strong evidence that important processes have been overlooked or the interrelationships misjudged. Agreement indicates that the analysis is correct or that errors balance out.

A kinetic model for the rumen has been developed (Baldwin et al., 1977) and offers a practicable means for quantitative integration of the innumerable activities of the rumen microcosm. A number of best estimates of rates have been made (Hungate et al., 1971) in order to represent the complete continuous fermentation system. Some simplification and generalizations have been necessary in developing the working model, but it now offers a framework on which modifications can be introduced as additional processes in the system are identified, and more rate measurements on the parts and on the total system are available.

The total activity of the rumen microbial ecosystem has been frequently measured, directly by in vivo measurements using markers such as polyethyleneglycol or lignin, by feed-feces difference, by total methane production and by zero time rate measurements of the chief waste products of the carbohydrate fermentation (namely, acetic, propionic and butyric acids, carbon dioxide and methane [Hungate et al., 1960, 1961]). Acetic acid is usually the most abundant, with less propionic and usually even less butyric. The acetate is readily interconvertible into butyrate (Leng and Leonard, 1965). These authors obtained evidence with $C^{14}T_3COOH$ that the hydrogens on the methyl group may exchange in reactions which have not been identified.

An equation to show the food disappearing and the products formed has been prepared from the results of analyses on a sheep fed equal portions of an alfalfa hay ration every two hours (Hungate et al., 1971). The food used was calculated from the elemental analyses of the feed and feces of the animal, on the assumption that the digestion occurred chiefly in the rumen. Zero time rate analyses were conducted with rumen contents to determine the rates of production of gases, acids and ammonia. The carbon not recovered in the waste products was assumed to have been assimilated into microbial cell bodies. Their average composition in hydrogen, nitrogen, and oxygen was estimated from values in the literature, and the following equation was obtained;

$$C_{20.08}H_{36.99}O_{17.406}N_{1.345} + 5.65\ H_2O \rightarrow C_{12}H_{24}O_{10.1} + 0.83\ CH_4 + 2.76\ CO_2 + 0.50\ NH_3 + C_{4.44}H_{8.88}O_{2.35}N_{0.785}\ \text{(microbial cells)} \quad (8)$$

This method of calculating the microbial cell yield gives a maximum value greater than that expected from the known biochemical mechanisms for gen-

TABLE III
Chief Carbohydrate Fermentation Products

Pure cultures	Rumen microbiota
Carbon dioxide	Carbon dioxide
Hydrogen gas	
Methane	Methane
Formic acid	
Acetic acid	Acetic acid
Propionic acid	Propionic acid
Butyric acid	Butyric acid
Lactic acid	
Ethanol	

erating ATP from the substrate used. But it does indicate that the rumen fermentation can accomplish an amount of growth work close to the theoretical maximum.

The Power Struggle

The carbohydrate fermentation products of pure cultures isolated from the rumen are shown in Table III. If these are compared with the fermentation products formed in the rumen, it is evident that although lactate, ethanol, succinate, formate, and H_2 are common products of many pure cultures, they do not accumulate in the rumen. Experiments were initiated to see whether they were not formed in the rumen or were formed and immediately used in further reactions. Were they intermediates? The results showed conclusively that lactate (Jayasuriya and Hungate, 1959) and ethanol (Moomaw and Hungate, 1963) were not important intermediates in the mixed rumen fer-

TABLE IV
Parameters for Possible Intermediates in the Rumen Fermentation

Intermediate	Concentration (nmoles/ml)	μ^a (min^{-1})	Flux (nmoles/ ml min)	Product formed	Rumen production accounted for (%)
Lactate	<12	0.03	0.36	Acetate	<1
Ethanol	Not detected	0.003		Acetate	
Succinate	4	10	40	Propionate	33
H_2	1	710	710	Methane	100
Formate	12	10	120	H_2 and CO_2	18

[a]The specific turnover rate constant for the pool of the intermediate.

mentation, whereas succinate (Blackburn and Hungate, 1963), hydrogen and formate (Hungate et al., 1970) were, as may be seen from Table IV.

ATP per Unit Carbohydrate. Since under anaerobic conditions fermentation is the mechanism by which the work of cell synthesis is accomplished, it seemed possible that reactions leading to ethanol and lactate were of minor importance in the rumen because they could not perform as much work as could be accomplished in reactions leading to succinate, H_2 and formate. One of the first products of cell synthesis is adenosine triphosphate (ATP). Only two ATP are formed from hexose when ethanol and lactate are produced, whereas four ATP are possible during formation of the other products. The pathways to these fermentation products are shown in Figures 3 and 4. The following equations show the stoichiometry of the conversion to acetate (Hungate, 1982):

$$C_6H_{12}O_6 + 4\ Pi + 4\ ADP \rightarrow 2\ CH_3COO^- + 2\ HCO_3^- + 4\ H_2 + 4\ H^+ + 4\ ATP \quad (9)$$

$$\Delta G_o' = -51\ kJ.$$

Although the bacterial strain isolated from a starving zebu at Archer's Post, Kenya (Margherita et al., 1964) exhibited almost exactly this fermentation with only a little ethanol formed, most of the isolated rumen bacteria producing acetate form also formate, lactate, ethanol, butyrate, propionate or succinate in pure culture.

In the rumen much of the H_2 of Eq. 9 is consumed by *Methanobrevibacterium ruminantium* or *Methanomicrobium mobile*. Their K_s for H_2 is about 10^{-6}/M and the concentration of H_2 dissolved in the rumen is of this same order of magnitude (Hungate, 1967; Hungate et al., 1970). When the H_2 is converted to methane, the free energy decrement is greater and more ATP is formed (Hungate, 1982).

$$C_6H_{12}O_6 + 4\ 1/2\ Pi + 4\ 1/2\ ADP \rightarrow 2\ CH_3COO^- + HCO_3^- + 3\ H^+ + CH_4 + 3\ 1/2\ H_2O + 4\ 1/2\ ATP,\ \Delta G_o' = -147\ kJ. \quad (10)$$

Succinate is one of the important primary products of fermentation in the rumen, formed chiefly by various species of *Bacteroides*. The stoichiometry of this fermentation is:

$$C_6H_{12}O_6 + 2/3\ CO_2 + 4\ Pi + 4\ ADP \rightarrow 4/3\ HOOCCH_2CH_2COO^- + 2/3\ CH_3COO^- + 2\ H^+ + 4\ 2/3\ H_2O + 4\ ATP,\ \Delta G_o' = -131\ kJ \quad (11)$$

The succinate is decarboxylated to propionate by *Veillonella* and *Selenomonas*, in a reaction that yields no ATP from oxidoreductions but may accomplish some chemiosmotic phosphorylation.

$$4/3\ HOOCCH_2CH_2COO^- \rightarrow 4/3\ CH_3CH_2COO^- + 4/3\ CO_2,\ \Delta G_o' = 0.3\ kJ \quad (12)$$

The net reaction for formation of propionate is the sum of these two equations,

$$C_6H_{12}O_6 + 4\ Pi + 4\ ADP \rightarrow 4/3\ CH_3CH_2COO^- + 2/3\ CH_3COO^- + 2\ H^+ + 2/3\ CO_2 + 4\ 2/3\ H_2O + 4\ ATP, \Delta G'_o = -131\ kJ \quad (13)$$

Propionate thus represents another route for disposition of H atoms, and as shown in Figure 4, four ATP can be formed also in this pathway of hexose fermentation, putting propionate in a competitive position with acetate. In butyrate production, three ATP can be made and its production entails less acid formed than is true of the other fermentations, but it is uncertain whether these features explain its formation in the rumen and other habitats.

All these products are formed in the normal rumen. An important feature of its normality is that substrate is limiting, and an advantage is possessed by organisms more efficient in substrate utilization, i.e., obtaining more ATP per hexose. Lactate and ethanol are very minor products, and do not serve for disposal of hydrogen; rather, it is stored in the propionic and butyric acids and in methane.

ATP per Unit Time. This efficiency in ATP synthesis per hexose is not necessarily of selective value when carbohydrate is not limiting. When starchy foods or other readily fermentable carbohydrate is fed in excess to ruminants, lactacidigenic bacteria rapidly outgrow the normal flora, producing so much lactic acid that the rumen contents become acid and the animal dies of lactacidosis (Hungate et al., 1951). It cannot metabolize the lactic acid, particularly the D-lactate (Dunlop, 1972), fast enough to keep pace with the production in the rumen. *Streptococcus bovis,* normally showing culture counts of 10^7/ml increases within 6 hours to 10^{10}/ml and the pH drops to almost 4.0. The *Streptococcus bovis* is then inhibited and is overgrown by the more aciduric lactobacilli. Unless cultures are taken prior to the acute illness the bloom of *S. bovis* may be missed.

These phenomena demonstrate that it is not ATP/hexose that is crucial but rather ATP/unit time. When carbohydrate is limiting, ability to obtain more ATP/hexose provides more ATP per unit time, but, when carbohydrate is in high concentration, the streptococci can process it faster and obtain more ATP/unit time than can the normal flora. If the change to a ration of readily available carbohydrate is made gradually, substrate remains limiting as the normal population increases rapidly enough to keep the substrate concentration low. The population remains stable. The premium remains on ATP/hexose, and the lactacidigenic bacteria cannot gain the ascendancy. Leedle et al. (1982) found little difference in the microflora of cattle fed maintenance rations of forage versus concentrates, though the proportions of cellulolytic, pectinolytic and xylan-digesting bacteria were somewhat less on the latter.

Since ATP represents the potential for chemical work of cell synthesis, the unit, work/time, represents power. The microbes compete for power much as do the organisms studying them! Scheifinger et al. (1975) were able to demonstrate a shift from a lactacidigenic to a propionigenic fermentation in

Selenomonas reminantium by decreasing the dilution rate in a continuous culture. This would decrease the concentration of substrate.

This principle of competition to obtain maximum power presumably is applicable to many anaerobic habitats. Rarely in anaerobic habitats is readily available substrate in excess, and we can expect acetate, propionate, and butyrate as the principal carbohydrate fermentation products. It would be interesting to test whether these are formed when nutrients other than energy sources limit growth.

The situation is different among aerobes. They can obtain up to 38 ATP per hexose. With this increased energy they can produce several times more cells per hexose than can the anaerobes, and nutrients besides energy sources more often limit their growth. The concept of selection for maximum power probably applies even less to photosynthetic organisms. They have sufficient energy to elaborate products only indirectly concerned with growth. This may be a factor in the relatively greater incidence of pigments, antibiotics and other metabolites among aerobes as compared to anaerobes.

Alimentary Tracts of Marine Animals

Most of the world's ocean areas are aerobic, and methanogenesis by free-living bacteria in sea water seems unlikely. However, with the advent of sensitive methods for the detection of methane, the biotic zone of the world's oceans has been found to contain a higher concentration of methane than is found at deeper layers (below 200 meters) or at the surface. Production of methane in the alimentary tracts of the animals living in this zone is postulated to explain this phenomenon. It seems likely to occur in animals such as fishes and marine mammals, and a microbial fermentation in the adult green turtle has been demonstrated (Fenchel, 1980). In view of methanogenesis in small termites, it seems possible that methane may be formed also in small marine animals, including invertebrates. The concentration of fermentable food in their gut may create sufficient O_2 demand to create adioxic conditions in which methanogenesis could occur.

MODERN NONALIMENTARY ANAEROBIC HABITATS

Aquatic Habitats

Some bodies of fresh water such as Lake Kivu and the Rio Negro of the upper Amazon contain so much hydrogen sulfide that it reduces all the oxygen. The depths of the Black Sea are an example of an anaerobic marine habitat. Some of the sulfide may be geothermal, arising from fissures in the earth but in other cases it is generated by sulfate-reducing bacteria, as discussed elsewhere in this volume. Some lakes in dry parts of the world become anaerobic during warm weather following cooler periods of extensive algal growth. The algae die and in their decomposition by euryoxic microbes all of the O_2 is consumed. Anaerobes then take over the decomposition processes, and in some instances *Clostridium botulinum* grows enough to produce toxins capable of killing ducks (Quortrup and Holt, 1940), as well as livestock drinking the water.

In some lakes in Northern Africa, sulfate-reducing bacteria attack the algae, producing H_2S, which kills all but the lungfishes. Dioxic conditions are restored by a bloom of purple sulfur bacteria which convert the sulfide again to sulfate.

Like the oceans, the waters of most lakes contain O_2, and any methanogenesis in them probably occurs in the alimentary tracts of animals. Dihydrogen has been detected in the waters of a salt pond (Scranton et al., 1983) by using an extremely sensitive method involving reduction of mercurous oxide by H_2 at high temperature, releasing the mercury as vapor measured by atomic spectrometry. The concentration of H_2 is small, a few nanomolar, but diurnal changes in its production indicate it plays a role in the microbial interactions in the various depths of the water. The concentration at the bottom is almost nil when the hypolimnion is anaerobic. The H_2 in the anaerobic layers is consumed in methanogenesis or sulfate reduction (Conrad et al., 1983).

Soils and Sediments

Soils are perhaps the most important, most interesting and most difficult microbial habitat to study. Factors such as light, moisture, temperature, and nutrients are so variable that it is extremely difficult to identify ecosystems for study, systems to which the investigator can return again and again as ideas and hypotheses for modeling develop. Soils contain an extremely varied biota including plants and animals as well as microbes. The variability in environmental factors has selected for resting stages in many soil microbes, notably the endospores of bacteria and a variety of spores in the actinomycetes and fungi. Organic matter, particularly plant cell wall material, added to soils undergoes successive stages of degradation leading ultimately to "humus," a material more poorly defined than cellulose, hemicellulose, or lignin, consisting of wall materials that are not readily degradable, together with the bodies of the microorganisms surviving the degradation. The proportion of wall materials in the "humus" of an arable fertile soil in a subtropical or tropical climate is less than in most soils in colder climates.

The extent to which anaerobic processes contribute to organic transformations in soils varies with the water content. The low solubility of O_2 in aqueous media allows most of it to be utilized in waterlogged soils more rapidly than it enters by diffusion. Numerous species of *Clostridium* as well as many euryoxic microbes grow in and degrade soil organic matter, including the bodies of microbes. Most types of organic matter can be attacked anaerobically, with the exception of the natural polymerized lignin of woody plants. The almost completely intact wood of Viking boats buried hundreds of years in anaerobic environments attests to the difficulty of anaerobic degradation of this material.

Undisturbed soils differ from aquatic sediments in having a greater penetration of O_2, a wider temperature range, and a wider range of moisture content. Both tend to be stratified according to depth, with the most ancient layers deepest. There is more communication between layers in soils because of the percolation of water, some of which can occur also in sediments. In soils in intertidal zones, water movement can be a major factor, but sediments deeply enough submerged that waves, tides and currents do not disturb them can exhibit a fairly reproducible ecosystem with a gradient toward decreasing organic matter with depth.

The gradient in O_2 concentration is somewhat steeper in most marine sediments than in soils, O_2 penetrating only a few millimeters below the water-sediment interface.

In a sense, the process of sedimentation resembles the plug-flow model in a vertical direction in that there is a linear stratification according to age, with new sediments occupying a position comparable to the mixed inoculum-substrate in the plug-flow model. The length of time involved is vastly greater, and the rate of the fermentation processes correspondingly slow.

The transformations of sulfur become important in sediments, particularly in those underlying seawater, which are much richer in sulfate than most fresh waters. Sulfate is a better H_2 acceptor than is carbon dioxide. Its reduction to H_2S by H_2 is exergonic to the extent of -152 kJ under standard conditions, as compared to the -135.6 kJ in the reduction of CO_2 to methane. This provides sulfate reduction with greater work capacity and enables it to support a larger population than can methanogenesis in the same habitat. Numerous studies show that in sedimentary layers containing sulfate, the oxidation of hydrogen by sulfate considerably exceeds the oxidation by CO_2 (Lovley and Klug, 1983). In deeper layers where the sulfate has been reduced, most of the H_2 produced is used in methanogenesis, but the flux of H_2 is much slower than in the upper layers of sediment containing more organic matter. A considerable part of the organic matter in these undisturbed sediments, particularly in the deeper layers, must be in the form of microbial cell bodies. The initial population near the surface of the sediment is the first stage of a food chain that proceeds through successive populations, each significantly smaller than the preceding population that was its food. The chief organic product of these fermentations is acetate. It can be oxidized by sulfate reducers if sulfate is available (Pfennig and Biehl, 1976; Widdel and Pfennig, 1981a,b) and caused to undergo an internal oxidation-reduction by the acetitrophic methanogens (Mah et al., 1978). Interestingly, Zinder and Koch (1983) have found that conversion of acetate to methane can be accomplished by a different method; in a thermophilic coculture, one organism oxidizes the acetate completely to CO_2, the second reducing CO_2 with H_2.

The ultimate fate of carbohydrate in the marine sediments, where time is unlimited (we hope), is the conversion to CO_2 and either CH_4 or H_2S.

Anaerobic Digestors

Toward the turn of the last century, anaerobic tanks were constructed for disposal of human excrements accumulated in quantity with the adoption of water as a means for transporting them. The enclosed digestors prevented escape of offensive odors and killed some of the microbial pathogens. The methane generated in the disposal plants was used as an energy source in the plant, and in a few communities was fed back into the municipal gas lines. Subsequently, with energy sources cheap and abundant, anaerobic digestion was largely replaced with aerated tanks. More recently the interest in renewable energy resources has led to further exploration of the possibilities for obtaining clean fuel by anaerobic digestion, as well as disposing of a nuisance and conserving plant nutrients.

To the extent that sulfate is in the digestor, it is reduced; the black color of digestor sludge is due chiefly to the metal sulfides it contains. Occurrence of these sulfides indicates anaerobic conditions; they oxidize spontaneously with O_2. The sulfide precipitation of copper in piggery wastes reduces copper toxicity.

The transformations of organic matter in the digestors are essentially the same as those in the anaerobic freshwater sediments, but they occur more rapidly due to higher temperature and better mixings. Plant cell wall materials constitute the quantitatively most important part of the organic matter fed into most digestors, and in this respect resemble the other anaerobic habitats we have discussed.

DIGESTION OF PLANT CELL WALLS

We have reviewed in an earlier section a rationale explaining the abundance of cellulose as a photosynthetic product in the cell walls of land plants. It is relatively rare in a pure state in nature, the closest approach being possibly the beta-1-4-glucan in the extracellular capsule of *Acetobacter xylinum* (Colvin and Leppard, 1977). Defatted cotton fiber, often used as a standard source of pure cellulose, consists of only 95% glucose.

The hemicellulose, cellulose, and lignin fractions obtained from plant cell walls by successive treatments with various solvents do not correspond to specific chemical entities in the natural polymer, as evidenced by the fact that enzymes of various muralytic bacteria do not remove exclusively hemicellulose, cellulose, or lignin. The lack of these materials in a pure state prevents adoption of a standard substrate for assaying "hemicellulose," "cellulose," or "lignin." It is thus a problem to select a substrate for comparisons of cellulases, and investigators are beginning to report enzymatic activity according to the particular substrate employed, i.e., as carboxymethylcellulase, pebble-milled cotton cellulase, crystalline cellulose cellulase, or cellulase as measured from other substrates or procedures.

New insight into plant cell wall structure may be obtained by using as substrate the wall itself. This involves a reverse of the usual procedure in which a mixture of enzymes is added to a single substrate; instead, a pure enzyme is added to the complex polymer composing the walls. It is necessary to ascertain the bonds being split by the pure enzyme, and for this the purified known polymers are essential. The linkages exposed to attack in the wall itself are divulged by the action of the pure enzyme on the natural polymer.

Increased complexity in cell wall components in plants presumably occurred because it increased the resistance of the wall to attack by muralytic microbes. Instead of a single type of enzyme, a battery of enzymes was needed. Particularly in the case of lignins (the wall components insoluble in 72% sulfuric acid), their inclusion in the wall polymer decreases the susceptibility to digestion by anaerobic microbes. We have already noted this in the almost perfect preservation of the wood of Viking boats, and modern experiments on digestion of plant materials support the view that native wood lignin undergoes little transformation in the absence of O_2.

The purified polymeric components of plant cell walls are extremely useful not only in identifying the nature of the bond split by a pure enzyme, but they are also helpful in isolating and identifying the bacteria elaborating the enzymes. Cellulose is digested by *Ruminococcus* species, *Bacteroides succinogenes,* some strains of *Butyrivibrio fibrisolvens, Eubacterium cellulosolvens, Clostridium lochheadii, C. longisporum, C. cellobioparum,* and *C. thermocellum.*

Pure cultures of these species vary in the digestion of pebble-milled cellulose. *Bacteroides succinogenes* is unique among the anaerobes in digesting only immediately adjacent cellulose, a feature shared with aerobic cellulolytic cytophagas. When it is grown in cellulose agar, the appearance of zones of cellulose digestion is very slow at agar concentrations above 1.2%, whereas it can occur within a few days if the concentration is below this level. I have found it helpful in isolating pure cultures from natural habitats to prepare the medium with 1.5% or more agar but without any cellulose, solidified as a long slant in an anaerobic tube, and then spreading a thin layer of 0.4–0.5% agar containing the cellulose and a dilution of the inoculum on the surface of the slant cooled to ca. 15°C. This temperature is high enough to allow spreading of the cellulose layer before it solidifies, and cold enough that solidification occurs before the cellulose flocculates too much. Through this weaker agar the cellulolytic organisms are freer to seek closer propinquity to the substrate and their enzymes, released as small vesicles, can better diffuse (Forsberg et al., 1981).

Even *Ruminocuccus albus,* reported not to attach closely to fibrous material in the rumen, apparently must maintain some sort of preferred position with respect to insoluble substrates. Rumen contents, diluted up to 10^{-5} and inoculated into cellulose agar roll tubes, fail to show growth of *R. albus, R.*

flavefaciens, and *Butyrivibrio*, and show only very small zones of cellulose digestion in higher dilutions. In the higher dilutions they can recover enough of their digestion products to become a small colony, but in the low dilutions noncellulolytic bacteria are so close they use all the soluble products before the clearing can be detected. In liquid cultures, on the contrary, digestion is most rapid in the lowest dilution. *Bacteroides succinogenes* can digest the cellulose, producing visible clearings in low dilutions, if the concentration of agar is low enough to allow it to move through the solidified medium. *Clostridium lochheadii* is so actively cellulolytic that it is able to grow in low dilution agar tubes in spite of the competition from noncellulolytic organisms.

The heterogeneity of the polymer of plant cell walls necessitates a multiplicity of enzymes for its successful digestion. This multiplicity might be achieved through a multitude of enzymes, each formed in a separate species, or by numerous enzymes produced by a single species. Both situations appear in the rumen. *Lachnospira multiparus* produces predominantely polygalacturonase enzymes and appears to be a specialist for digestion of the pectin in the middle lamella of plant cell walls. It grows in culture as long linear strands of cells, and apparently follows this same growth habit within the plant tissue, extending in long threads between the cell walls, seemingly accomplishing penetration of the tissue by rapid growth.

In contrast, *Ruminococcus albus* forms a variety of enzymes, including alpha-arabinosidase, polygalacturonase, xylanases, and cellulases, and is able alone to digest a significant fraction of the fiber in forage, as shown in Table V.

Even though *Bacteroides succinogenes* habitually attaches or is very close to the fiber it is digesting, studies of co-cultures with a noncellulolytic *Selenomonas* (Scheifinger and Wolin, 1973) showed that growth of the *Selenomonas* accounted for one third of the cellulose fermented.

The factors making lignin so indigestible are not well understood. Even under aerobic conditions it does not ordinarily decompose rapidly, and most of the digestion is by fungi. Some of them possess enzymes allowing the fungal filament to penetrate directly through a lignified cell wall, yet the entire carcass of the plant disintegrates only slowly over the years. Some of this can be

TABLE V
*Digestion by Ruminococcus of "Fiber" Components**

Component	Alfalfa hay (%)	Grass hay (%)
Hot water insoluble	2	17
2% H_2SO_4 soluble, "hemicellulose"	50	21
70% H_2SO_4 soluble, "cellulose"	55	28
Insoluble in 70% H_2SO_4, "lignin"	41	17

* Average values for four strains of ruminococci.

accounted for by the lack of non-energy-yielding nutrients. The wood of gymnosperms contains very little nitrogen, less than 0.05%, and is probably deficient in a number of other essential nutrients. The fungal habit of growing through the wood, transporting protoplasm to the growing point, may be a factor in fungal success in decomposing wood containing a paucity of non-energy nutrients, including nitrogen (Hungate, 1940; see Chapter 8, this volume).

Whereas the aerobic fungi can decompose more than 800 parts by weight of carbohydrate per single part of nitrogen available, bacteria require a greater proportion of nitrogen and probably also other nutrients. It should be mentioned that within recent years some species of anaerobic chytridiaceous fungi have been found in the rumen of cattle and sheep (Orpin, 1977), and appear to be more abundant in animals on a rather poor ration in which fiber digestion would seem to be particularly important (Bauchop, 1979). They can digest cellulose and it will be of considerable interest to learn whether they possess greater abilities to digest the lignified parts of plant cell walls than have been found for the bacteria (Hackett et al., 1977).

The flagellate protozoa in the hind-gut of termites can digest the cellulose in untreated wood (Hungate, 1938), whereas the rumen microbiota can digest it readily only after delignification, steam explosion, alkali treatment or massive high voltage X-irradiation (Lawton et al., 1951).

A colony of *Micromonospora propionici* isolated from the gut of *Amitermes wheeleri* (Hungate, 1946) was able to digest cellulose and capable of growing also on the wood from which the colony was collected. Further work on the muralytic bacteria in the Termitidae (termites containing bacteria but no protozoa), might yield additional species of bacteria growing on untreated wood. The high degree of comminution by the termite mandibles may be a factor in their ability to digest wood, but even extracts of the protozoa are able to digest as much as 20% of wood filed off with a wood rasp (Hungate, 1938).

The insolubility of native lignin appears to be an important factor limiting its anaerobic digestibility. Monomeric derivatives formed by chemical treatments to dissolve lignin are susceptible to methanogenesis by agnotobiotic enrichments (Healy and Young, 1979), but the insoluble polymeric lignin is not appreciably fermented.

The hemicellulose in forages is digested by ruminants to about the same extent as the cellulose. Some of the products of hemicellulose digestion by pure cultures are not fermented by them, and appear to be solubilized gratuitously in order to effect solubilization of other components which they can ferment (Dehority, 1973).

TRANSFORMATIONS OF NITROGENOUS MATERIALS

The nitrogenous materials of plants, though less abundant than the carbon-, hydrogen-, and oxygen-containing components of the plant skeleton, are nonetheless very important in anaerobic transformations. Proteins and nucleic acids compose a substantial part of the nonskeletal cell constituents, and together with other nitrogen-containing molecules, they contain sulfur, potassium, iron, and many trace mineral elements. The amino acids composing the proteins are synthesized by addition of ammonia or amines to intermediary metabolites of carbohydrates. When the amino acid is used as a source of energy under anaerobic conditions, the nitrogen is liberated as ammonia, again available for amino acid synthesis.

It might seem more efficient for proteolytic organisms to assimilate the amino acids directly into their own proteins rather than transforming them for energy, and presumably this does occur in nature, since amino acids have been found to be required growth factors for a great many microbes. But in the rumen, the amino acids introduced experimentally are chiefly fermented as a source of energy, and assimilated to only a very slight extent (Portugal, 1963). Peptides containing a number of amino acids seem to be more readily absorbed than are single amino acids (Wright, 1967).

The carbon-, hydrogen-, and oxygen-containing skeleton of amino acids left after deamination is fermented to acids, alcohols, and gases. The energy available in the fermentations of this skeleton is usually less than for an equal weight of carbohydrate because the carbons are not uniformly at the intermediate state of oxidation, some of them having been reduced or oxidized in the process of forming the carbohydrate-derived precursor of the amino acid.

The waste ammonia formed during amino acid fermentation is more than sufficient to neutralize the acid wastes, due in part to the fact that carbon-1 being already an acid, can contribute to CO_2 production through decarboxylation, but conversion of C-2 to a carboxyl group does not constitute a net increase in total acidity. In contrast, ammonia formation results in a net increase in alkalinity. Note that the peptide linkage in proteins effectively neutralizes the amino and carboxyl groups of the amino acids; its digestion recreates them.

The nitrogen is not converted to ammonia in fermentation products such as indole, skatole, and various organic amines. These, together with mercaptans and other compounds arising from sulfur, impart an unpleasant odor to fermentations of proteinaceous materials, in contrast to the pleasant aromas characteristic of most predominantly carbohydrate fermentations. The term putrefaction was early applied to the anaerobic decomposition of organic materials containing nitrogen, and fermentation was used to describe anaerobic carbohydrate transformations. The two processes are essentially similar in that they involve anaerobic rearrangement of atoms to yield energy, but the greater diversity in the elements composing the nitrogenous substrates leads to a greater variety of intermediate reactions and to additional fermentation products. The intermediary metabolism of fermentations of organic nitrogenous compounds has been nicely reviewed by Barker (1961).

Nitrogenous compounds are fermented in almost all habitats in which carbohydrate fermentations occur, since even if little nitrogenous material is in the habitat when the initial attack occurs, the bodies of the microbes fermenting the carbohydrate contain proteins and nucleic acids. The bodies not undergoing predation or scavenging can be used as a fermentation substrate. The ammonia released in this way is important as a nutrient for the microbes fermenting the carbohydrate. If the carbohydrate is in large excess all the ammonia will be assimilated into the microbial bodies decomposing the carbohydrate, but with greater proportions of nitrogenous substrates ammonia is a final product.

Some of the nitrogen transformations in the rumen have been elucidated through the extensive studies of Bryant on the nutritional requirements of rumen bacteria (1974) and of Allison on the fermentation of amino acids (1978). Ammonia is by far the most important source of nitrogen for the growth of rumen bacteria, and much of the amino acids in the proteins of the feed are fermented as a source of energy rather than assimilated into microbial cells. All the amino acids except leucine, isoleucine and valine can singly support an anaerobic fermentation; ammonia is released, and various acids are derived, according to the carbon skeletons of the amino acids, respectively. Thus isobutyric acid is produced from valine, isovaleric acid from leucine, 2-methyl-butyric acid from isoleucine, and phenylacetic acid from phenylalanine. These acids are required for the growth of many rumen bacteria which use them as precursors for synthesizing the amino acids from which they were derived for fermentation. The acid groups are carboxylated reductively and aminated in the exact reverse of their formation in fermentation. Both fermentation and synthesis require reduction or oxidation, respectively, of other substrates. These other substrates can include other amino acids, and classic studies of Stickland (1935) showed that although leucine, isoleucine, and valine were not fermentable as a single substrate, they could be fermented if one of a number of other amino acids was added as a second substrate; one was oxidized and the other reduced in what has subsequently been termed the Stickland reaction. This is true of many amino acids, and they have been classified as hydrogen donors or acceptors or either according to their tendency to give off or take up electrons, respectively (Barker, 1961).

This participation of more than one molecular substrate in an anaerobic fermentation by a single organism greatly increases the variety of substrates usable in anaerobic fermentation reactions. Synergisms between more than one organism, as discussed in the previous section, increases still more the metabolic mechanisms by which anaerobes can obtain energy. Our emphasis thus far on single substrates as supporting fermentations does not reflect the many mixed substrate reactions that occur in natural ecosystems. All members of the biota in a given community are exposed to similar electron and proton pressures in the environment, making possible an intercellular transfer of protons and/or hydrogen between separate members of the microbial community. Examples of these have already been mentioned in the section on carbohydrate transformations because the identification of the hydrogen-donating species was possible only in co-culture, but the phenomenon of one species feeding another is widespread. The reducing tendency of the rumen microcosm makes possible many of the detoxifying reactions in which hydrogen is substituted for the chlorine in many biocides. With the great diversity of substrates in ruminant feeds many specific metabolic reactions supporting growth are possible. Russell and Smith (1968) showed that heliotrin, an alkaloid in *Heliotropium europaeum,* was detoxified by a rumen bacterium that reduced it with H_2 or formate. A similar organism was described by Lanigan (1976) as *Peptococcus heliotrinreducans.*

Studies on another nitrogen-containing compound, mimosine, have provided an instance of a restricted distribution of a rumen microorganism. *Leucaena leucocephala* is a shrubby legume with a high protein content that grows well in the tropics and could be an important ruminant feed except that its mimosine content proves toxic to cattle and sheep in Australia when fed as the major source of protein. Raymond Jones, (1981), working at Townsville, Australia, found that when *Leucaena* was fed to cattle in Hawaii no toxicity was observed, and postulated that the rumen microbes detoxifying mimosine in Hawaii were absent from Australian animals. Transfer of rumen contents confirmed the hypothesis, and it appears that the responsible microbe was absent from Australia. This is a striking example on a continental scale that microbes are discontinuously distributed. The concept that a better microbe will evolve if a suitable environment is provided is not necessarily true.

Purines and pyrimidines are readily fermented by *Clostridium acidiurici, C. tetanomorphum,* and several other bacterial species (Barker, 1961). Many of these have been isolated from San Francisco Bay mud, and it seems possible that they use as substrate the nucleic acids derived from the microbial bodies in the sediment, but information on this point is lacking.

Information on purine metabolism in termites shows that uric acid is excreted by the Malphigian tubules, the excretory organ of termites, into the hind-gut at its junction with the mid-gut. The uric acid is fermented by bacteria, with production of ammonia that is then assimilated by other members of the gut microbiota (Potrikus and Breznak, 1980a,b). Termites are remarkably efficient in their assimilation of the sparse nitrogen in their food (Hungate, 1944a), and this recycling of uric acid is an important factor in this economy.

Another instance of the importance of a nitrogenous substrate was reported by Helge Larsen (personal communication). Large quantities of methane are released when salted fish are stored after the catch. Conditions become anaerobic and trimethylene oxide, the fish equivalent of urea, is converted to trimethylamine, from which certain methanogenic bacteria can form methane from each methyl group (Hippe et al., 1979).

These examples indicate the plethora of specific niches which may exist in diverse environments where anaerobic transformations of Earth's biotic complex occur.

ULTIMATE EVENTS IN ANAEROBIC TRANSFORMATIONS

The major substrate in anaerobic transformations is carbohydrate, and in an earlier section we described a number of the types of fermentations of carbohydrates by pure [axenic (*no stranger*)] microbial cultures. We have also analyzed in part the rumen fermentation to determine its carbohydrate fermentation patterns, i.e., the patterns in an open or agnotobiotic system. The pattern found in the rumen is probably that of a great many natural anaerobic habitats (Holmes and Freischel, 1978); in most of them substrates are chiefly insoluble and limiting, leading to competition for pathways by which the most anaerobic work can be accomplished. Since substrate is limiting the pathways yielding the most cell material per unit of substrate will have a competitive advantage, and therefore will tend to predominate.

This leads to formation of acetic, propionic, and butyric acids as waste products, much as in the rumen, because in the transformations producing them more energy is conserved in a high form, i.e., in microbial cells, than in fermentations forming other waste products. But in sediments and anaerobic digestors the decomposition process is carried beyond these acids to the ultimate conversion into CO_2, CH_4, and H_2S. This extended transformation is possible because of the greater time during which the organic materials are subject to microbial attack.

The average residence time of organic substrates in the rumen is perhaps a few hours in very small ruminants (Hungate et al., 1959), ranging up to perhaps two days in large ruminants on a poor ration. The general experience with municipal anaerobic digestors is that retention times of less than 10 days are difficult to maintain indefinitely; in sediments the complete conversion takes months to years.

The advantages of carbohydrate as a source of energy anaerobically are largely exploited in the processes leading to the rumen acids. In general, though not uniformly, the rate of microbial growth corresponds to the metabolic energy available, and in consequence, the rumen microbes, using the energy-rich early steps in carbohydrate breakdown, can on the average grow faster than the organisms depending on the later transformations less capable of metabolic work. The standard free energies of some of the most important transformations are shown in Table VI. Since in the anaerobic digestors it is desired to obtain as complete a transformation as is possible, the overall process must include also the anaerobic dehydrogenation (proton reduction) steps which are relatively energy poor, and will therefore support less rapid growth of the microbes effecting them. In a continuous fermentation, such as the anaerobic digestor, the residence time must equal the reciprocal of the net specific growth rate of the microorganisms; the slower the growth rate the longer the residence time.

From Table VI it is evident that the conversions of propionate and butyrate to methane, acetate and CO_2 are particularly poor in accomplishing

TABLE VI
Possible Work in Selected Fermentation Reactions

Reaction	Free energy change (kJ)
glucose → 2 lactate$^-$ + 2 H^+	−133
glucose + 4 H_2O → 2 acetate$^-$ + 2 HCO_3^- + 4 H^+ + 4 H_2	−206
glucose + H_2O → 2 acetate$^-$ + HCO_3^- + 3 H^+ + CH_4	−342
glucose + 3 H_2O → 3 HCO_3^- + 3 H^+ + 3 CH_4	−404
adenosine diphosphate + P_i → adenosine triphosphate + H_2O	+45
acetate$^-$ + H_2O → HCO_3^- + CH_4	−31
propionate$^-$ + 3/4 H_2O → 1/4 HCO_3^- + 1/4 H^+ + 3/4 CH_4 + acetate$^-$	−26
butyrate$^-$ + 1/2 H_2O + 1/2 HCO_3^- → 1/2 H^+ + 1/2 CH_4 + 2 acetate$^-$	−20

the work of cell growth. Starting with the pure substances only a minute conversion can occur before the concentration of H_2 reaches an equilibrium concentration, and H_2 consumption proceeds as rapidly as H_2 formation. But measurements of H_2 concentration in methanogenic habitats (Hungate, 1967; Scranton et al., 1983) show that the concentration of H_2 is kept below the equilibrium concentrations, and the anaerobic oxidation of butyrate and propionate, via dehydrogenation, hydration, and decarboxylation, proceeds as far as acetate, which is then converted to methane and carbon dioxide by acetitrophic methanogens, alone or in conjunction with proton reducers.

The conversion of propionate to methane, CO_2 and acetate has been accomplished in a two-member (monoxenic) culture by Boone and Bryant (1980) and of butyrate by McInerney et al. (1981). A methanogen, *Methanospirillum hungatei,* was seeded in excess in tubes of melted agar medium containing propionate or butyrate, and a series of these tubes were inoculated with dilutions of a digestor (enrichment) fed the corresponding acid. The methanogen was so abundant that its innumerable colonies were too small to distinguish, but the fewer and larger proton-reducing colonies, decreasing in number according to the dilution, could be picked and subcultured. Thus far it has not been possible to grow the proton reducer, i.e., the one removing the electron from the acid to H^+, in other than a two-member culture. It is difficult to maintain such low pressures of dihydrogen except with an accompanying organism having a great H_2 affinity.

The butyrate-oxidizing organism has been named *Syntrophobacter wolfii* and the propionate oxidizer *Syntrophomonas wolinii.* Additionally Mountfort and Bryant (1983) have demonstrated that an anaerobic rod in coculture with *Methanospirillum hungatei* or *Desulfovibrio* can oxidize benzoate to acetate. In this case a division time of 132 hr was necessary with a sulfate-reducing organism in the coculture, whereas with the methanogen it was 166 hr because of the greater energy available in the oxidation of H_2 by sulfate (−152.2 kJ) as compared to CO_2 (−135.6 kJ).

These various developments in our knowledge of anaerobic transformation of organic matter to products with the lowest energy state, i.e., CO_2, CH_4, and H_2S, can explain many of the previously little understood anaerobic transformations of organic matter in nature. With the availability of reliable methods for pure cultivation, including inhibitors of cell wall synthesis by eubacteria (Godsy, 1980), many methanogenic bacteria have been isolated from varied habitats. Undoubtedly, a great many additional strains will be found, differing in their nutrient requirements, substrates, substrate affinities and growth rates. The biochemical apparatus that evolved in very ancient times (Fox et al., 1980) has been preserved over the span of life on Earth and provides us with some intriguing clues on the early evolution of living things. The methanogen (Leigh and Jones, 1980) from the geothermal fissure grows optimally at 86°C, and some of the population in that habitat can grow well above 100°C, a temperature much below boiling at that water depth.

The pathways and the catalysts and coenzymes of methanogenesis are so unique that they seem worth cataloguing here (Table VII). Barker (1940) emphasized that four two-electron steps were concerned in the reduction of carbon dioxide to methane, and the efforts of Wolfe and his students at Illinois (Romesser, 1978) have developed a model on which many of the findings in his laboratory and others can be hinged. This is shown in Figure 6.

The anaerobic transformations discussed thus far are summarized in Figure 7. The processes in the upper part of the figure portray the fermentations of carbohydrates; the lower portion represents the processes of proton reduction and methanogenesis, the terminal stages in the anaerobic stabilization of organic substrates.

In the anaerobic oxidation of fats, the glycerol is readily fermented to

TABLE VII

Enzymes and Coenzymes Concerned in Methanogenesis

CoM mercaptoethanesulfonic acid, HS-C(H)(H)-C(H)(H)-SO_3H

Methyl CoM, H_3C-S-C(H)(H)-C(H)(H)-SO_3H

F_{420}, 8-hydroxy 5-deazaisoalloxazine

F_{342}, methanopterin, dihydromethanopterin, and carboxydihydromethanopterin (yellow fluorescent compound)

F_{430}, nickel tetrapyrrole, gives yellow color to methyl-CoM reductase

Methyl-CoM reductase, molecular weight 300,000, composed of 3 subunits

Membrane-linked hydrogenase

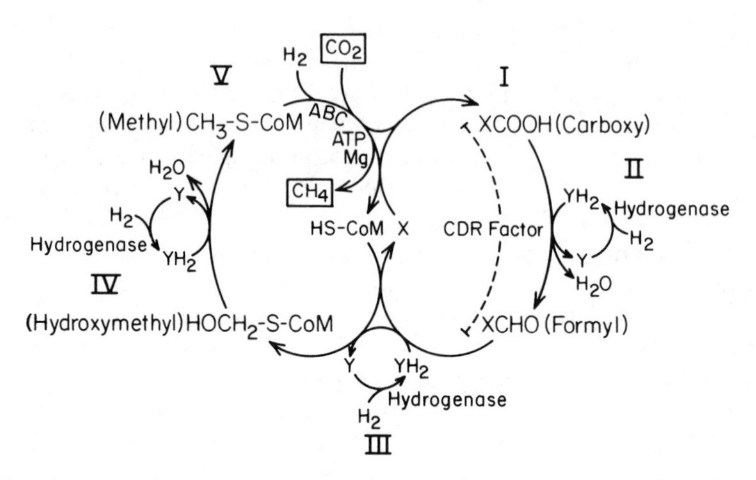

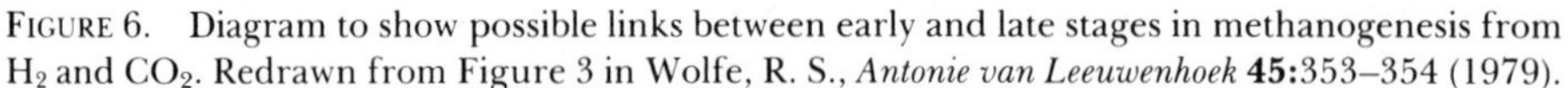
FIGURE 6. Diagram to show possible links between early and late stages in methanogenesis from H_2 and CO_2. Redrawn from Figure 3 in Wolfe, R. S., *Antonie van Leeuwenhoek* **45:**353–354 (1979).

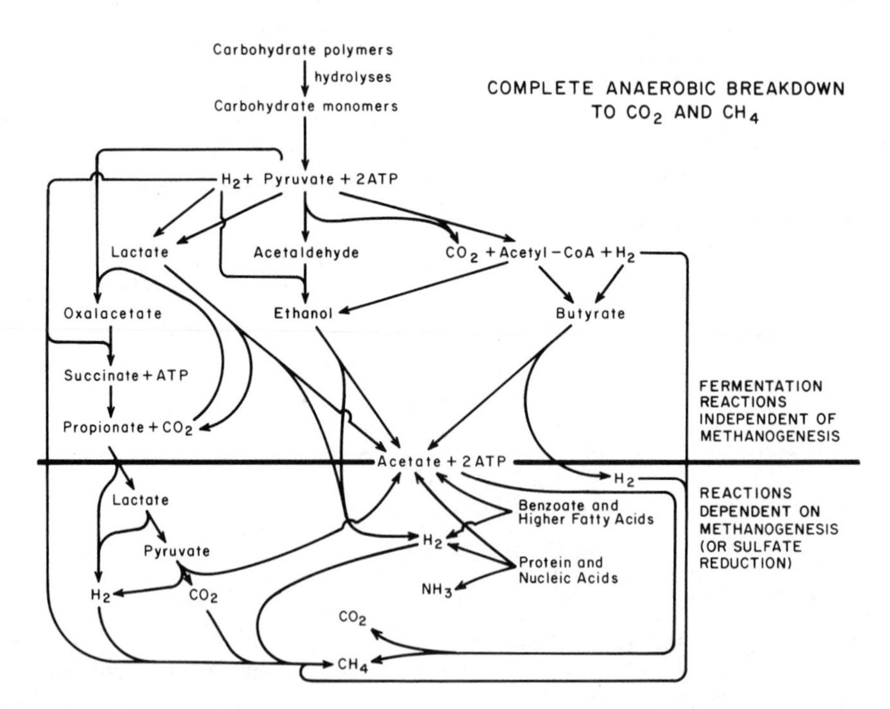

FIGURE 7. Diagram to show steps concerned in the conversion of carbohydrate to methane and carbon dioxide.

form propionate. Both being at the same average oxidation level, the fermentation involves an internal oxidation reduction similar to that in the homolactic fermentation. The long-chain fatty acids are oxidized by the same beta-oxidation process that occurs in aerobes, the difference being that the electrons are transferred to form methane or to reduce sulfate rather than combining with O_2. It had long been believed that compounds containing phenyl and other carbon ring groups required the energy available through oxidation with O_2 in order to break the ring, but Buswell early demonstrated that benzoate was converted to methane and CO_2 in anaerobic municipal digestors. Also, the soluble ring groups obtained from lignin have been found fermentable to methane, but as yet, good evidence for anaerobic cleavages of the bonds in natural insoluble lignin has not been presented.

On the basis of the low energy level represented anaerobically by carbon conversions to methane, a logical extension would be that the higher alkanes will tend to generate methane. The formation of methane from higher hydrocarbons is exergonic if H_2 is available as a source of the needed hydrogen. Muller (1957) has demonstrated the generation of methane from certain petroleum and wax samples he incubated for periods as long as 56 months. The maximum CH_4 production was from a sample of moderately purified paraffin wax which yielded 188.4 liters (corrected for the inoculum control) during 29 months of incubation with a sludge inoculum. In the five most productive cultures, the ratio of CH_4/CO_2 was 4.2, higher than the 3.0 ratio expected if the paraffin were reduced to methane with the H_2 obtainable by dehydrogenation and hydration. Some CO_2 must have dissolved or precipitated. It is tempting to suggest that a source of hydrogen was available in some samples and not in others. Dehydrogenation of some hydrocarbons to obtain hydrogen for reduction of others to methane is thermodynamically possible but may be difficult to catalyze anaerobically.

$$4\ C_2H_6 + 2\ H_2O \rightarrow CO_2 + 7\ CH_4,\ \Delta G_o' = -135\ \text{kJ} \qquad (14)$$

Reactions of this sort might explain the common occurrence of methane in petroliferous formations.

SOLUBLE NUTRIENTS IN ANAEROBIC TRANSFORMATIONS

Quantitatively, the most important soluble nutrients are the carbohydrates. Although the vast majority of carbohydrate plant cell wall material is insoluble, only the soluble molecules from digestion can diffuse into the protoplast of the microbe. The capacity for metabolizing the soluble carbohydrate exceeds the rate at which it is solubilized, and the affinity of the cells for the dissolved molecules is generally high. In consequence, the concentration of the soluble sugars is low. Russell and Baldwin (1978) have determined the K_s for several sugars in some species of rumen bacteria but did not measure the concentration of the sugars in the rumen. One of the great problems of autecology is to understand what a particular organism is doing in its natural habitat. A determination of its K_s for soluble substrates present in the habitat should permit estimation of the rate at which a given substrate could be used at its ambient concentration. Presuming that no other factors limited the rate of growth, determination of the rate at which the substrate can be taken up from the ambient concentration should permit estimation of the rate of growth and therefore of the productivity of the particular microbe concerned. This is, at the least, a challenge to ecologists.

Thus far, we have focused attention on the energy-containing substrates for accomplishing the work of synthesis in anaerobic transformations. We mentioned earlier the advisability of using some of the habitat material in designing media for isolation of new organisms with unknown growth requirements. The fluid from a successful enrichment culture can also be the habitat material utilized.

There are many nutritional factors besides the energy source that can limit the work done by anaerobes; *Bacteroides amylophilus* requires only starch and inorganic nutrients, and *Methanobacterium formicicum* and *M. thermoautotrophicum* need no organic nutrients, but this is exceptional. Most anaerobic bacteria require or are stimulated by various organic nutrilites. Marvin Bryant's extensive nutritional studies (1974) show that many rumen bacteria require acetate in the culture medium even though many of them produce it. Presumably during long evolution in a habitat containing ca. 0.3 % acetate, they have found it advantageous to absorb rather than to synthesize it. They cannot efficiently assimilate it at the low concentrations which they initially produce in a new culture lacking acetate. The optimal concentration for *Ruminococcus albus* strain 8 can be as much as 0.1 % (Hungate and Stack, 1982).

Similarly, as already mentioned, isobutyrate, isovalerate, and 2-methylbutyrate are used by many species for the synthesis of valine, leucine and isoleucine, respectively. Phenylacetic acid is a precursor of phenylalanine (Allison, 1965) and phenylpropionic acid is required (Hungate and Stack, 1982) for synthesis of capsular material in strain 8 of *Ruminococcus albus* (Stack and Hungate, 1984). 1,4-Naphthoquinone is required by *Succinivibrio dextrinosolvens* (Gomez-Alarcon et al., 1982).

The branched chain and phenylacetic acids are formed by deamination and decarboxylation of the corresponding amino acids (Allison, 1969) in reactions in which some ATP is formed, but it is only a supplementary source of energy for *Bacteroides ruminicola* (Russell, 1983) and carbohydrate is required, as is the case with the vast majority of microbes in the rumen. *Bacteroides ruminicola* utilizes peptides more readily than amino acids (Bryant, 1974).

Ammonia, probably the chief source of nitrogen for anaerobes in nature, is formed by deamination of amino acids, by nitrogen fixation, and by reduction of nitrate. Ammonia is supplied to the rumen also by urea diffusing from the blood and entering with the saliva. It rapidly splits into ammonia and CO_2. Hemin is a required growth factor for *Bacteroides ruminicola; Fusi-*

formis nigrescens needs vitamin K (Lev, 1959), and many rumen species require or are stimulated by one or more of the following vitamins: thiamin, riboflavin, biotin, folic acid, p-aminobenzoic acid, pyridoxamine, pantothenic acid, and B_{12}. Coenzyme M (mercaptoethanesulfonic acid) is required by *Methanobrevibacter ruminantium,* and an as yet unidentified factor is needed by *Methanogenium mobile.* Undoubtedly additional growth factors are important in the rumen microcosm but have not yet been identified.

In the case of *Ruminococcus albus* strain 8, it has been possible to obtain as rapid growth and digestion of cellulose on a chemically defined medium as on a medium containing rumen fluid. By inoculating this strain into a gnotobiotic sheep it should be possible to obtain a rate of cellulose utilization equal to that in a conventional animal, provided all the nutrients required by the strain are given in the proper concentrations. A challenge for microbial ecologists is to identify the kinds of bacteria essential to the sheep or other ruminants, identify their required nutrients, and simulate in a gnotobiotic animal the nutritional features characteristic of the domesticated animal (Lysons et al., 1977). This holds also for the biodigestor and other nonalimentary anaerobic ecosystems.

LEAKAGE OF METABOLITES FROM ANAEROBIC DIGESTERS OF WALL MATERIALS

The leakage of cellulose digestion products from the cellulose digester to noncellulolytic commensals has been described in previous paragraphs. The phenomenon is probably widespread. The factors concerned occur in many natural habitats, both aerobic and anaerobic. Leakage can maintain a highly heterogeneous and abundant population under conditions in which the muralytic species would not grow alone or would be much less productive. The commensal organisms may manufacture nutrilites needed by the muralytic species, absorb toxic traces of O_2, or metabolize waste products that might become inhibitory, as described already in the discussion of the removal of H_2 from the environment of proton-reducing anaerobic oxidizers of organic compounds.

A striking example of leakage was reported by Baresi et al. (1978). An acetate enrichment culture receiving no organic food except acetate elicited the development of *Methanosarcina barkeri* as the only organism capable of growing on the acetate; however, also within the culture were at least three other bacterial species, each showing a colony count equal to or greater than that of the methanogen, each producing H_2 when grown apart from the methanogen, and among them showing requirements for several vitamins and amino acids. The results indicate that the methanogen was unable to retain within itself all of the products it synthesized. Even on the very little energy the acetate provided, it synthesized and lost enough of its products to support the commensal growth of several other species.

Many of the nutrients in an anaerobic ecosystem are waste products excreted by some members of the community, and used by others. The branched-chain acids, waste products from amino acid utilization used by other species for synthesis of the amino acids, are an example. Nutritional interrelationships of this sort are widespread among anaerobes, and may explain the difficulty in reconstituting a formal gut flora with pure cultures of the bacteria isolated from the gut. The alimentary tract contains hundreds, possibly thousands, of distinct strains, each accomplishing some transformation by which it grows, and each excreting or secreting and leaking the diversity of nutrients which we find essential or stimulatory to the community as a whole.

The reasons that some members of the population have very complex nutritional requirements whereas others are prototrophic are not clear. One can reason that with the limited energy supplies available anaerobically it is more economical for an organism to absorb a nutrient from the environment than to generate it internally, and that, if the environment is such that availability of the nutrient is assured, members of the biota will become adapted to its use. It may be argued that leakage will be more prominent in anaerobes than in aerobes because less energy is available to maintain differences in concentrations of given substances within and without the cell. Whatever the basic explanations, exceedingly complex nutritional interrelationships can develop in anaerobic ecosystems.

The many factors concerned in the actual survival of species in nature are not well understood, but multiplicity of species in a community appears to confer stability on an ecosystem, whereas communities of few species are more fragile. Evolution of a few species which perform a multiplicity of functions might seem to be a possible course of evolution, but the expense of replicating metabolic machinery that would only rarely be used is probably too great a burden. Specialists integrated with other specialists seems to have been the more common course of evolution.

CONCLUSION

This brief eclectic consideration of anaerobic biotransformation cannot be an exhaustive account of anaerobic microbes and their activities in nature. It is intended rather to review some of the relationships impressing the author as generally applicable to problems of studying anaerobes in nature. Natural habitats are an endless challenge to the curious ecologist. Their analysis can endlessly enrich our fund of knowledge and breadth of understanding.

REFERENCES

Akin, D. E., Amos, H. E., Barton, F. E., II, and Burdick, D., 1973, Rumen microbial degradation of grass tissue revealed by scanning electron microscopy, *Agron. J.* **65:**825–828.

Akin, D. E., Burdick, D., and Amos, H. E., 1974, Comparative degradation of coastal bermudagrass, Coastcross-1 bermudagrass, and Pensacola bahiagrass by rumen microorganisms revealed by scanning electron microscopy, *Crop Sci.* **14:**537–541.

Allison, M. J., 1965, Phenylalanine biosynthesis from phenylacetic acid by anaerobic bacteria from the rumen, *Bioch. Biophys. Res. Comm.* **18:**30–35.

Allison, M. J., 1969, Biosynthesis of amino acids by ruminal micro-organisms, *J. Anim. Sci.* **29:**797–807.

Allison, M. J., 1978, Production of branched-chain volatile fatty acids by certain anaerobic bacteria, *Appl. Environ. Microbiol.* **35:**872–877.

Aranki, A., Syed, S. A., Kenney, E. B., and Freter, R., 1969, Isolation of anaerobic bacteria from human gingiva and mouse caecum by means of a simplified glove box procedure, *Appl. Microbiol.* **17:**568–576.

Balch, W. E., Schoberth, S., Tanner, R. S., and Wolfe, R. S., 1977, *Acetobacterium* a new species of hydrogen-oxidizing, carbon dioxide-reducing anaerobic bacteria, *Intern. J. Syst. Bact.* **27:**355–361.

Baldwin, R. L., Koong, L. J., and Wyatt, M. J., 1977, A dynamic model of ruminant digestion for evaluation of factors affecting nutritive value, *Agric. Systems* **2:**255–288.

Baresi, L., Mah, R. A., Ward, D. M., and Kaplan, I. R., 1978, Methanogenesis from acetate: enrichment studies, *Appl. Environ. Microbiol.* **36:**186–197.

Barker, H. A., 1936a, On the biochemistry of the methane fermentation, *Arch. Mikrob.* **7:**404–419.

Barker, H. A., 1936b, Studies upon the methane-producing bacteria, *Arch. Mikrob.* **7:**420–438.

Barker, H. A., 1940, Studies on the methane fermentation: IV. The isolation and culture of *Methanobacterium omelianskii, Antonie van Leeuwenhoek* **6:**201–220.

Barker, H. A., 1961, Fermentations of nitrogenous organic compounds. Chap. 3, in *The Bacteria,* Vol. II. (I. C. Gunsalus and R. Y. Stanier, eds.), Academic Press, New York, pp. 151–207.

Barker, H. A., Ruben, S., and Kamen, M. D., 1940, The reduction of radioactive carbon dioxide by methane-producing bacteria, *Proc. Natl. Acad. Sci.* **26:**426–430.

Barker, H. A., Kamen, M. D., and Haas, V., 1945, Carbon dioxide utilization in the synthesis of acetic and butyric acids by *Butryribacterium rettgeri, Proc. Natl. Acad. Sci.* **31:**355–360.

Bauchop, T., 1960, Studies on fermentative mechanisms in *Zymosarcina ventriculi,* Ph.D. Thesis, University of Glasgow.

Bauchop, T., 1967, Inhibition of rumen methanogenesis by methane analogs, *J. Bacteriol.* **94:**171–175.

Bauchop, T., 1979, Rumen anaerobic fungi of cattle and sheep, *Appl. Environ. Microbiol.* **38:**148–158.

Bauchop, T., and Elsden, S. R., 1960, The growth of micro-organisms in relation to their energy supply, *J. Gen. Microbiol.* **23:**457–469.

Bauchop, T., and Martucci, R., 1968, Ruminant-like digestion in the langur monkey, *Science* **161:**698–699.

Blackburn, T. H., and Hungate, R. E., 1963, Succinic acid turnover and propionate production in the bovine rumen, *Appl. Microbiol.* **11:**132–135.

Boone, D. R., and Bryant, M. P., 1980. Propionate-degrading bacterium, *Syntrophobacter wolinii* (sp. nov., gen. nov.) from methanogenic ecosystems, *Appl. Environ. Microbiol.* **40:**626–632.

Breznak, J. A., 1982, Intestinal microbiota of termites and other xylophagous insects, *Ann. Rev. Microbiol.* **36:**323–343.

Bryant, M. P., 1956, The characteristics of strains of *Selenomonas* isolated from bovine rumen contents, *J. Bacteriol.* **72:**162–167.

Bryant, M. P., 1974, Nutritional features and ecology of predominant anaerobic bacteria of the intestinal tract, *Am. J. Clin. Nutr.* **27:**1313–1319.

Bryant, M. P., and Burkey, L. A., 1953, Cultural methods and some characteristics of some of the more numerous groups of bacteria in the bovine rumen, *J. Dairy Sci.* **36:**205–217.

Bryant, M. P., Wolin, E. A., Wolin, M. J., and Wolfe, R. S., 1967, *Methanobacillus omelianskii,* a symbiotic association of two species of bacteria, *Arch. Microbiol.* **59:**20–31.

Buchner, E., 1897, Alkoholische Gärung ohne Hefezellen, *Ber. d. deutsch. chem. Ges.* **30:**117–124, 1110–1113.

Burri, R., 1902, Zur Isolierung der Anaëroben. *Centralbl. Bakteriologie II* **8:**533–537.

Buswell, A. M., and Sollo, F. W., 1948, The mechanism of the methane fermentation, *J. Am. Chem. Soc.* **70:**1778–1780.

Cahen, F., 1887, Ueber das Reductionsvermögen der Bacterien, *Zentralbl. f. Hygiene* **2:**386–396.

Calloway, D. H., 1968, The use of expired air to measure intestinal gas formation, *Ann. N. Y. Acad. Sci.* **150:**82–95.

Chung, K.-T., and Hungate, R. E., 1976, Effect of alfalfa fiber substrate on culture counts of rumen bacteria, *Appl. Environ. Microbiol.* **32:**649–652.

Colvin, J. R., and Leppard, G. G., 1977, The biosynthesis of cellulose by *Aceotobacter xylinum* and *Acetobacter acetigenum, Can. J. Microbiol.* **23:**701–709.

Conrad, R., Aragno, M., and Seiler, W., 1983, Production and consumption of hydrogen in a eutrophic lake, *Appl. Environ. Microbiol.* **45:**502–510.

Counotte, G. H. M., 1981, Regulation of lactate metabolism in the rumen. Ph.D. Thesis, University of Utrecht.

Counotte, G. H. M., and Prins, R. A., 1979, Regulation of rumen lactate metabolism and the role of lactic acid in nutritional disorders of ruminants, *Veterin. Sci. Commun.* **2:**277–303.

Counotte, G. H. M., Prins, R. A., Janssen, R. H. A. M., and DeBie, M. J. A., 1981, Role of *Megasphaera elsdenii* in the fermentation of DL-[2-^{13}C]lactate in the rumen of dairy cattle, *Appl. Environ. Microbiol.* **42:**649–655.

Dehority, B. A., 1973, Hemicellulose degradation by rumen bacteria, *Fed. Proc.* **32:**1819–1825.

Doddema, H. J., and Vogels, G. D., 1978, Improved identification of methanogenic bacteria by fluorescent microscopy, *Appl. Environ. Microbiol.* **36:**752–754.

Dunlop, R. H., 1972, Pathogenesis of ruminant lactic acidosis, *Adv. Vet. Sci. Comp. Med.* **16:**259–302.

Edwards, T., and McBride, B. C., 1975, New method for the isolation and identification of methanogenic bacteria, *Appl. Microbiol.* **29:**540–545.

El-Shazly, K., and Hungate, R. E., 1965, Fermentation capacity as a measure of net growth of rumen microorganisms, *Appl. Microbiol.* **13:**62–69.

Esmarch, E. von, 1886, Ueber einen Modifikation des Kochschen Plattenverfahrens zur Isolierung und zum quantitativen Nachweis von Mikroorganismen, *Zeitschr. f. Hyg.* **I:**293–301.

Fenchel, T. M., 1980, The protozoan fauna from the gut of the green turtle, *Chelonia mydas* with a description of *Balantidium bacteriophorus, Arch. Protistenk.* **123:**22–26.

Fontaine, F. E., Peterson, W. H., McCoy, E., Johnson, M. J., and Ritter, G., 1942, A new type of glucose fermentation by *Clostridium thermoaceticum* n. sp., *J. Bacteriol.* **41:**701–715.

Forsberg, C. W., Beveridge, T. J., and Hellstrom, A., 1981, Cellulase and xylanase release from *Bacteroides succinogenes* and its importance in the rumen environment, *Appl. Environ. Microbiol.* **42:**886–896.

Fox, G. E., Stackebrant, E., Hespell, R. B., Gibson, J., Maniloff, J., Dyer, T. A., Wolfe, R. S., Balch, W. E., Tanner, R. S., Mangrum, L. J., Zablen, L. B., Blackmore, R., Gupta, R., Bonen, L., Lewis, B. J., Stahl, D. A., Luehrsen, K. R., Chen, K. N., and Woese, C. R., 1980, The phylogeny of prokaryotes, *Science* **209:**457–463.

Fränkel, C., 1888, Ueber die Kultur anaërober Mikroorganismen, *Centralbl. Bakteriologie Parasitenk* **3:**735–740, 763–768.

Gay-Lussac, J. L., 1810, Extrait d'un memoire sur la fermentation, *Ann. Chim.* **76:**245–259.

Genthner, B. R. S., and Bryant, M. P., 1982, Growth of *Eubacterium limosum* with carbon monoxide as the energy source, *Appl. Environ. Microbiol.* **43:**70–74.

Godsy, E. M., 1980, Isolation of *Methanobacterium bryantii* from a deep aquifer by using a novel broth antibiotic disk method, *Appl. Environ. Microbiol.* **39:**1074–1075.

Gomez-Alarcon, R. A., Dowd, C. O., Leedle, J. A. Z., and Bryant, M. P., 1982, 1,4 naphthoquinone and other nutrient requirements of *Succinovibrio dextrinosolvens, Appl. Environ. Microbiol.* **44:**346–350.

Gruby, D., and Delafond, O., 1843, Recherches sur des animalcules se developpant en grande nombre dans l'estomac et dans les intestine, pendant la digestion des animaux herbivores et carnivores, *C. rend Acad. Sci.* **17:**1305–1308.

Hackett, W. F., Connors, W. J., Kirk, T. K., and Zeikus, J. G., 1977, Microbial decomposition of synthetic ^{14}C-labeled lignins in nature: lignin biodegradation in a variety of natural materials, *Appl. Environ. Microbiol.* **33:**43–51.

Harden, A., and Young, W. J., 1905, The alcoholic fermentation of yeast juice. *J. Physiol.* 32, Proc. 12 Nov., 1904, pp. i–ii.

Healy, J. B., Jr., and Young, L. Y., 1979, Anaerobic biodegradation of eleven aromatic compounds to methane, *Appl. Environ. Microbiol.* **38:**84–89.

Herwig, R. P., Staley, J. T., Nerini, M. K., and Braham, H. W., 1984, *Appl. Environ. Microbiol.* **47:**421–423.

Hippe, H., Casparic, D., Fiebig, K., and Gottschalk, G., 1979, Utilization of trimethylamine and other N-methyl compounds for growth and methane formation by *Methanosarcina barkeri, Proc. Natl. Acad. Sci. USA* **76:**494–498.

Holmes, P., and Freischel, M. R., 1978, H_2-producing bacteria in digesting sewage sludge isolated on simple, defined media, *Appl. Environ. Microbiol.* **36:**394–395.

Hungate, R. E., 1938, Studies of the nutrition of *Zootermopsis.* II. The relative importance of the termite and the protozoa in wood digestion. *Ecology* **19:**1–25.

Hungate, R. E., 1939, Studies of the nutrition of *Zootermopsis.* III. The anaerobic carbohydrate dissimilation by the intestinal protozoa. *Ecology* **20:**230–245.

Hungate, R. E., 1944, Studies on cellulose fermentation. I. The culture and physiology of an anaerobic cellulose-digesting bacterium, *J. Bacteriol.* **48:**499–513.

Hungate, R. E., 1944, Termite growth and nitrogen utilization in laboratory cultures, *Proc. Texas Acad. Sci.* **27:**91–98.

Hungate, R. E., 1946, The symbiotic utilization of cellulose, *J. Elisha Mitchell Scientific Society* **62:**9–24.

Hungate, R. E., 1947, Studies on cellulose fermentation. III. The culture and ioslation of cellulose-decomposing bacteria from the rumen of cattle, *J. Bacteriol.* **53:**631–645.

Hungate, R. E., 1955, Why carbohydrates? in: *Biochemistry and Physiology of Protozoa,* Volume II (S. Hutner and A. Lwoff, eds.), Academic Press, New York, pp. 195–197.

Hungate, R. E., 1962, Ecology of bacteria, in: *The Bacteria,* Volume IV (I. C. Gunsalus and R. Y. Stanier, eds.), Academic Press, New York, pp. 95–119.

Hungate, R. E., 1966, *The Rumen and Its Microbes,* Academic Press, New York.

Hungate, R. E., 1967, Hydrogen as an intermediate in the rumen fermentation, *Arch. Mikrobiol.* **59:**158–164.

Hungate, R. E., 1976, Microbial activities related to mammalian digestion and absorption of food, in: *Fiber in Human Nutrition* (G. A. Spiller and R. J. Amen, eds.), Plenum Press, New York, pp. 131–149.

Hungate, R. E., 1982, Methane formation and cellulose digestion—biochemical ecology of the rumen ecosystem, *Experientia* **38:**189–192.

Hungate, R. E., and Stack, R. J., 1982, Phenylpropionic acid: a growth factor for *Ruminococcus albus, Appl. Environ. Microbiol.* **44:**79–83.

Hungate, R. E., 1984, Microbes of nutritional importance in the alimentary tract, *Proceedings of the Nutritional Society* **43:**1–11.

Hungate, R. E., Dougherty, R. W., Bryant, M. P., and Cello, R. M., 1951, Microbiological and physiological changes associated with acute indigestion in sheep, *Cornell Veterinarian* **42:**423–449.

Hungate, R. E., Phillips, G. D., MacGregor, A., Hungate, D. P., and Buechner, H. K., 1959, Microbial fermentation in certain mammals, *Science* **130:**1192–1194.

Hungate, R. E., Phillips, G. D., Hungate, D. P., and MacGregor, A., 1960, A comparison of the rumen fermentation in European and zebu cattle. *J. Agric. Sci.* **54:**196–201.

Hungate, R. E., Mah, R. A., and Simesen, M., 1961, Rates of production of individual volatile fatty acids in the rumen of lactating cows, *Appl. Microbiol.* **9:**554–561.

Hungate, R. E., Smith, W., Bauchop, T., Yu, I., and Rabinowitz, J. C., 1970, Formate as an intermediate in the rumen fermentation, *J. Bacteriol.* **102:**389–397.

Hungate, R. E., Reichl, J., and Prins, R., 1971, Parameters of rumen fermentation in a continuously fed sheep: Evidence of a microbial rumination pool, *Appl. Microbiol.* **22:**1104–1113.

Jayasuriya, G. C. N., and Hungate, R. E., 1959, Lactate conversions in the bovine rumen, *Arch. Bioch. Biophys.* **82:**274–287.

Karrer, P., 1930, The enzymatic decomposition of native and reprecipitated cellulose, artificial silk, and chitin, *Kolloid-Zeit.* **52:**304–349.

Kluyver, A. J., and Donker, H. J. L., 1926, Die Einheit in der Biochemie, *Chem. d. Zelle u. Gewebe* **13:**134–190.

Kursteiner, J., 1907, Beiträge zur Untersuchungstechnik obligat anaërober Bakterien, sourie zur Lehre von der Anaërobiose überhaupt, *Centralbl. Bakteriologie* **19:**1–26, 97–115, 202–220, 385–394.

Lanigan, G. W., 1976, *Peptococcus heliotrinreducans,* sp. nov., a cytochrome-producing anaerobe which metabolizes pyrrolizidine alkaloids, *J. Gen. Microbiol.* **94:**1–10.

Latham, M. J., Brooker, B. E., Pettipher, G. L., and Harris, P. J., 1978a, *Ruminococcus flavefaciens* cell coat and adhesion to a cotton cellulose and to cell walls in leaves of perennial ryegrass (*Lolium perenne*), *Appl. Environ. Microbiol.* **35:**156–165.

Latham, M. J., Brooker, B. E., Pettipher, G. I., and Harris, P. J., 1978b, Adhesion of *Bacteroides succinogenes* in pure culture and in the presence of *Ruminococcus flavefaciens* to cell walls in leaves of perennial ryegrass (*Lolium perenne*), *Appl. Environ. Microbiol.* **35:**1166–1173.

Lawton, E. J., Bellamy, W. D., Hungate, R. E., Bryant, M. P., and Hall, E., 1951, Some effects of high velocity electrons on wood. *Science* **113:**380–382.

Leedle, J. A. Z., Bryant, M. P., and Hespell, R. B., 1982, Diurnal variations in bacterial numbers and fluid parameters in ruminal contents of animals fed low- or high-forage diets, *Appl. Environ. Microbiol.* **44:**402–412.

Leigh, J. A., and Jones, W. J., 1983, A new extremely thermophilic methanogen from a hydrothermal vent, *Abst. Ann. Meeting Amer. Soc. Microbiology,* p. 142.

Leng, R. A., and Leonard, G. J., 1965, Loss of methyl tritium from (^{3}H) acetate in rumen fluid. *Nature* **207:**760–761.

Lev, M., 1959, The growth-promoting activity of compounds of the vitamin K group and analogues for a rumen strain of *Fusiformis nigrescens. J. Gen. Microbiol.* **20:**697–703.

Levitt, M. D., and Ingelfinger, F. J., 1968, Hydrogen and methane production in man, *Ann. N. Y. Acad. Sci.* **150:**75–81.

Liborius, P., 1886, Beiträge zur Kenntnis des Sauerstoffbedürfnisses der Bakterien. *Zeitschr. f. Hygiene* **1:**115–177.

Lochhead, A. G., 1952, The nutritional classification of soil bacteria, *Proc. Soc. Appl. Bact.* **15:**15–20.

Lovley, D. R., and Klug, M. J., 1983, Sulfate reducers can outcompete methanogens at freshwater-sulfate concentrations, *Appl. Environ. Microbiol.* **45:**187–192.

Lysons, R. J., Alexander, T. J. L., and Wellstead, P. D., 1977, Nutrition and growth of gnotobiotic lambs, *J. Agric. Sci. Camb.* **88:**597–604.

McBee, R. H., 1977, Fermentation in the hindgut, in: *Microbial Ecology of the Gut* (R. T. J. Clarke and T. Bauchop, eds.), Academic Press, New York, pp. 185–217.

McInerney, M. J., Mackie, R. L., and Bryant, M. P., 1981, Syntrophic assocation of a butyrate-degrading bacterium and *Methanosarcina* enriched from bovine rumen, *Appl. Environ. Microbiol.* **41:**826–828.

Macy, J. M., Ljngdahl, L. G., and Gottschalk, G., 1978, Pathway of succinate and propionate formation in *Bacteroides fragilis, J. Bacteriol.* **134:**84–91.

Macy, J., Probst, I., and Gottschalk, G., 1975, Evidence for cytochrome involvement in fumarate reduction and adenosine-5′-triphosphate synthesis by *Bacteroides fragilis* grown in the presence of hemin, *J. Bacteriol.* **123:**436–442.

Magendie, F., 1816, Note sur les gaz intestinaux de l'homme sain, *Ann. Chim. Physique Ser. 2,* **2:**292–296.

Mah, R. A., Smith, M. R., and Baresi, L., 1978, Studies on an acetate-fermenting strain of *Methanosarcina, Appl. Environ. Microbiol.* **35:**1174–1184.

Margherita, S. S., and Hungate, R. E., 1963, Serological analysis of *Butyrivibrio* from the bovine rumen, *J. Bacteriol.* **86:**855–860.

Margherita, S. S., Hungate, R. E., and Storz, H., 1964, Variation in rumen *Butyrivibrio* strains, *J. Bacteriol.* **87:**1304–1308.

Moomaw, C. R., and Hungate, R. E., 1963, Ethanol conversion in the bovine rumen, *J. Bacteriol.* **85:**721–722.

Mountfort, D. O., and Bryant, M. P., 1982, Isolation and characterization of an anaerobic syntrophic benzoate-degrading bacterium from sewage sludge, *Arch. Microbiol.* **133:**249–256.

Muller, F. M., 1957, On methane fermentation of higher alkanes, *Antonie van Leeuwenhoek* **23**:369–384.

Mylroie, R. L., and Hungate, R. E., 1954, Experiments on the methane bacteria in sludge, *Can. J. Microbiol.* **1**:55–64.

Odelson, D. A., and Breznak, J. A., 1983, Volatile fatty acid production by the hindgut microbiota of xylophagous termites, *Appl. Environ. Microbiol.* **45**:1602–1613.

Omelianski, W., 1902, Ueber die Gärung der Cellulose, *Centralbl. f. Bakteriologie II* **8**:193–201, 225–231, 257–263, 289–294, 321–326, 353–361, 385–391.

Orpin, C. G., 1977, On the induction of zoosporogenesis in the rumen phycomycetes *Neocallimastix frontalis, Piromonas communis,* and *Sphaeromonas communis, J. Gen. Microbiol.* **101**:181–190.

Pasteur, L., 1858, Memoire sur la fermentation appelée lactique. Mémoires de la Société des sciences, de l'agriculture et des arts de Lille, 2nd ser., **5**:13–26.

Paterek, R., and Smith, P. H., 1983, Isolation of halophilic methanogenic bacterium from the sediments of Great Salt Lake and a San Francisco Bay saltern, *Abst. Ann. Meeting Amer. Soc. Microbiology,* p. 140.

Pfennig, N., and Biehl, H., 1976, *Desulfovibrio acetoxidans* gen. nov. and sp. nov., a new anaerobic sulfur-reducing, acetate-oxidizing bacterium, *Arch. Microbiol.* **110**:3–12.

Portugal, A. V., 1963, Some aspects of amino acid and protein metabolism in the rumen of the sheep. Ph.D. Thesis, Aberdeen University.

Potrikus, C. J., and Breznak, J. A., 1980a, Uric acid-degrading bacteria in guts of termites [*Reticulitermes flavipes* (Kollar)], *Appl. Environ. Microbiol.* **40**:117–124.

Potrikus, C. J., and Breznak, J. A., 1980b, Anaerobic degradation of uric acid by gut bacteria of termites, *Appl. Environ. Microbiol.* **40**:125–132.

Prévot, A. R., 1966, *Manual for the Classification and Determination of the Anaerobic Bacteria* (V. Fredette, ed. and transl.), Lea and Febiger, Philadelphia, 402 pp.

Prim, P., and Lawrence, J. M., 1975, Utilization of marine plants and their constituents by bacteria isolated from the gut of echinoids (Echinodermata), *Marine Biol.* **33**:167–173.

Quortrup, E. R., and Holt, A. L., 1940, Detection of potential botulinus-toxin-producing areas in western duck marshes with suggestions for control, *J. Bacteriol.* **41**:363–372.

Romesser, J. A., 1978, The activation and reduction of carbon dioxide to methane in *Methanobacterium thermoautotrophicum,* Ph.D. thesis, University of Illinois, Urbana.

Romesser, J. A., Wolfe, R. S., Mayer, F., Spies, E., and Walter-Mauruschaf, A., 1979, *Methanogenium,* a new genus of marine methanogenic bacteria, and characterization of *Methanogenium cariaci* sp. nov. and *Methanogenium marisnigri* sp. nov, *Arch. Microbiol.* **121**:147–153.

Roux, E., 1887, Sur la culture des microbes anaerobies. *Ann. Inst. Pasteur* **1**:49–62.

Russell, G. R., and Smith, R. M., 1968, Reduction of heliotrine by rumen microorganism, *Austral. J. Biol. Sci.* **21**:1277–1290.

Russell, J. B., 1983, Fermentation of peptides by *Bacteroides ruminicola* $B_1$4, *Appl. Environm. Microbiol.* **45**:1566–1574.

Russell, J. B., and Baldwin, R. L., 1978, Bacterial competition in the rumen, *Fed. Proc.* **37**:410.

Russell, J. B., Cotta, M. A., and Dombrowski, D. B., 1981, Rumen bacterial competition in continuous culture: *Streptococcus bovis* versus *Megasphaera elsdenii, Appl. Environ. Microbiol.* **41**:1394–1399.

Salyers, A. A., 1979, Energy sources of major intestinal fermentative anaerobes, *Am. J. Clin. Nutr.* **32**:158–163.

Scheifinger, C. C., and Wolin, M. J., 1973, Propionate formation from cellulose and soluble sugars by combined cultures of *Bacteroides succinogenes* and *Selenomonas ruminantium, Appl. Microbiol.* **26**:789–795.

Scheifinger, C. C., Latham, M. J., and Wolin, M. J., 1975, Relationship of lactate dehydrogenase specificity and growth rate to lactate metabolism by *Selenomonas ruminantium, Appl. Microbiol.* **30**:916–921.

Schnellen, C. G. T. P., 1957, Onderzoekingen over de methaangistung, Dissertation, Delft.

Scott, R. I., Williams, T. N., and Lloyd, D., 1983, Oxygen sensitivity of methanogenesis in rumen and anaerobic digester populations using mass spectrometry, *Biotechnol. Letters* **5**:375–380.

Scranton, M. I., Novelli, P. C., and Loud, P. A., 1983, Concentration and variability of hydrogen gas in an anoxic salt pond, *Abst. Ann. Meeting Amer. Soc. Microbiol.,* New Orleans, p. 275.
Smith, P. H., and Hungate, R. E., 1958, Isolation and characterization of *Methanobacterium ruminantium,* n. sp, *J. Bacteriol.* **75:**713–718.
Sprengel, C., 1832, Chemie für Landwirthe, Forstmänner und Cameralisten. II Theil. 1-699, Vandenhoeck and Ruprecht, Göttingen.
Stack, R. J., and Hungate, R. E., 1984, Effect of 3-phenylpropanoic acid on capsule and cellulases of *Ruminococcus albus* 8, *Appl. Environ. Microbiol.* **48:**218–223.
Stetter, K. O., Thomm, M., Winger, J., Wildgruber, G., Huber, H., Zillig, W., Jane-Covic, D., Konig, H., Palm, P., and Wunderl, S., 1981, *Methanothermus fervidus* sp. nov. a novel extremely thermophilic methanogen isolated from an Icelandic hotspring, *Centralbl. Bakteriologie* I Abt. Orig. **2:**166–178.
Stickland, L. H., 1935, Studies on the metabolism of the strict anaerobes. III. The oxidation of alanine by *Cl. sporogenes.* IV. The reduction of glycine by *Cl. sporogenes,* Biochem. J. **29:**889–898.
Taylor, E. C., 1982, Role of aerobic microbial populations in cellulose digestion by desert millipedes, *Appl. Environ. Microbiol.* **44:**281–291.
Thauer, R. K., Jungermann, K., and Decker, K., 1977, Energy conversion in chemotrophic anaerobic bacteria, *Microbiol. Rev.* **41:**100–180.
Thunberg, T., 1930, The hydrogen activating enzymes of the cells, *Quart. Rev. Biol.* **5:**318–347.
Thurston, J. P., Noirot-Timothée, C., and Arman, P., 1968, Fermentative digestion in the stomach of *Hippopotamus amphibius* (Artiodactyla: Suiformes) and associated ciliated protozoa, *Nature* **218:**882–883.
Thurston, J. P., and Noirot-Timothee, C., 1973, Entodiniomorph ciliates from the stomach of *Hippopotamus amphibius,* with descriptions of two new genera and three new species, *J. Protozool.* **20:**562–565.
van Niel, C. B., 1931, On the morphology and physiology of the purple and green sulphur bacteria, *Arch. f. Mikrobiol.* **3:**1–112.
Veillon, A., 1983, Sur un microcoque strictement anaérobie trouvé dans les suppurations fétides, *Compt. rendu. et Memoirs Soc. de biologie* **45:**807–809.
Volta, A., Lettres sur L'Air Inflammable des Marais auquel on a ajouté trois Lettres du neme Auteur tirees du Journal de Milan. Traduites de L'Italièn avec permission. Strasbourg, J. H. Hertz, Imprimeur de l'Universite 1778. From the first letter to Father Campé, written from Come 14 Nov. 1767.
Warburg, O., 1924, Über Eisen, den sauerstoffübertragenden Bestandteil des Atmungsferments, *Biochem. Zeitschr.* **152:**479–494.
Waterbury, J. B., Calloway, C. B., and Turner, R. D., 1983, A cellulolytic nitrogen-fixing bacterium cultured from the gland of Deshayes in shipworms (Bivalvia: Teredinidae), *Science* **221:**1401–1403.
Weiss, J. E., and Rettger, L. F., 1937, The gram negative *Bacteroides* of the intestine, *J. Bacteriol.* **33:**423–434.
Widdel, F., and Pfennig, N., 1981a, Studies in dissimilatory sulfate-reducing bacteria that decompose fatty acids. I. Isolation of new sulfate-reducing bacteria enriched with acetate from saline environments. Description of *Desulfobacter postgatei* gen. nov. sp. nov. *Arch. Microbiol.* **129:**395–400.
Widdel, F., and Pfennig, N., 1981b, Sporulation and further nutritional characteristics of *Desulfotomaculum acetoxidans, Arch. Microbiol.* **129:**401–402.
Wiegel, J., and Ljungdahl, L. G., 1981, *Thermoanaerobacter ethanolicus* gen. nov., spec. nov., a new extreme thermophilic anaerobic bacterium, *Arch. Microbiol.* **128:**343–348.
Wiegel, J., Brown, M., and Gottschalk, G., 1981, *Clostridium thermoautotrophicum* species novum, a thermophile producing acetate from molecular hydrogen and carbon dioxide, *Curr. Microbiol.* **5:**255–260.
Wieland, H., 1913, Über den Mechanismus der Oxydationsvorgänge, *Ber. Deutsch. chem. Ges.* **46:**3327–3342.

Wieringa, K. T., 1940, The formation of acetic acid from carbon dioxide and hydrogen by anaerobic sporeforming bacteria, *Antonie van Leeuwenhoek* **6:**251–262.

Wimpenny, J. W. T., and Samah, O. A., 1978, Some effects of oxygen on the growth and physiology of *Selenomonas ruminantium, J. Gen. Microbiol.* **108:**329–332.

Wood, W. A., 1961, Fermentation of carbohydrates and related compounds, in: *The Bacteria* Vol. II (I. C. Gunsalus and R. Y. Stanier, eds.), Academic Press, New York.

Wright, D. E., 1967, Metabolism of peptides by rumen microorganisms, *Appl. Microbiol.* **15:**547–550.

Wright, J. H., 1900, A simple method for anaerobic cultivation in fluid media. *Centralbl. Bakteriologie I* **27:**74–75.

Yamin, M. A., 1981, Cellulose metabolism by the flagellate *Trichonympha* from a termite is independent of endosymbiotic bacteria, *Science* **211:**58–59.

Yarlett, N., Lloyd, D., and Williams, A. G., 1982, Respiration of the rumen ciliate *Dasytricha ruminantium* Schuberg, *Biochem. J.* **206:**259–266.

Zehnder, A. J. B., Huser, B. A., Brock, T. D., and Wuhrmann, K., 1980, Characterization of an acetate-decarboxylating, non-hydrogen-oxidizing methane bacterium, *Arch. Microbiol.* **124:**1–11.

Zeikus, J. G., and Henning, D. L., 1975, *Methanobacterium arboriphilicus* sp. nov. an obligate anaerobe isolated from wetwood in trees, *Antonie van Leeuwenhoek* **41:**171–180.

Zinder, S. H., and Brock, T. D., 1978, Methane, carbon dioxide and hydrogen sulfide production from the terminal methiol group of methionine by anaerobic lake sediments, *Appl. Environ. Microbiol.* **35:**344–352.

Zinder, S., and Koch, M., 1983, Acetate oxidation by a thermophilic methanogenic syntrophic coculture, *Abst. Annual Meeting Amer. Soc. Microbiol,* p. 147.

3

THE MINERALIZATION OF ORGANIC MATERIALS UNDER AEROBIC CONDITIONS

George Hegeman

INTRODUCTION

Mineralization is the process by which the carbon and associated nitrogen, phosphorus, and sulfur of organic matter are returned to inorganic form. Under aerobic conditions, this means their most oxidized form (CO_2, NO_3^-, and SO_4^{-2}). Because the origin and general composition of organic carbon and the early stages of decomposition of organic polymers have been discussed recently (Burns, 1982; Krumbein, 1983; Ericksson and Johnsrud, 1982; Bolin, 1973), and because the discussion of cycling the organic forms of nitrogen and sulfur are best treated in other contexts, these will not be discussed here.

In what follows, an attempt is made neither to review exhaustively the tremendous number of transformations that comprise, in aggregate, what is known of the carbon mineralization process nor to list the bacteria that have been shown to carry out specific steps. Instead, it provides an historical and general overview of the major aspects of knowledge about bacterial oxidative metabolism functional in the aerobic mineralization process and how it developed, with some broad outlines of present knowledge and thoughts about future trends. References are provided to reviews to aid the reader in search of details.

HISTORY

The idea that all organic matter is fated to return eventually to its constituent mineral elements in a constant process of renewal has been, in broad

George Hegeman • Microbiology Group, Biology Department, Indiana University, Bloomington, Indiana 47405.

terms, a commonplace since antiquity. Recognition of the role of microbes in the cycling of organic materials and how the process occurs, however, came only after a conceptual background of considerable complexity had been laid.

Anthony van Leeuwenhoek, discoverer of microbes in the late 1600s, observed that microbes were most abundant and lively in a variety of infusions and other forms of organic matter actively undergoing spoilage (Dobell, 1932). Although Francesco Redi (probably around 1665) had shown that the doctrine of abiogenesis (spontaneous generation) did not apply in the classical case of the development of maggots in decaying meat, until the 1800s, the doctrine was still widely held to apply to microbes. The alternate notion that "germs" (the seeds of microbes) were at all times present in the environment (which Leeuwenhoek apparently believed) also did not permit a decision as to whether the growth of microbes in decomposing organic matter was a causal phenomenon or an incidental effect (Collard, 1976).

In the mid-1700s, Lazzaro Spallanzani provided strong evidence that decomposition was associated with the growth of microbes by showing that repeated heating can prevent spoilage of organic infusions. This evidence was less than completely convincing, however, since the time of heating required to prevent spoilage of different preparations varied and the treatment was consistently successful only when using containers that had been hermetically sealed to exclude air (Bulloch, 1960).

Chemistry

In the late 1700s, Priestley, Cavendish, Scheele, and Lavoisier engaged in experiments that laid the foundations for modern chemistry and demonstrated the presence and biological importance of oxygen in the air (Conant, 1950). Although the Swede Scheele had prepared and characterized oxygen by 1773, the publication of his findings was delayed until 1777. Therefore, Priestley's discovery in 1775 was that which had the most effect on others. In Priestley's communications with the French Academy (Easter meeting, 1775), he used as one test of the purity of his "air" its ability to support the life of a mouse (Conant, 1950). The life-sustaining property of air made it seem reasonable that the heating and hermetic sealing advised by Spallanzani (and, in the early 1800s, brought into general practice for preservation of foods by Francois Appert) was effective because it excluded oxygen (e.g., Gay-Lussac, 1810). It remained for Pasteur, with publication of his 1861 "Memoir on the Organized Bodies that Exist in the Atmosphere," to show in his famous and elegant experiments with swan-necked flasks that sterility, after heating, is due to effective protection from the germs in the air and not to the exclusion of oxygen (Pasteur, 1861).

The first side of Pasteur's contribution to the conceptual underpinnings of our present understanding of aerobic mineralization of organic compounds by microbes, the notion that they could *cause* chemical change, antedated his work on spontaneous generation. In this conviction he was preceded by Cagnard Latour (in 1836), Theodore Schwann (in 1837), and Friedrich Kutzing (also in 1837), all of whom concluded that yeast was causally responsible for alcoholic fermentation of sugar. This conclusion did not go undisputed. Great chemists of the time, J. Liebig and F. Wohler, ridiculed this notion and favored a nonbiological chemical mechanism. The second aspect of this side of Pasteur's contribution was the notion of *specificity* in microbiological transformations. This was drawn largely from his work on the butyric and lactic fermentations.

The scientific study of oxidative transformation of organic compounds by soil and water microbes can be argued to have begun with Schloesing and Muntz's study of the formation of nitrates in soil (1877). Davy, in 1814, had suggested that nitrates were formed from ammonia by oxidation in the soil by atmospheric oxygen. Liebig showed that atmospheric oxygen by itself did not effect "nitrification." Nitrates were prepared in France in "nitre gardens": manure piles with flat or bowl-shaped tops which, if regularly watered, yielded in time a crop of sodium and potassium nitrate crystals through the leaching of these soluble salts to the margins of the pile. Schloesing and Muntz showed that oxidation of amines and ammonia in "sewage water" in an experimental reactor developed with time as if dependent upon the growth of microbes and was halted by treatment of the column with disinfectants like chloroform. That nitrification was caused by microbes was confirmed by Warington, but it remained for Winogradsky to isolate the bacteria responsible for the process (Collard, 1976).

During the late 1800s, advances were being made in important areas of microbiological technique, especially methods of observation (staining, microscopy, photography) and in working with and obtaining pure cultures. It was again Winogradsky who—in his work with sulfur-oxidizing, nitrifying, and nitrogen-fixing bacteria—strongly reaffirmed the idea of specificity of microbiological activity and extended it to a new method for enriching in cultures, prior to isolation, organisms of specific physiological and nutritional types (Winogradsky, 1949). However, it was Beijerinck who first named, consciously employed, and systematically exploited the enrichment culture technique in the study of the oxidation of specific organic compounds by bacteria (van Iterson, 1940). The generality and success of the enrichment technique substantiates one of its underlying assumptions, namely that microbes are ubiquitous in the soil, so that a reasonably small sample (a gram or two) of any fertile soil should contain representatives of all major metabolic classes (Schlegel and Jannasch, 1967). Perhaps the most extensive application of the enrichment technique was that of L.E. den Dooren de Jong, Beijerinck's penultimate assistant, whose Ph.D. dissertation provides an extensive listing of bacterial abilities (den Dooren de Jong, 1926). Beijerinck also first demonstrated the existence of oligotrophic bacteria in the study, made with van Delden, in which they describe the isolation of "*Bacillus oligocarbophilus.*"

During a twenty-year period, beginning with the discovery of cell-free alcoholic fermentation by Buchner in 1897, the main outlines of glycolysis by muscle and of fermentation of glucose by yeast were established. Catabolism was appreciated as a series of distinct reactions, and students of energy-yielding metabolism in animals and microbes found unexpected common biochemical ground. Similarly, nutritionists and microbial biochemists during this time found that the vitamins needed for proper animal nutrition often had counterparts in, or were identical to, growth factors required by microbes. This coincidence is due to the fact that they are biosynthetic precursors of indispensable cofactors that function generally in metabolism. This was largely worked out with microbes during the period 1920 to 1935 (van Niel, 1949).

As early as 1924, A.J. Kluyver, Beijerinck's successor at the Microbiological Laboratories at Delft, saw that the apparent diversity of types of metabolism in microbes, plants, and animals was superficial and that all were, in fact, the reflection of the action in different combinations of fundamentally similar enzymes and coenzymes acting concertedly in similar sequences, so that a fundamental unity underlay the apparent great diversity (Kamp et al., 1959, van Niel, 1949). Kluyver conducted a wide-ranging study in microbial metabolism in what he later termed the field of comparative biochemistry, with the notion of unity as a guiding principle (Kluyver, 1924).

The actual mechanisms by which oxygen consumption is coupled to substrate oxidation remained uncertain until late in the development of biochemical understanding of metabolism (Mitchell, 1961; Fillingame, 1980). Progress was impeded for about 15 years in the 1920s and 1930s because of a division of the scientific community into two schools. One, impressed by Otto Warburg's successes in demonstrating the role of specific iron compounds in the activation of oxygen, stressed this aspect of respiratory metabolism. The other, with Thunberg and Wieland, were taken by the ubiquity and ease of demonstration of specific dehydrogenases capable of coupling substrate oxidation to the reduction of acceptors, such as quinones or methylene blue, and emphasized the role of these enzymes. It was not until 1926 when Fleisch, Szent-Gyorgyi, and Kluyver and Donker more or less simultaneously saw that these were *not* strictly incompatible views of generalized oxidative metabolism but that they could be made compatible if a natural "carrier" could take the place of the artificial acceptor of electrons from the substrate and if that carrier could be oxidized in turn by the "atmungsferment"—earlier shown to be an iron porphyrin by Warburg—that, in turn, could be oxidized by molecular oxygen to yield the oxidized substrate and water. This suggestion was verified by von Euler, Warburg, Kuhn, and others, who discovered and characterized the flavin nicotinamide nucleotide reduction-oxidation coenzymes during the 1930s. Roles and structures for other important cofactors (coenzyme A, ATP) were established shortly afterward.

In the 1920s and 1930s, a very active group of biochemists worked on bacterial metabolism at Cambridge. Using the manometric techniques invented in the early 1900s by Haldane and Barcroft and improved by Warburg to measure overall respiration, and the dye reduction technique of Thunberg to measure dehydrogenase activity, J.H. Quastel, T. Mann, D.M. Needham,

E.F. Gale, and Marjory Stephenson systematically studied the pathways functional in catabolism of numerous fatty acids, amino acids, alcohols, and carbohydrates (Collard, 1976). It also became clear during this period that nonfermentable substrates may be oxidized by bacteria in "anaerobic oxidations" that are formally respiration (a variant of normal electron transport comes into play) in which an alternate electron acceptor replaces oxygen. These studies both inspired the principle of "Unity in Biochemistry" and were impelled by it (Stephenson, 1949). Thanks to these workers and others, during the 1930s the early steps of bacterial catabolic metabolism and its overall major features became clear.

Regulation

In the early 1930s, the picture of aerobic mineralization by bacteria was far from clear. It was obvious, however, that a great many specific enzymes acting in concerted sequences were needed to explain the nutritional abilities of a single, even moderately versatile bacterium. A rough, minimal estimate could be made for some strains by reference to the work of, for instance, den Dooren de Jong (1926). Originally published in a relatively obscure thesis in Dutch, this became more generally known only when extensively quoted by Stephenson in her influential volume on bacterial metabolism (1949), first published in 1929. *Pseudomonas putida,* for example, could grow at the expense of 77 of 200 compounds tested when each was supplied as sole source of carbon in a synthetic medium. This implied that hundreds of specific enzymes must be able to function in the cell upon demand. Later studies revised this estimate upward (Stanier et al., 1967).

It was not surprising, then, when Karstrom (1930) discovered that the enzymes of bacteria fell into two broad classes: those that were present at all times ("constitutive") and those that were present in full amounts only when their substrate or a metabolic precursor of their substrate was present ("adaptive"). Changes in cellular enzyme complement by physiological mechanisms against a constant genetic background complemented genetic variation as a means for fitting the response of a population of bacteria to new circumstances. It was also not surprising that most of the catabolic pathways of *Pseudomonas* and similar versatile organisms active in mineralization of organic compounds in the soil proved to be "adaptive" or, as they were later termed, inducible. The mechanistic basis for control of "adaptive" enzyme synthesis had to await the operon model of Jacob and Monod (1961). However, the general occurrence of pathways composed of inducible enzymes provided a principle that could be used to test whether a given compound occurs as an intermediate in the pathway. According to the principle of "simultaneous adaptation" (Stanier, 1947), compounds that are intermediates in a pathway will be oxidized immediately by cells grown with (induced by) a precursor of that compound. Compounds oxidized via unrelated pathways or those that occur as intermediates earlier in the pathway are oxidized only after a lag that reflects the time needed to synthesize a new complement of enzymes. This generalization is usually applicable to analysis of inducible catabolic pathways but not universally so since, when individual enzymes in many pathways were studied, groups of enzymes were shown to be induced in blocks that sometimes included enzymes for earlier steps, with later intermediates sometimes functioning as inducers (Hegeman, 1966; Ornston, 1966, 1971).

The last forty years has been a time of considerable growth in knowledge about pathways for aerobic breakdown of natural compounds and their regulation (Clarke and Ornston, 1975; Ornston, 1971). This development has been somewhat uneven, however, with simple aromatic and hydroaromatic acids, amino acids, simple heterocyclic compounds, hydrocarbons, and simple sugars being best studied (Bull, 1980; Burns, 1982; Clarke and Ornston, 1975; Dagley, 1971, 1978; Dagley and Nicholson, 1970; Gottschalk, 1979; Krulwich and Pelliccione, 1979).

Pesticides and other man-made compounds which are widely dispersed in the environment in quantity pose disposal problems (polychlorinated and polybrominated biphenyls, etc.), as do crude oil components (Atlas, 1981); they have received considerable attention, but with uneven success.

Ironically, the quantitatively very important organic compounds, lignin and cellulose, have received great attention, but their breakdown process has resisted simple resolution. Because of their physical or chemical complexity in natural materials, simple, efficacious microbial processes that employ wood chips, sawdust, or the like as a basis for chemical production or single-cell protein manufacture are still not competitive with fossil fuel-based processes despite their reliance on inexpensive and renewable resources. The semicrystalline structure of unmodified cellulose makes it a difficult substrate for bacteria to attack, and the heteropolymeric ether-linked structure of lignin poses still other sorts of problems for bacteria (Crawford, 1981). In nature, fungi usually initiate attack on wood and insoluble plant products (Burns, 1982; see also Chapter 8, this volume). The bacteria enter the picture only as soluble compounds or considerably degraded lignin or lignocelluloses are formed.

It is likely that many processes that occur in soil are difficult to study because they are actually carried out by consortia of several types of microbes acting on concert (Bull and Slater, 1982). Because of the difficulties of working with mixed cultures, the actual processes that occur in the soil may be difficult or impossible to study in the laboratory unless a small number of organisms in unusually stable association (e.g., Bryant et al., 1967) are involved, or special techniques (e.g., continuous culture—see below) are employed. Yet a few such cases have been studied with success (e.g., Slater and Godwin, 1980).

THE BACTERIA

Beijerinck, Kluyver, their students, and many others isolated numerous specific types of bacteria throughout the first half of the 20th century (van

Niel, 1949). The aerobic bacteria that have been so far studied for their roles in the mineralization process exhibit great diversity of form and biology and often considerable specialization with respect to substrates. Some have such narrow nutritional niches, it is difficult to understand why they are so widespread in soils (e.g., *Bacillus fastidiosus;* van Iterson et al., 1940). In general, their dietary habits are only inferred from enrichment culture results and limited testing, and our understanding of the true pattern in the soil is no doubt far from complete. Actinomycetes (*Streptomyces*) can be enriched with gum rubber as the sole source of carbon, an example of the adaptation of mycelial organisms to polymers as substrates. Members of the *Corynebacterium, Mycobacterium,* and *Nocardia* groups can be enriched using hydrophobic, nonpolymeric materials (wax, mineral oil, cholesterol, etc.), suggesting their preferences for such materials (Krulwich and Pellicione, 1979; Perry, 1979). Other bacteria may specialize in one-carbon compounds as sole sources of carbon and energy for growth, even to the extent of using only methane and methanol (Colby et al., 1979). Still other microbes exhibit fairly catholic tastes in substrates but excel in using them when the substrates are present at only extremely low concentrations (Alexander, 1970; Kuznetsov et al., 1979).

STAGES OF MINERALIZATION

The events that ensue when organic material dies and falls into soil or water can be regarded to occur in roughly three stages:

1. Initiation of decomposition by the surface bacteria and any internal bacteria normally present, and soil organisms introduced through wounds. This primary attack is characterized by excretion of powerful proteases and other extracellular hydrolytic enzymes, possession of relatively inefficient (i.e., fermentative) energy metabolism, and a frequent requirement for one or more growth factors typically available in fresh but decomposing plant and animal matter. These organisms include, among many others, *Bacillus, Clostridium, Proteus,* and the lactic acid bacteria.
2. An intermediate stage then occurs in which fermentation end products and materials unused by the initial decomposers are attacked. This phase is, for the most part, aerobic since much of the available carbon that remains is in the form of fermentation end products that cannot be further converted under anaerobic conditions or materials refractory to attack anaerobically. The bulk of the carbon is released as CO_2 in this phase, and the food object is no longer recognizable. Organisms that participate in this process include members of the genera *Pseudomonas, Acinetobacter, Arthrobacter, Enterobacter,* and some of the more specialized members of the genus *Bacillus.*
3. A final stage is characterized by slow release of CO_2 aerobically from the most refractory of the organic remains. Here the one-carbon-

compound-oxidizing bacteria and other specialists come into play. This phase terminates with the ultimate disappearance of local high concentrations of organic matter through metabolism, outward diffusion of small molecules, and mechanical mixing (in surface soils) by the activities of insects and other animals. The refractory residue joins the other poorly attacked organic residues in the soil (humus) and continues slowly to feed a hardy oligotrophic population in the soil until the next "meal" arrives.

To the casual observer the mineralization process is apparently a seamless whole in which no easily recognizable stages occur. Different stages may occur simultaneously in the same area or in different zones of a gradient if the deposit is large. The overall rate at which it occurs depends on the deposit's situation with respect to key nutrients (H_2O, O_2, and the like) and the physical circumstances (particularly temperature). The process may halt temporarily until conditions change (pickled vegetables and silage may be thought of as mineralization "stuck" at the end of the first stage owing to the exclusion of air. Retting (of, e.g., flax and cocoa) is the controlled use of the earliest stages of mineralization to process crude agricultural products into a form more suitable for use by man. Composting and sewage treatment entail the application of various expedients (aeration, mixing) to speed the overall process of mineralization of organic material to yield a sanitized, stable product much reduced in bulk by the return of much of the initial carbon to the air as CO_2.

Knowledge of mineralization is dominated by studies performed with relatively few bacteria and pathways. For reasons indicated earlier, the mineralization of a relatively limited set of classes of chemical compounds has been studied in detail. For economic reasons, certain aspects of the process of mineralization have been studied in great detail (e.g., compost preparation for commercial mushroom production, conversion processes in the rumen). That only a relatively small subset of the bacteria acting in the mineralization process has been studied is due to more subtle reasons—partly historical precedent and partly microbiologists' desire to work with individual, tractable strains in pure culture.

The first step in the study of the breakdown of a given organic compound is usually the isolation of a bacterium capable of carrying out the process by means of the enrichment culture technique. Following Beijerinck, the cumulative liquid enrichment technique employs a simple neutral mineral (salts) medium with no added growth factors. The desired carbon and nitrogen source is added at relatively high concentrations (1-10 mmolar for the carbon source), a generalized inoculum (usually about a gram of fertile soil) is added, and incubation is carried out in the dark, aerobically without shaking as a shallow layer in a flask at about 30°C (den Dooren de Jong, 1926; van Iterson, 1940). This formula, applied to discover what bacteria use certain carbon sources as substrates, strongly favors the aerotactic soil bacteria like the pseudomonads, which quickly form a pellicle at the air–water boundary and exclude oxygen from the lower layers of the flask. It is at least partially for this reason that *Pseudomonas* has acquired its reputation for great nutritional versatility and that many degradative pathways have been studied in detail in *Pseudomonas* (Ornston, 1971). It is interesting that merely slightly acidifying the medium (to pH 6) and shaking an enrichment culture with a given carbon source can cause it to yield the immotile *Acinetobacter* rather than *Pseudomonas* as predominant organism (Baumann et al., 1968). However, an exhaustive study of the nutritional abilities of the aerobic pseudomonads (Stanier et al., 1967) did bear out den Dooren de Jong's (1926) conclusion. However, it should be noted that while species of *Pseudomonas* as a group are preeminently nutritionally versatile, that property does not necessarily extend to individual strains of the genus. It certainly seems true that the aerobic pseudomonads are more versatile than, say, *Salmonella* (Gutnick et al., 1969), a fact consistent with their role as intermediate-stage organisms in the mineralization process.

PATHWAY STRUCTURE

Catabolic pathways for degradation of related compounds that have been well studied have certain structural features that are of interest. Most obvious is their fanlike division into early compound-specific sequences and subsequent convergence upon common reaction sequences that, in turn, converge later on central catabolic cycles by providing intermediates (Table I) (Wheelis and Stanier, 1970; Ornston, 1971). The earliest stages often entail considerable modification of the primary substrate (insertion of molecular oxygen, net oxidation or reduction, cleavage of an aromatic ring, decarboxylation, etc.) with no net return to the bacterium. Kluyver termed this process "chip respiration" (Kamp et al., 1959) since the process is akin to the early stages in wood carving or sculpture where, by chipping, a crude block is prepared

TABLE I

Stages in the Mineralization of Organic Compounds by Pseudomonas

Primary substrates	Common intermediates	Central metabolic products
Quinic acid Shikimic acid 4-Hydroxyl benzoic acid	Protocatechuate ↓ 3-Oxo adipate	Acetyl-CoA and succinic acid
D-Mandelic acid L-Tryptophan Benzoic acid	Catechol ↓ 3-Oxo adipate	Acetyl-CoA and succinic acid
Camphor DL-Valine	Isobutyryl-CoA and isobutyrate	Propionyl-CoA

for finer work. Another more descriptive term originated with I.C. Gunsalus who termed this stage "peripheral metabolism" (personal communication) to distinguish it from the more central pathways by which the substrate carbon finds its way to the central reactions of intermediary metabolism. It is interesting that many peripheral sequences proceed without substrate activation with, for instance, coenzyme A, while in later stages, CoA esters or carboxylic acids are often intermediates. Peripheral sequences are dispersible, and their enzymes are often encoded on plasmids. Catabolic pathways for given compounds are usually very different from the biosynthetic pathways that give rise to those compounds.

GENETIC ORGANIZATION

The Chromosome

Genes for enzymes functional in many dissimilatory pathways in both *Pseudomonas aeruginosa* and *P. putida* have been mapped (Holloway, 1975). The less extensive catabolic enzyme genes in the enteric bacteria (Gutnick et al., 1969), *Salmonella* (Sanderson and Roth, 1983) and *Escherichia* (Bachman, 1983), and of *Bacillus* (Henner and Hoch, 1980) have also been mapped.

Those genes that specify a coordinately regulated group of enzymes are typically found to be linked (Wheelis and Stanier, 1970), as required by their participation in an operon (Jacob and Monod, 1961). However, both in the case of *P. aeruginosa* (Rosenberg and Hegeman, 1969) and *P. putida* (Leidigh and Wheelis, 1973), a degree of supraoperonic linkage far greater than expected on the basis of random gene distribution was found. The selective pressures that would tend to produce intense clustering of genes for functionally similar degradative pathways were discussed by Leidigh and Wheelis (1973), who speculated that genetic exchange mediated by processes that required close linkage (transduction or plasmid-mediated recombination) may have acted to produce this chromosomal specialization. Genetic material can be transferred between *P. aeruginosa* and *P. putida* (Chakrabarty and Gunsalus, 1970), and to other bacteria as well.

Plasmids

A number of pathways or key steps in pathways that can be recognized by a distinct nutritional phenotype are encoded by genes on plasmids in aerobic bacteria. These pathways include those functional in the catabolism of camphor, salicylate, toluate, naphthalene, octane and related *n*-alkanes, benzene sulfonate, and hydrogen (Clarke, 1978). Plasmid-encoded steps are sometimes partially redundant upon chromosomally encoded counterparts. The extent of the role of plasmids in effecting genetic exchange among bacteria in nature is difficult to assess. Although the analogy with infectious transfer of multiple drug resistance (Watanabe, 1963) is attractive and features of plasmid stucture suggest that they are well-designed to transfer genetic material (Campbell, 1981), Sanderson (1976) has concluded that the major lines of evolution in Enterobacteriaceae have not been greatly influenced by gene transfer and integration. Compared to parasitic or aquatic organisms soil organisms probably have little opportunity to mix or migrate, yet viruses, whose reproductive cycle depends on infectious transfer, are readily demonstrated for most soil bacteria. Reanny et al. (1982) cite a number of lines of evidence suggesting that genetic exchange in natural microbial communities does occur and may be an important evolutionary factor.

PROSPECTS AND TRENDS IN RESEARCH

Pathways

A large body of knowledge now exists about various specific pathways for mineralization of organic compounds by aerobes (Bull, 1980; Burns, 1982; Clarke and Ornston, 1975; Colby et al., 1975; Crawford, 1981; Dagley and Nicholson, 1970; Dagley, 1975, 1978; Gottschalk, 1979; and Perry, 1979, in part).

Xenobiotic Compounds as Pollutants

In the early 1960s, a new and urgent interest in the environment and man's effects upon it emerged. Carson (1962), in particular, sensitized the public to the dangers of chemical pollution. Alexander (1965) pointed out that the spreading of quantities of poorly degraded and sometimes toxic novel compounds into the environment creates new problems for the mineralizing bacteria. Dagley noted (1975) that technological man has struck a Faustian bargain in his growing dependence upon synthetic chemicals, many of which were originally designed to be nondegradable. Examples of such chemicals include DDT, Chlordane, Kepone, and pentachlorophenol. One may ask how long it will take bacteria in nature to "learn" to degrade these xenobiotic chemicals, if they ever do. In the evolution of pathways for catabolism of naturally occurring compounds, bacteria have had the advantage of a very long exposure to the substrates in question, together with their related biosynthetic intermediates or compounds of similar structure (Beam and Perry, 1973). This is not true of xenobiotic compounds, although some resemble familiar, degradable chemicals (e.g., the herbicide Dalapon is a chlorinated propionic acid).

It has been known for some time that addition of degradable organic material to soil contaminated with a poorly degraded organic compound stimulates degradation of this latter material in a nonspecific way. Physiological processes of the flora that develops at the expense of the degraded material presumably act nonspecifically upon the poorly degradable material. In 1970 (Horvath and Alexander), this phenomenon was named "cometabolism," and a number of instances were cited by Horvath (1972). Use of this term was criticized on a number of grounds (cf. Hulbert and Krawiec, 1977, Perry, 1979). It was pointed out that there is really no need to dignify nonspecific processes of metabolism with a separate name and that there is danger of confusion with processes like "cooxidation" [the process in which oxidation of one gaseous hydrocarbon by bacteria occurs only when another such compound is being oxidized as substrate for growth (Leadbetter and Foster, 1960)]. Nevertheless, it seems reasonable that nonspecific attack upon xenobiotic compounds by bacterial enzymes not evolved by selection to function in this way may play a role in pollutant transformation. It is noteworthy that some intermediates and products of these nonspecific conversions are even more toxic and less degradable than the compounds from which they are derived.

Some information is available about the way new enzyme activities and pathways are acquired in the laboratory (Clarke, 1978; Mortlock, 1982, 1983), and human intervention to speed this process in the case of obnoxious pollutants either by cut-and-splice methods of genetic engineering or in other ways (Kellog et al., 1981; Kibane et al., 1982; Senior et al., 1976) seems possible. However, whether the bacteria newly endowed with laboratory-acquired genetic equipment can survive and grow when reintroduced into the natural environment has yet to be determined; whether these survivors could accelerate mineralization as intended is still further from being established.

Another approach to this problem is to employ pesticides or other synthetic chemicals with a potential for pollution by taking into account the "tastes" of the bacteria—either by building upon known biodegradable molecules (e.g., sugar-based detergents, etc.) or by requiring that proposed new agents meet some standard for degradability or be restricted in use.

A problem of considerable magnitude is also presented by substances that are only slowly degraded (e.g., polychlorinated biphenyls, pentachlorophenol), but are degraded even more slowly or not at all under anaerobic conditions or in the relatively infertile deep layers of the soil. These may move with time in aquifers used as water sources for agriculture and direct human consumption. Considerable interest has been shown in treating point sources of such pollution (e.g., waste dump sites) by the addition to soil of growth-limiting nutrients (nitrogen, phosphorus, etc.) and "engineered" or adapted strains of bacteria that can presumably act on the pollutants *in situ*. Field trials of such techniques have so far met with only limited success.

Mixed Cultures

Compounds that are not degraded materially by individual strains of bacteria are, nonetheless, often mineralized at a measurable rate in the soil. Presumably, the combined efforts of a number of organisms are required to degrade such compounds. This is likely to be the situation for many recalcitrant compounds. It has been pointed out that continuous culture, rather than traditional batch culture, techniques are a valuable approach to the study of such processes (Veldkamp and Kuenen, 1973); the use of such methods for enriching functional communities of microbes has been reviewed (Parkes, 1982). Studies that employ chemostats to examine degradation of recalcitrant complex substrates have often revealed "core" communities of several microbes involved in the overall process. These typically are composed of one or more organisms involved in direct attack upon the compound supplied, and one or more that do not contribute directly to the mineralization process, but that either supply growth factors required by those involved directly in the primary attack or remove inhibitory products, thereby earning themselves a place in the consortium.

Although there is as yet little experience with this approach, it appears that continuous culture techniques can yield the same degradative communities from different soils (Senior, 1977) and therefore may be as reproducible as batch cultures. It is of particular interest that continuous systems can employ realistically low concentrations of toxic substrates and yet support populations of densities suitable for analysis; in addition, continuous removal of potentially toxic products overcomes a major shortcoming of batch culture approaches. Continuous systems can permit the complex physical structure of soils to be approximated, thereby accommodating phenomena, such as adsorption, that occur on surfaces, and physical association among members of consortia that is necessary for proper growth and for mineralization. Adaptive phenomena in communities may also be studied over time, and particularly effective strains can be selected (Senior et al., 1976).

A special form of the continuous flow culture system that has been applied to the evaluation of degradability and the environmental toxicity of poorly degraded pesticides, heavy metals, and like substances is the "microcosm." This employs living material taken by sampling directly from the environment (e.g., cores, mixed phytoplankton) and examines the changes that occur in the community structure during exposure to the test substrate or toxicant. Microcosms can provide valuable information but, as has been pointed out by Pritchard et al. (1979), the initial period of exposure is the most important to observe since adaptation can occur, and longer term measurements may reveal the ultimate potential of the community to resist and degrade an agent, but not the community's reaction to initial exposure.

CONCLUSIONS

At present, the complete oxidation of organic material is attributed largely to the activities of microorganisms; investigations during the past two centuries have provided the details of the existence and role of contributing environmental factors (e.g., O_2), of the biochemical mechanisms by which organic substrates are converted to CO_2, and—most recently—of the importance of mixed microbial communities in the mineralization of complex organic matter. The relatively recent introduction of nonnatural materials, "organic pollutants," into the biosphere presents a challenge to microbe and microbiologist alike. Whether the task of assuring conversion of nonnatural substrates to CO_2 is accomplished by continuing evolution of microorganisms or by recent methods such as genetic engineering remains for the future to reveal.

ACKNOWLEDGMENTS. Work in the author's laboratory is supported by research grants PCM 7812482 and PCM 8314087 from the U. S. National Science Foundation.

REFERENCES

Alexander, M., 1965, Biodegradation: problems of molecular recalcitrance and microbial fallibility, *Adv. Appl. Microbiol.* **7:**35–80.

Alexander, M., 1970, *Microbial Ecology,* John Wiley and Sons, New York.

Atlas, R., 1981, Microbial degradation of petroleum hydrocarbons: an environmental perspective, *Microbiol. Rev.* **45:**180–209.

Bachman, B., 1983, Linkage map of *Escherichia coli* K12, *Microbiol. Rev.* **47:**180–230.

Baumann, P., Doudoroff, M., and Stanier, R. Y., 1968, A study of the *Moraxella* group. II. Oxidase-negative species (genus *Acinetobacter*), *J. Bacteriol.* **95:**1520–1541.

Beam, H. W., and Perry, J. J., 1973, Microbial degradation of cycloparaffinic hydrocarbons via co-metabolism and commensalism, *J. Gen. Microbiol.* **82:**163–169.

Bolin, B., 1973, The carbon cycle, in: *Chemistry in the Environment: Readings from Scientific American* (C. L. Hamilton, ed.), W. H. Freeman and Co., San Francisco, pp. 53–61.

Bryant, M. P., Wolin, E. A., Wolin, M. J., Wolfe, R. S., 1967, *Methanobacillus omelianski,* a symbiotic association of two species of bacteria, *Arch. Microbiol.* **59:**20–31.

Bull, A. T., 1980, Biodegradation, in: *Contemporary Microbial Ecology,* (N. C. Ellwood, J. N. Hedges, M. J. Latham, J. M. Lynch, and J. H. Slater, eds.), Academic Press, New York, pp. 107–136.

Bull, A. T., and Slater, J. H., (eds.) 1982, *Microbial Interactions and Communities,* Volume 1. Academic Press, New York.

Bulloch, W., 1960, *The History of Bacteriology,* Oxford University Press, New York.

Burns, R. G., 1982, Carbon mineralization by mixed cultures, in: *Microbial Interactions and Communities,* Volume 1, Academic Press, New York, pp. 475–543.

Campbell, A., 1981, Evolutionary significance of accessory DNA elements in bacteria, *Annu. Rev. Microbiol.* **35:**55–84.

Carson, R. L., 1962, *Silent Spring,* Houghton-Mifflin, New York.

Chakrabarty, A. M., and Gunsalus, I. C., 1970, Transduction and genetic homology between *Pseudomonas* species *aeruginosa* and *putida, J. Bacteriol.* **103:**830–832.

Clarke, P. H., 1978, Experiments in microbial evolution, in: *The Bacteria,* Volume VI (L. N. Ornston and J. R. Sokatch; eds.), Academic Press, New York, pp. 137–218.

Clarke, P. H., and Ornston, L. N., 1975, Metabolic pathways and regulation, in: *Genetics and Biochemistry of Pseudomonas,* (P. H. Clarke and M. H. Richmond, eds.), John Wiley, New York, pp. 191–340.

Clarke, P. H., and Richmond, M. H. (eds.), 1975, *Genetics and Biochemistry of Pseudomonas,* John Wiley, New York.

Colby, J., Dalton, H., and Whittenbery, R., 1979, Biological and biochemical aspects of growth on C_1 compounds, *Annu. Rev. Microbiol.* **33:**481–518.

Collard, P., 1976, *The Development of Microbiology,* Cambridge University Press, Cambridge, England.

Conant, J. B. (ed.), 1950, *The Overthrow of the Phlogiston Theory,* Case 2, Harvard Case Histories in Experimental Science, Harvard University Press, Cambridge, Massachusetts.

Crawford, R. L., 1981, *Lignin Biodegradation and Transformation,* John Wiley, New York.

Dagley, S., 1971, Catabolism of aromatic compounds by microorganisms, *Adv. Microbiol. Physiol.* **6:**1–46.

Dagley, S., 1975, A biochemical approach to some problems of environmental pollution, in: *Essays in Biochemistry,* Volume 11, Academic Press, London, pp. 81–138.

Dagley, S., 1978, Pathways for the utilization of organic growth substrates, in: *The Bacteria,* Volume III, (L. N. Ornston and J. R. Sokatch eds.), Academic Press, New York, pp. 305–388.

Dagley, S. and Michelson, D. E., 1970, *An Introduction to Metabolic Pathways,* John Wiley and Sons, New York.

den Dooren de Jong, L. E., 1926, Bijdrag tot de kennis van het mineralsatieproces. Dissert. Rotterdam.

Dobell, C., 1932, *Antonie van Leeuwenhoek and His "Little Animals"* (reprinted as Dover 5594 in 1960, Dover Publications, New York). Staples Press, London.

Ericksson, K.-E., and Johnsrud, S. C., 1982, Mineralisation of carbon, in: *Experimental Microbial Ecology* (R. G. Burns and J. H. Slater, eds.), Blackwell Publications, Oxford, pp. 134–153.

Fillingame, R. H., 1980, The proton-translating pumps of oxidative phosphorylation, *Annu. Rev. Biochem.* **49:**1079–1113.

Gottschalk, G., 1979, *Bacterial Metabolism,* Springer-Verlag, New York.

Gutnick, D., Calvo, J. M., Klopotowski, T., and Ames, B. N., 1969, Compounds which serve as the sole source of carbon or nitrogen for *Salmonella typhimurium* LT-2, *J. Bacteriol.* **100:**215–219.

Hegeman, G.D., 1966, Synthesis of enzymes of the mandelate pathway by *Pseudomonas putida, J. Bacteriol.* **91:**1155–1160.

Henner, D. J., Hoch, J. A., 1980, The *Bacillus subtilis* chromosome, *Microbiol. Rev.* **44:**57–82.

Holloway, B. W., 1975, Genetic organization of *Pseudomonas,* in: *Genetics and Biochemistry of Pseudomonas* (D. H. Clarke and M. H. Richmond, eds.), John Wiley and Sons, New York, pp. 133–161.

Horvath, R. S., 1972, Microbial co-metabolism and the degradation of organic compounds in nature, *Bacteriol. Rev.* **36:**146–155.

Horvath, R. S., and Alexander, M., 1970, Cometabolism: a technique for the accumulation of biochemical products, *Can. J. Microbiol.* **16:**1131–1132.

Hulbert, M. H., and Krawiec, S., 1977, Cometabolism: a critique. *J. Theor. Biol.* **69:**287–292.

Jacob, F., and Monod, J., 1961, Genetic regulatory mechanisms in the synthesis of proteins, *J. Mol. Biol.* **3:**318–356.

Kamp, A. F., La Riviere, J. W. M., and Verhoeven, W. (eds.), 1959, *Albert Jan Kluyver: His Life and Work,* North-Holland Publishing Co., Amsterdam.

Karstrom, H., 1930, Uber die Enzymbildung in Bacterien. Thesis, Helsingfors.

Kellog, S. T., Charterjee, D. K., and Chakrabarty, A. M., 1981, Plasmid-assisted molecular breeding: a new technique for enhanced biodegradation of persistent toxic chemicals, *Science* **214:**1133–1135.

Kibane, J. J., Charterjee, D. K., Kans, J. S., Kellog, S. T., and Chakrabarty, A. M., 1982, Biodegradation of 2,4,5-T by a pure culture of *Pseudomonas cepacia, Appl. Environ. Microbiol.* **44:**72–78.

Kluyver, A. M., 1924, Eenheid en Verscheidenheid in de stifwisseling der microben, *Chem. Weekbl.* **21:**no. 22.

Krulwich, T. A., and Pelliccione, N. J., 1979, Catabolic pathways of coryneforms, nocardias, and mycobacteria, *Annu. Rev. Microbiol.* **33:**95–112.

Krumbein, W. E., 1983, *Microbial Geochemistry,* Blackwell Scientific Publications, Oxford.

Kuznetsov, S. I., Dubinia, G. A., and Lapteva, N. A., 1979, Biology of oligotrophic bacteria, *Annu. Rev. Microbiol.* **33:**377–388.

Lal, R., Saxena, D. M., 1982, Accumulation, metabolism, and effects of organochlorine insecticides on microorganisms, *Microbiol. Rev.* **46:**95–127.

Leadbetter, E. R., and Foster, J. W., 1968, Bacterial oxidation of gaseous alkanes, *Arch. Mikrobiol.* **35:**92–104.

Leidigh, B. J., and Wheelis, M. L., 1973, The clustering on the *Pseudomonas putida* chromosome of genes specifying dissimilatory function, *J. Mol. Evol.* **2:**235–242.

Mitchell, P., 1961, Coupling of phosphorylation to electron and hydrogen transfer by a chemiosmotic pump, *Nature (London)* **191:**144–148.

Mortlock, R. P., 1982, Metabolic acquisition through laboratory selection, *Annu. Rev. Microbiol.* **36:**259–284.

Mortlock, R. P., 1983, Experiments in evolution using microorganisms, *Bioscience* **33:**308–313.

Ornston, L. N., 1966, The conversion of catechol and protocatechuate to β-ketoadipate by *Pseudomonas putida.* IV. Regulation. *J. Biol. Chem.* **241:**3800–3810.

Ornston, L. N., 1971, Regulation of catabolic pathways in *Pseudomonas, Bacteriol. Rev.* **35:**87–116.

Parkes, R. J., 1982, Methods for enriching, isolating and analysing microbial communities in laboratory systems, in: *Microbial Interactions and Communities* (A. T. Bull and J. H. Slater, eds.), Academic Press, New York, pp. 45–102.

Pasteur, L., 1861, Memoire on spontaneous generation. in: *Pasteur, L.: Oeuvres Reunies,* 7 Volumes, Pasteur Vallery-Radot (ed.), Masson et Cie, Paris, pp. 1922–1939.

Perry, J. J., 1979, Microbial co-oxidations involving hydrocarbons, *Microbiol. Rev.* **43:**59–72.

Pritchard, P. H., Bourquin, A. W., Fredrickson, H. L., and Maziarz, T., 1979, System design factors affecting environmental fate studies in microcosms, in: *Microbial Degradation of Pollutants in the Marine Environment* U.S. Environmental Protection Agency, Gulf Breeze, Florida, pp. 251–272.

Reanny, D. C., Roberts, W. P., Kelly, W. J., 1982, Genetic interactions among microbial communities, in: *Microbial Interactions and Communities (T. A. Bull and J. H. Slater, eds.), Academic Press, New York, pp. 287–322.*

Rosenberg, S. L., and Hegeman, G. D., 1969, Clustering of functionally related genes in *Pseudomonas aeruginosa, J. Bacteriol.* **99:**353–355.

Sanderson, K. E., 1976, Genetic relatedness in the family *Enterobacteriaceae, Annu. Rev. Microbiol.* **30:**303–326.

Sanderson, K. E., and Roth, J. R., 1983, Linkage map of *Salmonella typhimurium, Microbiol. Rev.* **47:**410–553.

Schlegel, H. G., and Jannasch, H. W., 1967, Enrichment cultures, *Annu. Rev. Microbiol.* **21:**49–70.

Schloesing, T., and Muntz, A., 1877, Sur la nitrification par les ferments organisés, C. R. Acad. Sci. **84:**301–303.

Senior, E., 1977, Characterization of a microbial association growing on the herbicide Dalapon, Ph.D. Thesis, University of Kent.

Senior, E., Bull, A. T., and Slater, J. H., 1976, Enzyme evolution in a microbial community growing on the herbicide Dalapon, *Nature* (London) **263:**476–470.

Slater, J. H., and Godwin, D., 1980, Microbial adaptation and selection, in: *Contemporary Microbial Ecology* (N. C. Ellwood, J. N. Hedges, M. J. Latham, J. M. Lynch, and J. H. Slater, eds.), Academic Press, New York, pp. 137–160.

Stanier, R. Y., 1947, Simultaneous adaptation: a new technique for study of metabolic pathways, *J. Bacteriol.* **54:**339–348.

Stanier, R. Y., Palleroni, N. J., and Doudoroff, M., 1967, The aerobic pseudomonads: a taxonomic study, *J. Gen. Microbiol.* **43:**159–271.

Stephenson, M. J., 1949, *Bacterial Metabolism,* Longmans, Green, London.

van Iterson, Jr., G., den Dooren de Jong, L. E., and Kluyver, A. J., 1940, *Martinus Willem Beijerinck, His Life and Work,* M. Nijhoff, The Hague.

van Niel, C. B., 1949, The "Delft School" and the rise of general microbiology, *Bacteriol. Rev.* **13:**161–174.

Veldkamp, H., and Kuenen, J. G., 1973, The chemostat as a model system for ecological studies, *Bull. Ecol. Res. Commun. (Stockholm)* **17:**347–355.

Watanabe, T., 1963, Infective heredity of multiple drug resistance in bacteria, *Bacteriol. Rev.* **27:**87–115.

Wheelis, M. L., and Stanier, R. Y., 1970, The genetic control of dissimilatory pathways in *Pseudomonas putida, Genetics* **66:**245–266.

Winogradsky, S., 1949, *Microbiologie due sol, problems et methodes,* Masson et cie, Paris.

4

STAGES IN THE RECOGNITION OF BACTERIA USING LIGHT AS A SOURCE OF ENERGY

Norbert Pfennig

INTRODUCTION

In the beginning of the nineteenth century, de Saussure established that, for the formation of organic matter by green plants in the light, the amount of carbon dioxide assimilated was stoichiometrically related to the amount of cell material formed and molecular oxygen liberated. Since cell material was more reduced than carbon dioxide, it was generally believed that the oxygen produced originated from carbon dioxide. It was Ingenhousz who had shown in 1779 that only the green-pigmented parts, and not the colorless parts of the plants nor the animals, were capable of oxygen production in the light (Rabinowitch, 1945). We can understand, therefore, that the three characteristics of green color, carbon dioxide assimilation, and oxygen evolution conceptually became the fundamental properties of photoautotrophic organisms for the following 130 years.

Although the necessity of light for growth and oxygen production of green plants was established, it was J. R. Mayer (1845) who first recognized light as the form of energy (radiant energy) that green plants were able to convert into the chemical energy necessary for the synthesis of cell material. With this concept arose the recognition of radiant energy as the only possible source of the necessary chemical energy for autotrophic carbon dioxide assimilation. This idea became topical again when Winogradsky (1890) formulated his concept of the chemosynthetic mode of life of certain bacteria. He envisaged inorganic oxidation processes as sources of energy for autotrophic carbon dioxide assimilation. Van Niel (Kluyver and van Niel, 1956)

Norbert Pfennig • Faculty of Biology, Konstanz University, D-7750 Konstanz, Federal Republic of Germany.

recalled that it was Lebedeff in 1908 who clearly expressed that "there appears to be no difference in principle between the mechanism of CO_2 assimilation by green plants and by chemosynthetic bacteria."

THE STUDIES OF ENGELMANN ON THE RESPONSES OF GREEN ALGAE AND OF PURPLE BACTERIA TO LIGHT

The sensitive reaction of motile aerobic chemotrophic bacteria to air was first studied by Engelmann (1881, 1882). He recorded the characteristic accumulations of such bacteria in certain regions of the liquid under a coverslip. Engelmann observed what twenty years later Beijerinck called "Atmungsfiguren." The bacteria were taken from enrichments in different types of infusions of decaying plant material. Depending on their relationship to oxygen, the swarming bacteria accumulated in lines more or less distant from the edge of the liquid or from an air bubble under the cover slip. When in a moist chamber the air was replaced by hydrogen, the bacteria soon moved to the edge of the water, and after a while became immotile. Gassing the preparation with oxygen had the opposite effect: the bacteria remained highly motile in a zone a few millimeters away from the edge. The most sensitive reacting bacteria were usually vibrioid or spirilloid.

From these experiments, Engelmann developed the idea of using such motile spirilla as sensitive reagents for oxygen. Active motility at some distance from a green algal cell was an indication of a strong supply of oxygen, while cessation of motility indicated depletion or absence of oxygen. With the use of his indicator bacteria, Engelmann (1881, 1882) showed for the first time that the chloroplasts were the sites of oxygen evolution in the light. As a particularly suitable object, Engelmann used filaments of *Spirogyra,* a green alga containing spiral-shaped chloroplast bands. Algae and indicator bacteria were anaerobically sealed off under a cover slip, and the preparation was illuminated under the microscope. After some time in the light, the bacteria congregated around the algal filaments close to the green chloroplast bands. By changing either the light intensity of his white light source or the spectral region of a microspectrum that he alternatively used for illumination, Engelmann was able to reversibly increase or decrease the distance of the swarming bacteria from the green chloroplast. These reactions indicated stronger or weaker photosynthetic oxygen production.

When filamentous green algae were illuminated with the entire microspectrum, the indicator bacteria soon accumulated in two main regions, the blue and red parts of the spectrum. The numerical distribution of the bacteria in the illuminated region was taken as a relative measure for the intensity of photosynthetic oxygen production in the different spectral regions. In this way, Engelmann (1883a) obtained the first action spectra of photosynthesis, which on the whole agreed reasonably well with the absorption spectra of the corresponding algae.

The use of the microspectrum allowed Engelmann (1883a, 1884) for the first time to determine whether various light-absorbing pigments in different kinds of algae supported photosynthetic activity. Engelmann measured diatoms and brown, red, and blue-green algae. He discovered congregations of his swarming indicator bacteria not only in spectral regions of chlorophyll absorption but also in the absorption ranges of the accessory pigments. From these observations, he concluded that light absorbed by both the chlorophylls and the other pigments was effective in photosynthesis (Blinks, 1954).

Engelmann (1883b) also discovered that when he used certain motile pigmented bacteria, indicator bacteria did not respond. However, the motility of the pigmented bacteria themselves was sensitive to different regions of the microspectrum. With the red-pigmented motile bacteria that he named *Bacterium photometricum,* conspicuous cell accumulations occurred in the wavelength regions 800–900 nm, 580–610 nm, and 520–550 nm. Since the maxima of congregation of this bacterium coincided with the maxima of its light absorption, Engelmann recognized that the light-sensitive motile organisms by themselves formed an image of their absorption spectrum. He called this wavelength-dependent cell distribution a "bacteriospectrogramm." Engelmann concluded that the red pigment, which was called bacteriopurpurin by Lankester (1876), supported photosynthesis as did chlorophyll in the green-pigmented organisms. This was a bold and far-reaching conclusion, since at the time and for many years later it was generally believed that a photosynthetic organism was necessarily green, exhibited the absorption characteristics of green plant chlorophylls, and produced oxygen in the light. Engelmann's *Bacterium photometricum,* however, not only lacked the absorption maxima of chlorophyll and did not form oxygen in the light, it also exhibited a particularly strong phototactic activity and accumulation in the invisible infrared region between 800–900 nm, in which photosynthetic activity had not been recorded before.

In the year between 1887 and 1888, Engelmann (1888) resumed his studies on the purple bacteria and their relation to light. From different investigators, he received samples of purple bacteria collected from natural habitats, as well as a pure culture of *Spirillum rubrum* (later *Rhodospirillum rubrum*). All these different purple bacteria exhibited absorption spectra that were very similar to the one of *B. photometricum,* regardless of whether the species was able to form intracellular globules of elemental sulfur from sulfide. Engelmann proposed, therefore, to separate the light-sensitive purple sulfur bacteria from the likewise sulfur-globule-forming colorless sulfur bacteria (Winogradsky, 1888) and to group them together with the permanently sulfur-free purple bacteria under the name "purple bacteria." This arrangement was later again proposed by Molisch (1907), maintained by van Niel (1944), and is still in use today.

The different species of purple bacteria exhibited the same kind of sensitivity to large or small changes in light intensity both in space and in time. Engelmann (1888) coined the term "Schreckbewegung" (shock movement, phototactic reaction) for the sudden reversal of the direction of motion by individual cells when they moved from higher light intensity into an area of lower light intensity or into complete darkness. As a consequence of this kind of phobic reaction, the cells of a suspension became trapped in an illuminated area ("Lichtfalle," light trap) of an otherwise shaded cuvette.

Engelmann realized that it was the same kind of phobic reaction that caused the cells to congregate in certain bands of the spectrum. Only radiation that was absorbed by the bacteriopurpurin of the cells was capable of causing a response; the other spectral regions were "dark" to these bacteria. Engelmann was convinced by all these observations that the purple bacteria were photosynthetic organisms.

In order to test the light dependence of growth of the purple bacteria, Engelmann (1888) inoculated two tall glass cylinders with purple sulfur bacteria and some decaying seaweed. He observed development of the purple bacteria only in the cylinder which he incubated in diffuse daylight. No growth could be detected in the cylinder that was kept in the dark. However, when Engelmann removed the second cylinder to diffuse daylight, he noticed development of the purple bacteria in this vessel, as well.

Following the ideas of his time as well as the results of his own experiments with green-colored algae, Engelmann expected that the photosynthetic activity of the purple bacteria would be accompanied by oxygen evolution. In his experiments with *Bacterium photometricum* in 1883, he had been unable to detect oxygen formation with his indicator bacteria. When he repeated the search for oxygen evolution with other types of purple sulfur bacteria in 1888, he obtained results which he interpreted as indicative of weak oxygen formation.

Engelmann (1888) tried to understand this very weak reaction as the result of an intracellular oxygen consumption for the oxidation of sulfide and stored sulfur to sulfate. The latter oxidation process had just been discovered and described for the colorless sulfur bacteria by Winogradsky (1887). The question remained open, however, why oxygen evolution could not be observed in the case of the purple bacteria that did not oxidize reduced sulfur compounds.

WINOGRADSKY'S WORK ON THE SULFUR BACTERIA

At about the same time as Engelmann was pioneering experiments on the phototactic reactions of the purple bacteria, Winogradsky studied the colorless and red sulfur bacteria (Winogradsky 1887, 1888). He collected the filamentous forms *Thiothrix* and *Beggiatoa* from various sulfur springs and muddy ponds in southern Germany and Switzerland. Winogradsky recognized that these bacteria thrived only in places where both hydrogen sulfide and oxygen were available. Under such conditions, the cells formed globules of elemental sulfur, which by their strong light refraction caused the striking whitish color of these mat-forming bacteria. The elemental sulfur was further oxidized to sulfate, whereby the cells became free of sulfur globules when they were deprived of hydrogen sulfide. In considering the physiological significance of this oxidation process, Winogradsky developed the new idea that these bacteria obtained chemical energy for growth by the oxidation of sulfur. This capacity would allow the organisms to use the traces of organic substances in the natural habitat exclusively for the formation of cell material. On the basis of his later studies with the first pure cultures of nitrifying bacteria in mineral media, Winogradsky (1890) extended the idea to the concept of the chemosynthetic mode of life of certain bacteria: the energy conserved during the oxidation of inorganic compounds such as sulfide and sulfur, or ammonia and nitrite, is used by the chemoautotrophic bacteria to assimilate CO_2 into cell material in the same way as the energy of light is used by the photoautotrophic organisms.

During his studies on the colorless sulfur bacteria, Winogradsky pursued the morphological and taxonomic investigations of the purple sulfur bacteria. By the microscopic observation of different morphological types of sulfur purple bacteria collected from nature and maintained in crude cultures with mud and decaying plant material in tall glass cylinders in the laboratory, Winogradsky established (1888) the first detailed systematic survey of these bacteria. He showed that the pleomorphistic views on the group, expressed by Lankester (1876), could not be upheld. On the basis of morphological characteristics, Winogradsky established reproducibly recognizable species and genera, most of which were later confirmed by pure culture studies (Pfennig, 1967).

The formation of intracellular globules of elemental sulfur from hydrogen sulfide and their disappearance by further oxidation to sulfate is a conspicuous feature common to both the colorless and the purple bacteria. Since Winogradsky had just recognized the oxidation of sulfide to sulfate as a physiological process providing energy for growth, he thought of the purple sulfur bacteria as being chemosynthetic bacteria in the same way as the colorless sulfur bacteria. In his monograph on the sulfur bacteria, Winogradsky remarked that, in contrast to *Beggiatoa,* he met the pink or reddish patches of the purple bacteria in the deeper, more quiet and anoxic parts of the sulfide-containing waters of sulfur spring effluents and ponds. He assumed that the oxygen required by the purple sulfur bacteria in nature or in his slide preparations was provided by photosynthesis of the small green organisms which he more or less regularly encountered amidst the groups of purple bacteria.

Concerning the function of the red pigment bacteriopurpurin in the purple bacteria, Winogradsky felt that it remained completely obscure. He was not convinced of Engelmann's conclusion that the phobophototactic behavior of the purple bacteria implied a photoautotrophic mode of life. Not only did Winogradsky expect green pigmentation and oxygen evolution in photosynthetic organisms, but he had also made different observations on the effect of light on the motility of the purple sulfur bacteria in his glass cylinder cultures (Winogradsky, 1888; Pfennig, 1962, 1967). While Engelmann observed a stimulatory action of illumination on the motility of his purple bacteria, Winogradsky observed that at the side of his cylinders which faced the window, the pink and purple red patches of sulfur bacteria consisted almost exclusively of nonmotile cells in aggregates. He found actively motile cells in samples taken from the sediment or from the shaded side of the cylinder.

On the basis of concepts and experimental results existing at that time, the problems posed by the conflicting ideas expressed by Engelmann and Winogradsky could not be resolved.

THE CONTRIBUTION OF MOLISCH TO THE KNOWLEDGE OF THE PURPLE BACTERIA

Molisch (1907) published the results of his experiments with purple bacteria in a monograph. He used complex organic media for the enrichment and pure cultures of his newly described species. While several of the strains were facultatively aerobic and could be grown on agar medium in the dark like ordinary aerobic bacteria, growth of a few of his strains depended on light and the exclusion of air. Molisch recognized that under anaerobic conditions, all his purple bacteria grew only in the light. Because he observed good growth only in the presence of peptone and an additional carbon source, e.g., glycerol, he described their metabolism as a type hitherto unknown, namely, as photoassimilation of organic substances. None of the strains used sulfide or formed intracellular globules of sulfur. Accordingly, Molisch established a separate family, the Athiorhodaceae, for his purple nonsulfur bacteria. He grouped this family together with Winogradsky's purple sulfur bacteria, Thiorhodaceae, in the order Rhodobacteria, purple bacteria.

Molisch substantiated the separation of the purple sulfur bacteria from the colorless sulfur bacteria and the new assemblage with the purple nonsulfur bacteria by pointing out that both purple groups contained two types of pigments. Ewart (1897) and later Molisch (1907) had independently discovered that Lankester's (1876) bacteriopurpurin consisted of a green pigment, bacteriochlorin (later, bacteriochlorophyll a), and reddish pigments, bacterioerythrins (later identified as carotenoids).

Molisch made strong efforts to test his purple bacteria for the production of oxygen in the light. He applied four different methods, among these Beijerinck's (1901) most sensitive and specific test with luminous bacteria. However, oxygen formation could not be detected in illuminated suspensions of his purple bacteria. On the basis of his experiments, Molisch rejected Engelmann's concept of the photoautotrophic nature of the purple bacteria. His studies did not contribute to a solution of the problems posed by Winogradsky's purple sulfur bacteria. However, several investigators realized that these problems could only be solved by the study of pure cultures.

PROGRESS IN THE STUDY OF THE SULFUR BACTERIA

As far as the chemosynthetic colorless sulfur bacteria were concerned, considerable progress was achieved. Keil (1912) was able to cultivate the aerobic filamentous colorless sulfur bacteria *Thiothrix* and *Beggiatoa* for some time in mineral media with sulfide as energy source and CO_2 as sole carbon source. The experimental setup was, however, so complicated that no further physiological studies could be done with these bacteria. More successful was the demonstration of the existence of Winogradsky's chemoautotrophic mode of life with reduced sulfur compounds as energy source for bacteria of the *Thiobacillus* group (Nathanson, 1904; Beijerinck, 1904; Waksman and Joffee, 1922).

Of the attempts before 1930 to cultivate species of the purple sulfur bacteria under controlled conditions, the work of two authors must be mentioned. Skene (1914) improved the culture conditions and culture media, but did not obtain a pure culture. In his experiments, the purple sulfur bacteria grew only in the presence of sulfide and light; they were autotrophic, because they grew with CO_2 as carbon source in the absence of organic substances. The role of oxygen for his bacteria remained obscure because Skene was unable to exclude green-colored organisms from his cultures.

Bavendamm (1924) continued the experiments of Skene and achieved the first pure culture of *Lamprocystis roseopersicina* and *Chromatium warmingii.* With these bacteria he obtained autotrophic growth under controlled anaerobic conditions in the light. However, because he used test tubes with soft agar containing 5–10% $CaCO_3$, Bavendamm was unable to carry out any quantitative determinations with these cultures. For the interpretation of his results, Bavendamm followed a concept of the physiology of the purple bacteria that was advanced by Buder in 1919.

THE WORK OF BUDER WITH PURPLE SULFUR BACTERIA

In 1915, Buder reported that he became interested in the purple sulfur bacteria as objects for the study of the irritability of unicellular organisms by light. Buder stated that since the work of Engelmann (1888), little research had been done in this field; he also emphasized that he did not intend to solve the metabolic problems raised by the observations and theoretical conclusions of Engelmann and Winogradsky. Buder collected the purple sulfur bacteria *Thiospirillum jenense* and *Chromatium okenii* from natural habitats and maintained them, as had Winogradsky, for several months in the laboratory in mud-containing tall glass cylinders exposed to diffuse daylight. These raw cultures represented the source of the actively motile cells that Buder required for his phototaxis experiments.

On the basis of his own careful studies, Buder (1915, 1918) fully appreciated the original observations on the phobic reactions of purple bacteria reported by Engelmann. This was probably one of the reasons why Buder (1919) tried to reconcile the conflicting ideas of Engelmann and Winogradsky regarding the metabolism of the purple sulfur bacteria. Buder considered the capacity for autotrophic CO_2-assimilation confirmed by the experiments of Skene (1914). Because oxygen was not evolved during metabolism in the light, Buder supposed that in these bacteria the capacities for photosynthesis and for chemosynthesis were either present side by side, or even that they might be closely linked together. In this case, the bacteria would consume the photosynthetically-produced oxygen for the chemosynthetic oxidation of hydrogen sulfide and sulfur to sulfate, and hence no oxygen would be detectable. Buder's suggestion provided an interesting working hypothesis; however, experimental tests were not carried out. As was pointed out later by van Niel (1932), Buder did not discuss the possibility of a purely photosynthetic metabolism in the absence of sulfide. Such a metabolism should have been possible in principle, since it was assumed that sulfide and sulfur represented the oxidizable substrates only for the chemosynthetic metabolism.

In fact, as Skene and Bavendamm had shown, autotrophic development in the light was fully dependent on the availability of hydrogen sulfide and, therefore, photosynthesis and chemosynthesis could not be thought to function independently of each other. How the two processes could be thought to be linked together remained to be solved.

VAN NIEL'S BROADENED CONCEPT OF PHOTOSYNTHESIS

At the time when Buder developed his reconciling hypothesis for the photometabolism of the purple sulfur bacteria, it was generally believed that the oxygen produced in green plant photosynthesis originated from the CO_2 that became assimilated into cell material. This implied, of course, that any kind of photoautotrophic CO_2 assimilation had to be accompanied by oxygen evolution. Also, it is reasonable to say that at that time the oxidation of sulfide and sulfur to sulfate was generally imagined to require the participation of molecular oxygen; this was certainly so in case of most chemosynthetic colorless sulfur bacteria. Consequently, there was no other way to understand the formation of sulfate from reduced sulfur compounds by the sulfur bacteria than by thinking that the necessary oxygen must be provided either by air or by photosynthesis in the absence of air.

In his fundamental paper on the physiology of the purple and green sulfur bacteria, van Niel (1932) discussed the theoretical background of his experimental work. He pointed out that Kluyver and Donker (1926) in their important contribution "Die Einheit in der Biochemie" provided a new possibility regarding the sulfide and sulfur oxidation of the purple sulfur bacteria.

These authors introduced into the understanding of microbial biochemistry Wieland's concept of hydrogen transfer mechanisms as a means of biological oxidations and reductions. In particular, Kluyver and Donker suggested that hydrogen acceptors other than oxygen could well be involved in the dehydrogenation of sulfide or sulfur. Van Niel (1932) states that his experiments with pure cultures of purple and green sulfur bacteria were carried out "with a view to examine this possibility."

Van Niel adopted Beijerinck's glass-stoppered bottle method for the cultivation of the purple sulfur bacteria in defined mineral media under anaerobic conditions in the light. He determined the stoichiometric relationship between the amount of hydrogen sulfide oxidized to sulfate and the amount of cell material formed. The results were incompatible with the assumption of a chemosynthetic mode of growth because the yields were considerably higher than those obtained for the same amount of sulfide oxidized by true aerobic chemoautotrophic bacteria.

In fact, van Niel's quantitative experiments could best be understood in terms of the molar relationships of a light dependent chemical reduction-oxidation equation. The amount of carbon dioxide reduced to cell material was in stoichiometric proportion to the amount of sulfide that was oxidized to sulfate. Van Niel also showed that the equation for the photosynthesis of the purple sulfur bacteria was in complete agreement with the generally accepted overall equation for green plant photosynthesis. The difference was that in the latter process, carbon dioxide was reduced to cell material with water as the hydrogen donor being oxidized to molecular oxygen, while in case of the purple sulfur bacteria sulfide or sulfur served as hydrogen donor being oxidized to sulfate. The recognition of photosynthetic carbon dioxide assimilation as a photochemical oxidation-reduction process allowed van Niel to formulate the electron donor in general terms so that any other inorganic or organic compound could be imagined to serve in its place:

$$2\,H_2A \quad + \quad CO_2 \quad \xrightarrow{\textit{light}} \quad 2\,A \quad + <CH_2O> \quad + \quad H_2O$$

2 H_2A	CO_2	2 A	$<CH_2O>$ + H_2O
reduced hydrogen donor	hydrogen acceptor	oxidized hydrogen donor	reduced acceptor

Examples for the use of several different organic compounds by purple sulfur bacteria anaerobically in the light were given by Müller (1933). In later years, when van Niel (1944) carried out his comprehensive studies on the purple nonsulfur bacteria, it became evident that these bacteria were also potentially photoautotrophic. They could be grown anaerobically in the light in growth factor-supplemented mineral media with carbon dioxide as carbon source, and either hydrogen or thiosulfate as photosynthetic reductant.

OXYGENIC AND ANOXYGENIC PHOTOSYNTHESIS: SOLUTIONS TO CLASSICAL PROBLEMS

Van Niel's broadened concept of photosynthesis greatly stimulated biochemical research on the light and dark reactions involved. When in the years between 1950 and 1960, all enzymatic steps of autotrophic carbon dioxide assimilation in green plants were established (Bassham and Calvin, 1957), what Lebedeff had anticipated in 1908 was confirmed, namely, that the pathway could be the same in most photoautotrophic organisms and in the chemoautotrophic bacteria. The whole of autotrophic carbon metabolism comprised light-independent, "dark" reactions.

During the same time, it was also established experimentally that the role of light is to provide energy that is converted into chemical energy by the chlorophyll-containing reaction centers. Via the photosynthetic electron transport system, chemical energy becomes available for metabolism in the form of adenosine triphosphate and reduced pyridine nucleotides (Arnon et al., 1954; Frenkel, 1954).

When water is the electron donor in oxygenic photosynthesis of green plants and cyanobacteria, one light reaction in each of two consecutively functioning photosystems is required to bring the electrons to $NAD(P)^+$ (Duysens and Amesz, 1962). In contrast, when hydrogen, reduced sulfur compounds, or organic compounds serve as electron donors in anoxygenic photosynthesis of purple or green bacteria, only one photosystem is involved, with bacteriochlorophyll a in most cases (Frenkel, 1959). Overall energetic considerations would indicate that the latter type of photosynthesis using low potential electron donors could be more efficient than oxygenic photosynthesis with water. However, the photosynthetic electron transport system of the purple bacteria does not involve reactions in which the energy of, e.g., molecular hydrogen could be conserved. The dry weight yields of both types of photosynthesis are quite similar. This was experimentally demonstrated by determination of the mol quanta consumed per gram of cell dry weight formed in monochromatic light-limited continuous cultures (quantostat; Göbel, 1978). In *Anabaena cylindrica* 7120, the quantum requirement was 0.62 Einstein per gram dry weight (λ = 628 nm; 33°C and 12.8 hr doubling time; Göbel, 1976). When *Rhodopseudomonas acidophila* was grown photoautotrophically with hydrogen as electron donor, the quantum requirement was 0.69 Einstein per gram dry weight (λ = 860 nm, 30°C and 7 hr doubling time).

Not only the biochemical problems of photosynthesis were resolved in detail. The phototactic reactions of the purple bacteria, first revealed by Engelmann and further studied by Buder, attracted renewed interest in the light of the newly developed concepts of photosynthesis. We owe to Clayton (1959, 1964) the most detailed experimental studies and a comprehensive review on the phototaxis of the purple nonsulfur bacteria.

STRUCTURE OF THE PHOTOSYNTHETIC APPARATUS IN PURPLE AND GREEN BACTERIA

In the colorless cytoplasm of eukaryotic uni- or multicellular algal species, the green pigmented chloroplasts are easily recognized by ordinary light microscopy. As early as 1882, Engelmann demonstrated with his indicator bacteria method that the green chloroplasts are the sites of oxygen evolution in the light. Engelmann concluded from this that the chloroplasts were the organelles of photosynthetic carbon dioxide assimilation. Neither in the blue-green algae nor even in the largest species of the purple and green bacteria were comparable pigmented subcellular organelles discernible; the cells of these prokaryotic phototrophs appeared more or less uniformly pigmented in the light microscope. On the basis of light microscopic studies, it was therefore generally assumed that the photosynthetic pigments of phototrophic bacteria and many blue-green algae were dispersed throughout the cytoplasm or chromatoplasm. Not until special methods were developed to prepare ultrathin sections of cells for the study by electron microscopy was it possible to raise and answer the question of the structure of the photosynthetic apparatus in phototrophic bacteria.

Purple Bacteria

When French in 1938 determined the absorption characteristics of a suspension of living cells and of a purple-red cell-free extract of the purple nonsulfur bacterium *Rhodospirillum rubrum,* he obtained virtually identical absorption spectra. The purple-red pigment complex could be separated from the cell-free extract like a protein, by precipitation with ammonium sulfate. However, this kind of fractionation method was not suited for a separation of the blue bacteriochlorophyll a from the red carotenoids of the complex. Both pigments appeared to be bound to the same soluble protein which was presumed to be evenly distributed in the cytoplasm.

Progress in the study of the subcellular components associated with photosynthesis in the purple bacteria was initiated by the experiments of Schachman and co-workers (1952). Like French in 1938, these authors prepared pigmented extracts of *Rhodospirillum rubrum* by grinding the cells with alumina. Using high-speed centrifugation, they were able to sediment and purify from the cell-free extract a purple-red particulate fraction with a high sedimentation constant (190–200 S). When shadowed preparations of this fraction were studied under the electron microscope, disk-shaped particles of relatively uniform size with an apparent diameter of 110 nm were discovered. The shape of these particles was reminiscent of that of red blood cells. No such particles could be obtained from aerobically grown, colorless cells of the same bacterium. Schachman et al. (1952) called the disk-shaped structures carrying the native photosynthetic pigment system of the cells "chromatophores"; the authors supposed that these structures were the site of the pigment-associated photosynthetic reactions in the cells. The first experimental evidence for photochemical activity of isolated "chromatophores" was obtained two years later. Using a cell free "chromatophore" fraction, Frenkel (1954) demonstrated the light-dependent synthesis of adenosine triphosphate from adenosine diphosphate and inorganic phosphate. This energy-conserving process was called photophosphorylation.

Considerable progress in the understanding of the actual arrangement of the photosynthetic pigment-bearing structures within the cells was achieved by electron microscopic investigations of ultrathin sections of phototrophic bacteria. Vatter and Wolfe (1958) published the first cytological studies on *Rhodospirillum rubrum* grown anaerobically in the light and aerobically in the dark. Only in the red-pigmented, light-grown cells did the cytoplasm contain numerous roundish membrane structures 50–100 nm in diameter, the interior of which appeared less electron dense than the surrounding cytoplasm. It was concluded that the membrane rings originated from vesicular membrane structures that were identical to the isolated "chromatophores" of Schachman et al. (1952). In addition to *R. rubrum,* Vatter and Wolfe studied the fine structure of *Rhodopseudomonas sphaeroides* and *Chromatium vinosum* strain D. In the latter two species membrane-bounded vesicles of similar appearance as in *R. rubrum* were present, although of somewhat smaller size (about 50 nm in diameter). All these results favoured the view that the "chromatophores" represented the structural units of photosynthetic activity in the purple bacteria.

Subsequent investigations with *R. rubrum* and other purple bacteria showed that the in vivo structure of the photosynthetic apparatus was actually more complex than originally thought. Working with *R. rubrum,* Tuttle and Gest (1959) used mild osmotic lysis to open the cells and liberate the cytoplasm. These authors did not obtain "chromatophores" in large amounts. Rather, they found that the red-pigmented fraction with photochemical activity sedimented at low centrifugal forces, unlike the "chromatophore" fraction of Schachman et al. Tuttle and Gest concluded that the membrane vesicles were actually part of a reticulum of vesicular membrane extensions that penetrated from the cytoplasmic membrane into the cytoplasm. The "chromatophores" would then originate by comminution of the in vivo membrane system during disruption of cells by abrasion or sonication.

The fine structure studies carried out in later years on various purple bacteria fully confirmed the concept of Tuttle and Gest along two lines of evidence. First, electron microscopy of thin sections of *R. rubrum* and *Rhodopseudomonas sphaeroides* revealed that the intracytoplasmic membranes originated from and were continuous with the cytoplasmic membrane of the cells (Cohen-Bazire and Kunisawa, 1963; Drews and Giesbrecht, 1963; Boatman 1964; Holt and Marr, 1965). The development of the vesicles was studied in cells that were shifted from aerobic, chemotrophic to anaerobic, phototrophic growth conditions. The vesicles first appeared as indentations of the cytoplasmic membrane, which then invaginated and became constricted, resulting in spherical vesicles that were open at the cell membrane pole. In later stages, vesicular or tubular structures extended into the cytoplasm in the form of connected vesicles or bulged tubes. The second line of evidence was based on ultrastructural studies of other species and genera of the purple bacteria. Thin sections of *Rhodospirillum molischianum* (Drews, 1960; Giesbrecht and Drews, 1962; Gibbs et al., 1965; Hickman and Frenkel, 1965) and of *Rhodomicrobium vannielii* (Boatman and Douglas, 1961; Conti and Hirsch, 1965; Trentini and Starr, 1967) revealed that different types of photosynthetic membrane systems existed among the purple bacteria. The cells of *Rhodospirillum molischianum* were found to contain several stacks of short lamellae that originated from the cytoplasmic membrane as infolded flattened disks. Characteristically, the stacks were not parallel to the cytoplasmic membrane, but formed a sharp angle to it. In *Rhodomicrobium vannielii,* the intracytoplasmic membrane system consisted of many layers of paired lamellae peripherally arranged in parallel to the cytoplasmic membrane. The membrane layers usually appeared open in one or both ends of the cells.

From these observations, it became clear that the vesicular membrane system discovered by Vatter and Wolfe (1958) was not a universal structural feature of the purple bacteria. Consequently, the concept of the "chromatophores" as independent structural and functional units of photosynthesis in purple bacteria could not be upheld.

During the following fifteen years, fine structure studies on many species of the purple nonsulfur bacteria (Rhodospirillaceae) confirmed that various types of intracytoplasmic photosynthetic membrane systems originated from the cytoplasmic membrane and were continuous with it (Cohen-Bazire, 1963; Cohen-Bazire and Sistrom, 1966; Remsen, 1978).

In the purple sulfur bacteria (Chromatiaceae), the situation was less clear and remains so. Most species of this family possess a vesicular type of intracytoplasmic membrane system, and the cells appear filled with vesicles (Cohen-Bazire, 1963; Remsen, 1978). It was pointed out by Remsen (1978) that the vesicles almost certainly arise from the cytoplasmic membrane, but that, even so, part of the vesicles may be separated from it. Remsen suggested that such "free vesicles" may be associated with the formation of membranes that enclose intracellular sulfur globules.

Green Bacteria

The first fine structure studies of phototrophic bacteria by Vatter and Wolfe (1958) included thin sections of the green sulfur bacterium *Chlorobium limicola.* No vesicular membranous structures comparable to those of *Chromatium vinosum* could be detected within the cytoplasm. Instead, the authors observed electron-opaque granules, 15–25 nm in diameter, within the cells. These granules were tentatively considered to represent the "chromatophores" of the green bacteria.

Three years later, Bergeron and Fuller (1961) studied thin sections of a thiosulfate-utilizing strain of *Chlorobium limicola.* Like Vatter and Wolfe (1958), the authors detected numerous particles of 15 nm width, which, together with polyphosphate granules, were the only unusual structural elements in the cytoplasm. After disruption of the *Chlorobium* cells with ultrasonic treatment or Hughes press, Bergeron and Fuller (1961) obtained the bulk of the photosynthetic pigment system in the cell-free extract. During ultracentrifugation, a green-pigmented fraction with particles about 15 nm in diameter sedimented together with ribosomes. From these results, the authors assumed that the sedimented particles were identical with those observed in the cytoplasm of thin sections. The authors considered the particles to be the structural units of photochemical activity in the green sulfur bacteria.

New observations and concepts on the photosynthetic apparatus of the green bacteria were established when more refined methods were applied both for electron microscopy and for cell disruption. Cohen-Bazire (1963) and Cohen-Bazire et al. (1964) studied the fine structure of six different strains of *Chlorobium limicola* and *C. thiosulfatophilum* containing either bacteriochlorophyll c or d. Thin sections of all these strains revealed large oblong vesicular bodies adjacent to the cytoplasmic membrane over the entire cortical region of the cells. These bodies were 30–40 nm wide and 100–150 nm long and were named "chlorobium vesicles." Each vesicle was completely bounded by a thin (3 nm), electron-dense membrane that was closely appressed to the cytoplasmic membrane on one side. The assumption that the chlorobium vesicles were firmly attached to the cytoplasmic membrane was corroborated using cell material disrupted by mild osmotic lysis. In such lysates, all the photosynthetic pigments sedimented at low centrifugal forces in association with the spheroplast fragments; the supernatant was colorless. When *Chlorobium* cells were disrupted in a French pressure cell, low speed centrifugation sedimented only residual intact cells and large cell fragments; however, the supernatant remained green pigmented. After high-speed centrifugation and purification on a linear sucrose gradient, Cohen-Bazire (1963) obtained a deeply green-pigmented particle fraction with a considerably higher specific bacteriochlorophyll content than the original cells. Negatively stained preparations of this particle fraction revealed that the bulk of the material consisted of elongated bodies of similar dimensions as the chlorobium vesicles of the thin sections.

Cohen-Bazire concluded from these results that the cytoplasmic membrane-attached chlorobium vesicles were the structures that carried the bulk of the bacteriochlorophyll c or d of the green sulfur bacteria. The 15-nm particles, observed by Bergeron and Fuller (1961), must therefore have originated from the content of the chlorobium vesicles by comminution.

Holt and co-workers (1966) confirmed the observations and interpretations of Cohen-Bazire (1963) by fine structure studies of green sulfur bacteria from a *Chloropseudomonas* culture. A fine structure of the photosynthetic apparatus comparable to that of *Chlorobium* was later detected in the green sulfur bacteria *Pelodictyon clathratiforme* (Pfennig and Cohen-Bazire, 1967), *Prosthecochloris aestuarii* (Gorlenko and Zhilina, 1968), *Chlorobium phaeobacteriodes,* and *C. phaeovibrioides* (Pfennig, 1967). The thermophilic gliding, filamentous *Chloroflexus aurantiacus* was recognized by Pierson and Castenholz (1974) as a member of the phototrophic green bacteria on the basis of its possession of bacteriochlorophyll c and chlorobium vesicles. In later years, Staehelin and co-workers (1978) coined the name "chlorosomes" for the unique cytoplasmic membrane-attached vesicular bodies of both families of the green bacteria, the Chlorobiaceae and Chloroflexaceae.

No definite and detailed solution has so far been published for the structural and functional organization of the photosynthetic apparatus in the green bacteria. We still rely on a model that was developed by Olson and co-workers (1977) on the basis of evidence accumulated by several authors. According to this model, the chlorosomes contain only the accessory bacteriochlorophylls c, d or e which serve exclusively in light harvesting. The chlorosomes are firmly bound to the cytoplasmic membrane by bacteriochlorophyll a–protein complexes. In the attachment area of the chlorosomes to the cytoplasmic membrane, the latter contains the photochemical reaction centers with bacteriochlorophyll a, as well as the functionally associated cytochromes and carotenoids (Pierson and Castenholz, 1978; Staehelin et al. 1978).

In retrospect, it is interesting to note that the original concepts that assumed a fundamental difference between the photosynthetic apparatus of the purple and green bacteria (e.g., Bergeron and Fuller, 1961) had to be modified gradually. Today, there is little doubt that the intrinsic photochemical reactions of all kinds of photosyntheses are membrane-bound processes. From a cytological point of view, it appears quite reasonable that the chlorosomes of the green bacteria have been compared with the phycobilisomes of the blue-green algae (Pierson and Castenholz, 1978). Both types of vesicular bodies are located on the surface of typical membranes that carry the photochemical reaction centers, and both types of bodies contain accessory photosynthetic pigments with no function other than light harvesting.

The cytoplasm of eukaryotic phototrophic organisms harbors the separately membrane-bounded chloroplasts as organelles of photosynthesis. It is now certain that such separated units do not exist in the phototrophic bacteria. The original concept of the "chromatophores" as the structural and functional units of photosynthesis in the purple bacteria was most likely formed in analogy to the chloroplasts of the eukaryotic cells. This concept had to be modified when it became apparent that the membrane vesicles are parts of an intracytoplasmic membrane system that is continuous with the cytoplasmic membrane. The fundamental difference in cellular organization between eukaryotic and prokaryotic cells was clearly expressed by Stanier and van Niel as early as 1962: "In eucaryotic cells, respiration and photosynthesis take place in specific membrane-bounded organelles or plastids, the mitochondria and chloroplasts respectively." "In the procaryotic cell, there is no equivalent structural separation of major sub-units of cellular function. . . . In fact one can say that no unit of structure smaller than the cell in its entirety is recognizable as the site of either metabolic unit process." Progress in research on the cytology of the prokaryotic phototrophic bacteria over the past twenty years has fully confirmed these concepts, established in the early days of fine structure research.

CONCLUSION

In the century that has elapsed since bacteria were first suspected of possessing photosynthetic capabilities, studies of the purple and green bacteria have contributed concepts and insights regarding photosynthesis that would otherwise have been unavailable. Because of their unique ability to employ electron donors other than water, the intimate association of their photosynthetic apparatus with the cell membrane, and their possession of pigments that allow photosynthesis employing light of wavelengths not used by oxygenic phototrophs, these bacteria have revealed steps in the history of photosynthesis itself. We are now in a position to fully appreciate the original observations and the spirited interpretations of the pioneers who studied the phototrophic bacteria.

REFERENCES

Arnon, D. I., Allen, M. B., and Whatley, F. R., 1954, Photosynthesis by isolated chloroplasts, *Nature (London)* **174**:394–396.

Bassham, J. A., and Calvin, M., 1957, The Path of Carbon in Photosynthesis, Prentice Hall, Englewood Cliffs, N. J.

Bavendamm, W., 1924, Die farblosen und roten Schwefelbakterien des Süß- und Salzwassers, Gustav Fischer, Jena.

Beijerinck, M. W., 1901, Photobacteria as a reactive in the investigation of the chlorophyll function, *Proc. Akad. Sci. Amst.* **4**:45–49.

Beijerinck, M. W., 1904, Ueber die Bakterien welche sich im Dunkeln mit Kohlensäure als Kohlenstoffquelle ernähren können, *Centralbl. Bakt.*, Abt. 2, **11**:592–599.

Bergeron, J. A., and Fuller, R. C., 1961, The photosynthetic macromolecules of *Chlorobium thiosulfatophilum*, in: *Biological Structure and Function* (T. W. Goodwin, O. Lindberg, eds.), Vol. II, Academic Press, New York, pp. 307–324.

Blinks, L. R., 1954, The photosynthetic function of pigments other than chlorophyll, *Annu. Rev. Plant Physiol.* **5**:93–114.

Boatman, E. S., 1964, Observations on the fine structure of spheroplasts of *Rhodospirillum rubrum*, *J. Cell. Biol.* **20**:297–311.

Boatman, E. S., and Douglas, H. C., 1961, Fine structure of the photosynthetic bacterium *Rhodomicrobium vannielii*, *J. Cell. Biol.* **11**:469–483.

Buder, J., 1915, Zur Kenntnis des *Thiospirillum jenense* und seiner Reaktionen auf Lichtreize, *Jb. Bot.* **56**:529–584.

Buder, J., 1918, Bakteriospektrogramme von Purpurbakterien, *Ber. dtsch. bot. Ges.* **36**:103–104.

Buder, J., 1919, Zur Biologie des Bacteriopurpurins und der Purpurbakterien, *Jb. Bot.* **58**:525–628.

Clayton, R. K., 1959, Phototaxis of purple bacteria, in: *Handbuch der Pflanzenphysiologie* (W. Ruhland, ed.), Band 17, Teil 1, Springer-Verlag, Berlin, pp. 371–387.

Clayton, R. K., 1964, Phototaxis in microorganisms, in: *Photophysiology,* Vol. 2, (A. C. Giese, ed.), Academic Press, London, pp. 51–77.

Cohen-Bazire, G., 1963, Some observations on the organization of the photosynthetic apparatus in purple and green bacteria, in: *Bacterial Photosynthesis* (H. Gest, A. San Pietro, and L. P. Vernon, eds.), Antioch Press, Yellow Springs, Ohio, pp. 89–114.

Cohen-Bazire, G., and Kunisawa, R., 1963, The fine structure of *Rhodospirillum rubrum, J. Cell. Biol.* **16:**401–419.

Cohen-Bazire, G., Pfennig, N., and Kunisawa, R., 1964, The fine structure of green bacteria, *J. Cell. Biol.* **22:**207–225.

Cohen-Bazire, G., and Sistrom, W. R., 1966, The procaryotic photosynthetic apparatus, in: *The Chlorophylls* (L. P. Vernon and G. R. Seely, eds.), Academic Press, New York, pp. 313–341.

Conti, S. F., and Hirsch, P., 1965, Biology of budding bacteria. III. Fine structure of *Rhodomicrobium* and *Hyphomicrobium* spp., *J. Bacteriol.* **89:**503–512.

Drews, G., 1960, Untersuchungen zur Substruktur der "Chromatophoren" von *Rhodospirillum rubrum* and *Rhodospirillum molischianum, Arch. Mikrobiol.* **36:**99–108.

Drews, G., and Giesbrecht, P., 1963, Zur Morphogenese der Bakterien-"Chromatophoren" (Thylakoide) und zur Synthese des Bakterio-chlorophylls bei *Rhodopseudomonas spheroides* und *Rhodospirillum rubrum. Zentbl. Bakteriol. Parasit. kde. Infektionskr. Hyg.,* Abt. 1: Orig. Reihe A 190, 508–536.

Duysens, L. N. M., and Amez, J., 1962, Function and identification of two photochemical systems in photosynthesis. *Biochim. Biophys. Acta* **64:**243–260.

Engelmann, Th. W., 1881, Neue Methode zur Untersuchung der Sauerstoffausscheidung pflanzlicher und tierischer Organismen, *Pflügers Arch. ges. Physiol.* **25:**285–292.

Engelmann, Th. W., 1882, Zur Biologie der Schizomyceten, *Bot. Ztg.* **40:**321–325; 337–341.

Engelmann, Th. W., 1883a, Farbe und Assimilation, *Bot. Ztg.* **41:**1–13, 17–29.

Engelmann, Th. W., 1883b, *Bacterium photometricum.* Ein Beitrag zur vergleichenden Physiologie des Licht- und Farbensinnes, *Pflügers Arch. ges. Physiol.* **30:**95–124.

Engelmann, Th. W., 1884, Untersuchungen über die quantitative Beziehung zwischen Absorption des Lichtes und Assimilation in Pflanzenzellen, *Bot. Ztg.* **42:**81–93; 97–105.

Engelmann, Th. W., 1888, Die Purpurbakterien und ihre Beziehungen zum Licht, *Bot. Ztg.* **46:**661–669; 677–689; 693–701; 709–720.

Ewart, A. J., 1897, On the evolution of oxygen from coloured bacteria, *J. Linn. Soc. Bot.* **33:**123–155.

French, C. S., 1938, The chromoproteins of photosynthetic purple bacteria, *Science* **88:**60–62.

Frenkel, A. W., 1954, Light induced photophosphorylation by cell free preparations of photosynthetic bacteria, *J. Am. Chem. Soc.* **76:**5568–5569.

Frenkel, A. W., 1959, Light-induced reactions of bacterial chromatophores and their relation to photosynthesis, *Annu. Rev. Plant Physiol.* **10:**53–70.

Gibbs, S. P., Sistrom, W. R., and Worden, P. B., 1965, The photosynthetic apparatus of *Rhodospirillum molischianum, J. Cell. Biol.* **26:**395–412.

Giesbrecht, P., and Drews, G., 1962, Elektronenmikroskopische Untersuchungen über die Entwicklung der "Chromatophoren" von *Rhodospirillum molischianum* Giesberger, *Arch. Mikrobiol.* **43:**152–161.

Göbel, F., 1976, Der Quantenbedarf des Wachstums phototropher blaugrüner Bakterien, in: *Jahresbericht,* Gesellschaft für Strahlen- und Umweltforschung, München, pp. 70–71.

Göbel, F., 1978, Quantum efficiencies of growth, in: *The Photosynthetic Bacteria* (R. K. Clayton and W. R. Sistrom, eds.), Plenum Press, New York, pp. 907–925.

Gorlenko, V. M., and Zhilina, T. N., 1968, Study of the ultrastructure of green sulfur bacteria, strain SK-413, *Mikrobiologiya* **37:**1052–1056.

Hickman, D. D., and Frenkel, A. W., 1965, Observations on the structure of *Rhodospirillum molischianum, J. Cell. Biol.* **25:**261–278.

Holt, S. C., and Marr, A. G., 1965, Location of chlorophyll in *Rhodospirillum rubrum, J. Bacteriol.* **89:**1402–1412.

Holt, S. C., Conti, S. F., and Fuller, R. C., 1966, Photosynthetic apparatus in the green bacterium *Chloropseudomonas ethylicum, J. Bacteriol.* **91:**311–322.

Keil, F., 1912, Beiträge zur Physiologie der farblosen Schwefelbacterien, *Beitr. Biol. Pflanz.* **11:**335–372.

Kluyver, A. J., Donker, H. J. L., 1926, Die Einheit in der Biochemie, *Chemie der Zelle und Gewebe* **13:**134–190.

Kluyver, A. J., and van Niel, C. B., 1956, *The Microbes' Contribution to Biology,* Harvard University Press, Cambridge, Mass.

Lankester, R., 1876, Further observation on a peach- or red-coloured bacterium—*Bacterium rubescens, Quart. J. microc. Sci.* N.s. **16:**27–40.

Mayer, J. R., 1845, *Die organische Bewegung in ihrem Zusammenhang mit dem Stoffwechsel: Ein Beitrag zur Naturkunde,* Drechler'sche Buchhandlung, Heilbronn.

Molisch, H., 1907, Die Purpurbakterien nach neuen Untersuchungen, Gustav Fischer, Jena.

Müller, F. M., 1933, On the metabolism of the purple sulphur bacteria in organic media, *Arch. Mikrobiol.* **4:**131–166.

Nathanson, A., 1904, Über eine neue Gruppe von farblosen Schwefelbakterien und ihren Stoffwechsel, *Mitt. zool. Station,* Neapel **15:**655–680.

Olson, J. M., Prince, R. C., and Brune, D. C., 1977, Reaction-center complexes from green bacteria, *Brookhaven Symp. Biol.* **28:**238–246.

Pfennig, N., 1962, Beobachtungen über das Schwämen von *Chromatium okenii, Arch. Mikrobiol.* **42:**90–95.

Pfennig, N., 1967, Photosynthetic bacteria, *Annu. Rev. Microbiol.* **21:**285–324.

Pfennig, N., and Cohen-Bazire, G., 1967, Some properties of the green bacterium *Pelodictyon clathratiforme, Arch. Microbiol.* **59:**226–236.

Pierson, B. K., and Castenholz, R. W., 1974, A phototrophic gliding filamentous bacterium of hot springs, *Chloroflexus aurantiacus, Arch. Microbiol.* **100:**5–24.

Pierson, B. K., and Castenholz, R. W., 1978, Photosynthetic apparatus and cell membranes of the green bacteria, in: *The Photosynthetic Bacteria* (R. K. Clayton and W. R. Sistrom, eds.), Plenum Press, New York, pp. 179–197.

Rabinowitch, E. I., 1945, *Photosynthesis and Related Processes,* Vol. 1, Interscience Publ. Inc., New York.

Remsen, Ch. C., 1978, Comparative subcellular architecture of photosynthetic bacteria, in: *The Photosynthetic Bacteria* (R. K. Clayton and W. R. Sistrom, eds.), Plenum Press, New York, pp. 31–60.

Schachman, H. K., Pardee, A. B., and Stanier, R. Y., 1952, Studies on the molecular organization of microbial cells, *Arch. Biochem.* **38:**213–221.

Skene, M., 1914, A contribution to the physiology of the purple sulfur bacteria, *New Phytologist* **13:**1–17.

Staehelin, L. A., Golecki, J. R., Fuller, R. C., and Drews, G., 1978, Visualization of the supramolecular architecture of chlorosomes (Chlorobium type vesicles) in freeze-fractured cells of *Chloroflexus aurantiacus, Arch. Microbiol.* **119:**269–277.

Trentini, W. C., and Starr, M. P., 1967, Growth and ultrastructure of *Rhodomicrobium vannielii* as a function of light intensity, *J. Bacteriol.* **93:**1699–1704.

Tuttle, A. L., and Gest, H., 1959, Subcellular particulate systems and the photochemical apparatus of *Rhodospirillum rubrum, Proc. Natl. Acad. Sci. U.S.A.* **45:**1261–1269.

Vatter, A. E., and Wolfe, R. S., 1958, The structure of photosynthetic bacteria, *J. Bacteriol.* **75:**480–488.

Van Niel, C. B., 1932, On the morphology and physiology of the purple and green sulphur bacteria, *Arch. Mikrobiol.* **3:**1–112.

Van Niel, C. B., 1944, The culture, general physiology, morphology and classification of the non-sulfur-purple and brown bacteria, *Bacteriol. Rev.* **8:**1–118.

Waksman, S. A., and Joffee, J. S., 1922, Microorganisms concerned in the oxidation of sulfur in soil, *J. Bacteriol.* **7:**239–256.

Winogradsky, S. N., 1887, Über Schwefelbakterien, *Bot. Ztg.* **45**:489–496.

Winogradsky, S. N., 1888, *Beiträge zur Morphologie und Physiologie der Bakterien,* Heft 1, Leipzig, A. Felix.

Winogradsky, S. N., 1890, Recherches sur les organismes de la nitrification, *Ann. Inst. Pasteur (Paris)* **4**:257–275.

5

OXYGENIC PHOTOSYNTHESIS IN PROKARYOTES

Mary Mennes Allen

INTRODUCTION

Cyanobacteria (blue-green algae, blue-greens) are prokaryotic organisms that contain a photosynthetic apparatus similar in structure and function to that present in the chloroplast of the phototrophic eukaryotes (Stanier, 1977). They differ from *Prochloron,* a prokaryote that contains chlorophyll b instead of phycobiliproteins as light-harvesting pigments. They differ from the purple and green bacteria, particularly because they carry out oxygenic photosynthesis, but also because of differences in the ultrastructure, in the chemical composition of the photosynthetic apparatus, and in nutritional requirements and growth physiology. The mechanism of cyanobacterial photosynthesis is identical to that of photosynthetic eukaryotes. Accordingly, in this discussion of oxygenic prokaryotic photosynthesis, only those areas where cyanobacteria played a role in the understanding of the process or only those details which are characteristic of blue-greens will be reviewed.

All photosynthetic eukaryotes, as well as two groups of prokaryotes, the cyanobacteria and prochlorophytes, contain photosystems I and II (PS I and PS II) and perform oxygenic photosynthesis. The structure of the reaction center complexes appears to be conserved among all these organisms (Glazer, 1983), but there is diversity in their antenna complexes. The ability of cyanobacteria to grow anaerobically has been retained since the Precambrian, when their predecessors were probably the first organisms to produce oxygen. The mechanism of oxygenic photosynthesis seems not to have been altered during the evolutionary events that separated cyanobacteria from algae and higher plants.

Chlorophyll a is the photosynthetic pigment common in all organisms

Mary Mennes Allen • Department of Biological Sciences, Wellesley College, Wellesley, Massachusetts 02181.

capable of oxygenic photosynthesis. It absorbs only a small part of the light spectrum, but, under saturating light conditions, it alone is sufficient to drive photosynthesis. Cyanobacteria have evolved ancillary light absorbing pigment systems that enhance their growth and survival in their particular ecological domains. Accessory pigments extend the absorption spectrum and work with chlorophyll a to harvest additional light for photosynthesis. It is the oxygen-evolving photosynthetic mechanism that makes cyanobacteria unique among bacteria and places them in many of the varied ecological niches in which they are found. Since cyanobacteria absorb light in regions of the visible and near infrared complementary to other bacteria, there is little competition for the photosynthetic opportunities available.

Among oxygen-evolving organisms, the cyanobacteria have a versatile physiology and wide ecological tolerance that contribute to their great competitive success. Within this, the largest and most widespread group of photosynthetic prokaryotes, there are species able to grow aerobically as photoautotrophs, anaerobically using H_2S, photoheterotrophically, chemoheterotrophically, and as anaerobic or aerobic dinitrogen fixers. Although thermal springs are the only presently known environment where cyanobacteria are the sole oxygen-evolving organisms (Brock, 1967), "extreme" environments are typically colonized by cyanobacteria; photosynthetic eukaryotes are either absent from or infrequent in environments that select for halophiles, thermophiles, psychrophiles, or aerobic dinitrogen-fixing bacteria. The wide variety of illuminated habitats that cyanobacteria occupy includes arid desert soils deficient in combined nitrogen (Cameron and Fuller, 1960), regions recently exposed to volcanism (Shields, 1957), hot springs (Castenholtz, 1969), and Antarctic lakes (Fogg and Horne, 1970). The ability of cyanobacteria to endure extreme fluctuations in temperature is important for their survival in deserts, where they are the major photosynthetic microbe. Cyanobacteria, therefore, are pioneer species in newly created ecological niches as well as having been pioneers in the evolution of life as evidenced by the fossil record. Paleobiological data (Schopf, 1974; Schopf and Walter, 1982) indicate that oxygen-producing blue-greens were present in the Early Precambrian and that the development of an oxygenated environment was almost entirely the result of their photosynthesis.

When combined nitrogen is limiting and the environment is lighted, such as in tropical soils or rice paddies, nitrogen-fixing cyanobacteria are abundant. Many cyanobacteria fix atmospheric dinitrogen, a property that they share with a variety of other bacteria but which is absent from all eukaryotes except symbionts of nitrogen-fixing prokaryotes. Heterocyst-forming cyanobacteria are active nitrogen-fixers; heterocysts have no PS II activity and thus cannot produce the oxygen that rapidly and irreversibly denatures nitrogenase, the enzyme catalyzing the reduction of dinitrogen to ammonia. In many cyanobacteria, this cellular separation of two incompatible activities allows both processes to go in the same filament of cells. Many nonheterocystous filamentous blue-greens can fix dinitrogen anaerobically (Rippka and Waterbury,

1977), and the unicellular *Gloeothece* (Wyatt and Silvey, 1969) fixes dinitrogen aerobically.

FROM BLUE-GREEN ALGAE TO CYANOBACTERIA

Because cyanobacteria were green, microscopic in size, performed oxygenic photosynthesis, and often were found in the same habitats as green algae, they long were considered algae and had been studied largely by botanists. Until the early 1960s, most studies were carried out using blue-greens as "plant" representatives to approach certain problems for which they might be best suited because of their pigmentation or rapid growth. Contaminating bacteria were generally ignored, and the group was poorly studied microbiologically. Many aspects of plant photosynthesis were, however, elucidated using blue-greens as experimenal objects, as will be reviewed below. Only after blue-greens were recognized as prokaryotes could meaningful comparisons be made with eukaryotic phototrophs, other prokaryotic phototrophs, and later, with cyanelles and with *Prochloron.* This last, the only genus in Prochlorophyta, was described by Lewin and Cheng (1975) as 0_2-evolving, phototrophic extracellular symbionts of marine invertebrates (ascidians). Although indisputably prokaryotic with respect to cell ultrastructure, these organisms lack phycobilisomes (structures found throughout the cyanobacteria), and possess P700-chlorophyll a-protein and chlorophyll a/b-protein similar to those of green plants, as well as two chlorophylls (a and b). The group and its affinities have been reviewed recently (Lewin, 1981; Withers et al., 1978; Seewaldt and Stackebrant, 1982).

Similarities between the mechanism of photosynthesis, the nucleic acids, and the protein-synthesizing systems of the cyanobacteria and chloroplasts have led to hypotheses suggesting an endosymbiotic origin of chloroplasts (Mereschkowsky, 1905; Sagan, 1967). The close relationships between phycobilins of red algae and cyanobacteria led to the hypothesis of cyanobacterial origin for the rhodophytan chloroplast (Stanier and Cohen-Bazire, 1977). Although the presence of chlorophyll b in the light-harvesting pigment system of *Prochloron* has encouraged the proposal that *Procholoron* may be the evolutionary predecessor of green plant chloroplasts (Stanier and Cohen-Bazire, 1977), much more information is required to resolve this issue (Glazer, 1983). Comparison of 16 S ribosomal RNA sequences suggests no specific relationship between *Prochloron* and chloroplasts of green algae or higher plants, but shows highest sequence homology with cyanobacteria (Seewaldt and Stackebrandt, 1982).

History of the Relationship between Blue-Green Algae and Bacteria

The blue-green algae were recognized as an autonomous group of algae by Stizenberger (1860), who called them Myxophyceae and differentiated them from other algae by pigmentation. To emphasize their blue pigmentation, Sachs (1874) named the organisms Cyanophyceae. From his study of apochlorotic algae, which he grouped together with pigmented ones in his classification system, Cohn (1853) was the first to argue that bacteria and blue-greens were closely related. He later named blue-greens Schizophyceae and grouped them with the bacteria (Schizomycetes) in the division Schizophyta (Cohn, 1875). Pringsheim (1949) has reviewed the classical literature supporting and opposing Cohn's views on the similarity of bacteria and the blue greens.

Not until electron microscopic studies of fine structure were possible did the concept of the close relationshp between bacteria and blue-greens gain support, although Stanier and van Niel (1941) had proposed that bacteria and blue-greens originated from common ancestors, noting that they had as common features the absence of true nuclei, absence of sexual reproduction, and absence of plastids, and thus argued that on morphological bases alone the separation of bacteria and blue-greens was impossible. Stanier and van Niel argued that Copeland's kingdom of Monera (1938) be used for the two divisions of blue-greens and bacteria. In 1962, Stanier and van Niel developed a scientific definition of bacteria to allow their separation from other protists and from viruses, but, again they could not clearly separate blue-greens from bacteria purely on morphological grounds; physiological and biochemical criteria were also needed. Both groups of organisms have prokaryotic cellular organization—absence of internal membranes separating the nucleus from the cytoplasm and isolating the enzymatic machinery of photosynthesis and respiration in specific organelles; nuclear division occurs by fission, not by mitosis; and a unique mucopeptide is the chief strengthening agent in their cell wall. Added proof that blue-greens are prokaryotic came from the work of Taylor and Storck (1964), who showed that blue-greens shared with other bacteria 70 S ribosomes, while eukaryotic, cytoplasmic (nonplastid) ribosomes have a sedimentation value of 80 S.

Modern Methods for Establishment of Affinities among Organisms

Modern methods for establishing affinities among organisms give further evidence justifying classification of blue-greens as bacteria (proposal by Stanier et al., 1978). Unicellular blue-greens show a range of 35–71 moles percent G + C in their DNA (Stanier et al., 1978), while filamentous forms range between 39 and 52 moles percent (Edelman et al., 1967). A more recent study of the DNA base composition (Herdman et al., 1979a) of 176 strains showed that only two subgroups, the unicells and the filamentous nonheterocystous strains, have major DNA base compositional divergences; for all other strains, the overall range is 38 to 47 moles percent. The genome sizes of 128 pure strains of cyanobacteria were described by Herdman et al. (1979b). The majority of unicellular strains contained genomes of 1.6 to 2.7×10^9 daltons, comparable in size to other bacteria and interpreted as corresponding to multiples of two times a basic unit of 1.2×10^9 daltons. Most pleurocapsalean and filamentous strains possessed large genomes distributed into three distinct groups, interpreted as corresponding to multiples of three, four, and six times the basic unit. Herdman et al. (1979b) suggested that genome evolution in cyanobacteria occurred by a series of duplications of a small ancestral genome, thus increasing their morphological complexity.

Ribosomal RNA (rRNA) sequence homologies determined by comparisons of T1 oligonucleotide catalogs of ^{32}P-labeled 16 S rRNAs have been used to assess phylogenetic relationships with the cyanobacteria and to assess similarities with other bacteria (Bonen and Doolittle, 1976). Similarity coefficients (S values) which reflect the number of oligonucleotide sequences common to two catalogs are derived and from matrices of such coefficients, relatedness can be correlated.

These studies indicate that cyanobacteria as a group do not appear more distant from certain bacterial groups, such as *Escherichia coli, Bacillus subtilis,* and *Rhodopseudomonas sphaeroides,* than some of these groups are from each other (Bonen and Doolittle, 1976). For example, blue-greens on average show as much homology (86.6%) with *B. subtilis* as the latter does with *E. coli.* Each of the four cyanobacterial 16 S rRNAs is more closely related to *B. subtilis* 16 S rRNA than it is to *R. sphaeroides* 16 S rRNA and more closely related to the latter than to *E. coli* 16 S rRNA. This supports Bonen and Doolittle's theory that blue-greens are more closely related to bacilli than to enterics. The relationship extends also to post-transcriptional modification patterns: modified oligonucleotides were identical in all four blue-greens and more similar to those of bacilli than to those of enterics. The 16 S rRNA homologies among the four cyanobacteria are in agreement with the classification system of Stanier et al. (1971) (Rippka et al., 1979). Other similarities of transcriptional and translational machinery of cyanobacteria with other bacteria are reviewed by Doolittle (1979 and 1982) and Stanier and Cohen-Bazire (1977).

Recognition as Prokaryotae

In the eighth edition of *Bergey's Manual of Determinative Bacteriology* (Buchanan and Gibbons, 1974), cyanobacteria were for the first time made one of two divisions of the Prokaryotae, bringing them together taxonomically with other bacteria.

DISCOVERY OF THE ROLE OF PHOTOSYNTHETIC PIGMENTS AND ELECTRON CARRIERS IN CYANOBACTERIA

Accessory Pigments

That light absorbed by accessory pigments (i.e., those other than chlorophyll a) is utilized for photosynthesis was first suggested by the pioneering experiments of Engelmann (1883), who used motile bacteria as indicators of oxygen evolution from photosynthesis of algae of various colors. Action spectra (measured by the number of oxygen-sensitive bacteria accumulating along a filament of the blue-green *Oscillatoria* exposed to a microspectrum of light) compared to absorption spectra of cells indicated that the motile bacteria accumulated along the orange part of the spectrum and that the distribution corresponded to light absorption by phycocyanin. Nearly 60 years passed while his conclusions were disputed.

The relative amounts of light absorbed at different wavelengths were not quantitatively correlated with photosynthetic activity in different wavelengths until measurements of quantum yield (measured as 0_2-evolved per incident quantum as a function of wavelength of light) for various species of algae were made by Emerson and Lewis (1941a,b). In various regions of the spectrum, different plant pigments absorb different proportions of the total absorbed radiation. For any pigment photochemically inactive in photosynthesis, the observed quantum yield for the total radiation absorbed will decrease at the wavelength where that pigment is the principal absorbing molecule. Emerson and Arnold reasoned that a comparison of wavelength dependence of quantum yield with the fractions of total absorbed light absorbed by the different components would reveal not only which pigments were photochemically active in photosynthesis, but their relative activity as well. The quantum yield of the cyanobacterium *Chroococcus* was therefore measured at a succession of wavelengths throughout the visible spectrum, and the dependence of quantum yield on wavelength was compared to the proportion of light absorbed by each pigment. These comparisons demonstrated that light absorbed by the accessory pigment phycocyanin was utilized in photosynthesis with an efficiency equal to that of the light absorbed by chlorophyll, and that light absorbed by carotenoids was largely unavailable for photosynthesis (Emerson and Lewis, 1941b). Measurements by Haxo and Blinks (1950) of several marine blue-greens suggested that, as for the red algae they tested, there was low photosynthetic activity in regions of maximum chlorophyll absorption, but that light trapped by phycoerythrin and phycocyanin was utilized in photosynthesis with an efficiency approaching 100%. Emerson and Lewis (1943) found a sharp drop in the quantum yield above 685 nm in *Chlorella* (hence the term "red drop") and assumed a similar drop in *Chroococcus,* based on the 1941 data. The apparent efficiency of chlorophyll could be increased more than additively by simultaneous illumination using light of a wavelength absorbed by an accessory pigment (Emerson, 1958); it was later hypothesized that this enhancement effect was attributable to light absorption by phycobilins (Blinks, 1960). Blinks showed that the action spectrum for the Emerson effect was almost identical with that of chromatic transients—changes in oxygen evolution rate recorded upon altering the wavelength of incident light when intensities were adjusted to give equal steady rates of photosynthesis at both wavelengths.

Mechanisms of Energy Transfer

The mechanism of energy transfer was first examined by Arnold and Oppenheimer (1950), who concluded from absorbance spectra that resonance transfer of energy was the important mechanism for transfer of energy from phycocyanin to chlorophyll. They further estimated, from the frequency of energy transfer, that the probable distance between phycocyanin and chlorophyll molecules should be 40 Å.

The first evidence for a pigment, in small concentration, able to efficiently trap energy and transfer it to neighboring molecules of chlorophyll a was obtained in fluorescence studies in red and blue-green algae by Duysens (1951). In phycobiliprotein-containing organisms, quanta absorbed by the phycobilins excited chlorophyll a to a stronger fluorescence than did quanta absorbed by chlorophyll a itself. Duysens further concluded that chlorophyll a occurred in two different modifications, differing in fluorescence yield, and that energy is transferred from the phycobilins to the highly fluorescent part of chlorophyll. Low temperature fluorescence spectra of various blue-greens showed that light energy absorbed by phycobilins was more readily transferred to certain forms of chlorophyll than to others (Goedheer, 1968). Excitation energy was, therefore, effectively transferred from phycocyanin to chlorophyll. For efficient energy transfer, the fluorescence spectrum of the energy donor pigment must overlap with the absorption spectrum of the energy acceptor, and the pigments must be in close proximity to each other. However unlikely it is that the energy is transferred as light *per se,* the mechanism(s) of the transfer remain to be elucidated.

Existence of Two Photosystems

Studies of cyanobacteria provided significant experimental support for the hypothesis, based on the enhancement (or Emerson) effect, that oxygenic photosynthesis involved two photosystems. Using *Anacystis* and *Nostoc,* as well as green and red algae, Kok and Hoch (1961) obtained evidence for the existence of a pigment, P700, that could be bleached by photochemical events involving chlorophyll a, and its light-absorbing properties re-established by photochemical stimulation of the "accessory" pigment phycocyanin. Additional evidence for the occurrence of two different light-induced reactions came from the studies of Jones and Myers (1964). Using *A. nidulans,* they demonstrated two enhancement spectra: against background monochromatic light of 690 nm, known to be absorbed maximally by chlorophyll a, the enhancement spectrum corresponded to the absorption specrum of phycocyanin; against background monochromatic light of 620 nm, known to be absorbed maximally by phycocyanin, the enhancement spectrum corresponded to the absorption spectra of chlorophyll a and the carotenoids. These two systems have been designated "PS I" (the chlorophyll-carotenoid system) and

"PS II" (the accessory pigment system). The results of this type of study were especially clear when cyanobacteria were employed, because the absorption maxima of chlorophyll a and the accessory pigments (phycocyanins) are farther apart than in green algae (which contain chlorophyll b as the principal accessory pigment).

Amesz and Duysens (1962) had already reported evidence of two oxidation-reduction systems in *A. nidulans.* The two systems were distinguished as follows: one system effected cytochrome oxidation and NADP reduction, while the other system effected O_2 production and cytochrome reduction. The action spectra of the two systems were distinct. In that same study and in a related study with red algae (Duysens and Amesz, 1962), it was observed that 3-(3,4-dichlorophenyl)-1,1-dimethyl urea (DCMU), a known inhibitor of photoevolution of O_2 by green plants, inhibited O_2 evolution, but not NADP reduction, in the cyanobacterial and algal systems.

These studies were crucial to the developing notion that in oxygenic photosynthesis, the two distinguishable photosystems that existed acted in series. It was subsequently demonstrated, again largely from studies with cyanobacteria (see, e.g., Amesz, 1964), that electrons are transferred from PS II to PS I via a carrier system involving plastoquinone. The present view is that the accessory pigment (phycocyanin or chlorophyll b)-containing system mediates the light-dependent oxidation of water, freeing O_2 and yielding electrons that reduce the intermediate carrier system and cytochromes. The chlorophyll a-containing system, upon illumination, oxidizes the cytochromes and reduces pyridine nucleotides.

Other Components of Cyanobacterial Photosynthesis

Other components known to occur in plant photosynthetic reactions were later shown present in blue-greens as well: ferredoxin (Arnon, 1963, 1965), ferredoxin-reducing substance (Fujita and Myers, 1965), NADP: ferridoxin oxidoreductase (Susor and Krogmann, 1967), plastocyanin (Lightbody and Krogmann, 1967), membrane bound cytochrome 554 or cytochrome f (Katoh, 1969), and a low-potential soluble cytochrome C_{549} (Holton and Myers, 1963). C550 was found (Knoff, 1973) from spectroscopic observations to be a substance that underwent photoreduction even at liquid nitrogen temperatures. Knoff also noted that cytochrome f and cytochrome b_{558} were photoreduced by PS II and photooxidized by PS I in a cell-free preparation from *Nostoc muscorum,* consistent with the hypothesis that these two cytochromes are part of a chain of electron carriers linking PS II and PS I. Knoff also detected photoreduction of C550 by PS II at 77°K and concluded that C550 was closely associated with the primary electron acceptor for PS II; similarly, photosystem I photoxidation of P700 at 77°K was consistent with the notion that P700 served as the primary electron donor of PS I. Recent reviews on the mechanism of photosynthesis in cyanobacteria are available (Krogmann, 1973, 1977; Ho and Krogmann, 1982).

Fractionation of Photosynthetic Function

Fractionation of chloroplast (i.e., eukaryotic) photosynthetic apparatus into isolated pigment-protein complexes differing in chemical composition and in partial reactions of photosynthesis provided additional evidence for the existence of two photosystems. Similar results have also been obtained by cyanobacteria.

Ogawa et al. (1969) were the first to separate two photochemical systems in cyanobacteria; they employed Triton X-100 detergent treatment and sucrose density centrifugation of sonicated *Anabena variabilis* washed free of phycobiliproteins. Two colored bands were separated: a heavy, blue-colored band corresponding chemically and spectrally to PS I (enriched in long-wavelength chlorophyll a, P700, and with a higher β-carotene to total xanthophyll ratio) and a light, orange-colored band containing components expected for PS II (enriched both in short-wavelength chlorophyll a and carotenoids). The fragments were, however, inactive in $NADP^+$ photoreduction as well as O_2 evolution; it continues to be difficult to obtain active and stable thylakoids from cyanobacteria. Hill reaction and coupled photophosphorylation appear to be the most labile reactions (Binder et al., 1976); most workers have found that PS II activity is especially easily inactivated, and isolated particles, accordingly, are rarely able to evolve O_2.

The most promising PS II particles (showing the highest oxygen-evolving activity to date), prepared from a thermophile by lauryldimethylamine oxide (LDAO) solubilization, did exhibit oxygen evolution; however, the effect of DCMU was not reported, and some PS I activity remained in the particles (Stewart and Bendall, 1981). The only other method extant for preparing PS II particles that retain oxygen-evolving capacity was described by England and Evans (1981), who developed a rapid procedure employing LDAO with French pressure-cell breakage; they believe this may be more generally applicable to other cyanobacteria. Oxygen evolution by such preparations was substantially inhibited by DCMU.

Guilkema and Sherman (1982) and Sadewasser and Sherman (1981) have recently utilized a high resolution gel system to detect over 90 membrane proteins and six chlorophyll-containing bands in *Anacystis nidulans* preparations. These investigators (1981) also used 3,3,5,5-tetramethylbenzidine (TMBZ) with hydrogen peroxide to stain the gels; this revealed five polypeptides with heme-dependent peroxidase activity. Labeling with ^{59}Fe showed nine iron-containing bands, and lactoperoxide-catalyzed iodination defined at least 41 proteins with surface-exposed domains. This work, in conjunction with fluorescence and fractionation studies, allowed the first construction of a topological model of functional complexes for the photosynthetic lamellae of cyanobacteria in two closely related organisms; further comparable analyses of other cyanobacteria and photosynthetic mutants are necessary to ascertain comparative structure–function relationships in photosynthetic membranes among cyanobacteria and to allow comparison with chloroplast membrane models.

Photosynthetic Mutants of Cyanobacteria

Among the many techniques used to understand photosynthesis, an important approach has been to block certain steps of the process selectively. One approach is to isolate mutants unable to carry out photosynthesis and then to determine which steps of the process do or do not occur in each mutant. Prokaryotic organisms with only a haploid genome on which the information for the photosynthetic apparatus must be coded would seem well suited for the study of the genetics of photosynthesis. An analysis of mutants defective in electron transport activity would provide an approach to the relationship of membrane structure and function. Only recently (Astier et al., 1979) have a variety of mutants of cyanobacteria modified in photosynthetic function become available. Spontaneous pigment mutants were selected by growth in far-red light by Myers et al. (1980); Stevens and Myers (1975) mutagenized cells to obtain pigment mutants; and Astier et al. (1979) have isolated three DCMU-resistant mutants of a chemoheterotrophic strain. Because most cyanobacteria are obligate photoautotrophs, photosynthetic mutations tend to be lethal. Sherman and Cunningham (1977) therefore isolated temperature sensitive (ts) mutants by nitrosoguanidine mutagenesis and selection for high fluorescence or resistance to metronidazole (Guilkema and Sherman, 1980). Protein and fluorescence patterns of the mutants indicated that the mutations were distinct and affected numerous membrane functions (Sherman and Cunningham, 1977).

Path of Carbon in Photosynthesis

The path of carbon in cyanobaterial photosynthesis was first studied by Norris et al. (1955) as part of short term $^{14}CO_2$ incorporation experiments in 27 different phototrophs. Paper chromatography showed the majority of the label was found in hexose monophosphates and phosphoglyceric acid after 5 min of incorporation in each test species. A compound later identified as citrulline (Linko et al., 1957) became radioactive in appreciable amounts in several blue-greens, but not in other organisms. Detailed kinetic analyses of the products of $^{14}CO_2$ assimilation were done in several unicellular cyanobacteria (Pelroy and Bassham, 1972; Ihlenfeld and Gibson, 1975) and were consistent with the net fixation of CO_2 via the reductive pentose phosphate cycle.

PHOTOSYNTHETIC STRUCTURES IN CYANOBACTERIA

Two features of the cyanobacterial photosynthetic apparatus distinguish it from the chloroplast structures present in most eukaryotic plants. In cyanobacteria, photosynthetic thylakoids are present in varying unstacked arrangements in the cytoplasm, whereas in eukaryotes the thylakoids are within the chloroplast envelope membrane and are usually stacked as grana. Secondly, cyanobacteria contain phycobiliproteins organized into discrete phycobilisome structures on the surface of the thylakoids. Among eukaryotes, only the red algae contain phycobilisomes; other eukaryotes and *Prochloron* have a light-harvesting protein complex (CPII) containing chlorophyll b integrated into the chloroplast membrane.

Thylakoids

The earliest electron microscopic studies of cyanobacteria (Niklowitz and Drews, 1957; Ris and Singh, 1961) showed the cytoplasm to be traversed by a complex system of paired lamellae that were not separated from other parts of the cytoplasm by an enclosing membrane. After cell breakage, Petrack and Lipmann (1961) showed that phycocyanin-free fragments of the lamellar system contained all cell chlorophyll and could carry out photophosphorylation. They suggested that phycocyanin may function by protecting against photoxidation instead of being essential for photosynthesis. Shatkin (1960) compared thin sections of whole cells of *Anabaena variabilis* with thin sections of the 105,000*g* pellet from broken cells which Petrack and Lipmann had studied; he showed that the pellet consisted of flattened, smooth vesicles, 30–300 nm in diameter, derived from the peripheral lamellae. Calvin and Lynch (1952) had previously observed by electron microscopy a 36,000*g* particulate fraction from a unicellular blue-green; they described the particle as grana-like because they were of a similar size to chloroplast grana and contained all the chlorophyll and carotenoids of the cell. Cox et al. (1964) showed that a membrane fraction carried out photosynthetic oxygen evolution, and Allen (1968b) demonstrated that the amount of chlorophyll in *Anacystis nidulans* was directly proportional to the amount of thylakoid membrane present. Although Thomas and DeRover (1955) had shown loss of Hill reaction to be correlated with a loss of phycocyanin, Susor and Krogmann (1964) demonstrated that lamellae devoid of the major accessory pigment retained Hill reaction, although the quantum efficiency with water was low.

The chlorophyll and electron transport systems of cyanobacterial photosynthesis, therefore, are contained in an extensive system of flattened membranous sacs or thylakoids that are separated from a cell membrane that itself contains no component of the photosynthetic machinery. There is but one known exception: in unicellular *Gloeobacter violaceous,* the cell membrane is the location both of chlorophyll and the electron transport system (Rippka et al., 1974).

Phycobilisomes

Phycobiliproteins were found localized in structures termed phycobilisomes; these appear in most cyanobacteria as globular bodies 40 nm in diameter on the outer surface of photosynthetic lamellae (Gantt and Conti,

1966) when gentle preparative techniques are used. They can be isolated intact in concentrated salt solutions (Gantt and Lipschultz, 1972) after cell breakage, treatment with the nonionic detergent triton X-100, and sucrose gradient centrifugation. Isolated phycobilisomes contain only protein—both biliproteins (see below) and colorless linker polypeptides (Tandeau de Marsac and Cohen-Bazire, 1977). In *Gloeobacter,* however, phycobilisomes are present in a single cortical layer in contact with the inner surface of the cell membrane (Rippka et al., 1974).

Phycobiliproteins are a major constituent of cyanobacterial cells; in some species, such as *Anacystis nidulans,* they may represent up to 40% of the cell protein (Myers and Kratz, 1955). Phycocyanins usually have a broad *in vivo* absorption band (*ca.* 620 nm, phycoerythrins and phycoerythrocyanin *ca.* 565 nm, and allophycocyanin *ca.* 650 nm). All cyanobacteria studied contain allophycocyanin and C-phycocyanin and some contain C-phycoerythrin; their fluorescence maxima show that a stepwise energy transfer is possible from phycoerythrin or phycoerythrocyanin to phycocyanin to allophycocyanin to chlorophyll a (Gantt and Lipschultz, 1973). The function of allophycocyanin in light harvesting and energy transfer was demonstrated by Lemasson et al. (1973) who found a distinct peak in the action spectra of several cyanobacteria at 650–655 nm; the height of the peak correlated with allophycocyanin level. The small allophycocyanin peak at 650 nm is completely masked by phycocyanin and chlorophyll a absorption in this spectral region, but allophycocyanin appears to be the most efficient light-harvesting pigment. The close proximity of pigments in the phycobilisome and the close proximity of phycobilisomes to thylakoids had been predicted earlier by Arnold and Oppenheimer (1950) as a way to maximize energy transfer. Current knowledge and hypotheses on phycobilisomes and their structure and function have been reviewed (Gantt, 1980; Glazer, 1982, 1983; and Cohen-Bazire and Bryant, 1982).

NUTRITIONAL ADAPTATIONS IN THE PHOTOSYNTHETIC APPARATUS

Many useful insights into problems of microbial ecology can be derived from controlled experimenal studies on pure cultures (Stanier, 1977). However, comparative experimental studies became possible only relatively late in the elucidation of the biochemistry of photosynthesis. The first consistent purification of strains was done by "pour plates" (Allen, 1952) and by repeated transfer in liquid medium (Gerloff et al., 1950). Application of traditional bacteriological techniques (Allen, 1968a; Stanier et al., 1971) finally allowed purification and characterization of strains at will. Many of the physiological and metabolic factors accounting for the wide natural distribution of cyanobacteria are now being understood, and much of this understanding comes from studies of nutritional adaptation of pure cultures in the laboratory.

Light Quality and Quantity

The synthesis of the biliproteins C-phycoerythin and phycoerthrocyanin is regulated by the wavelength of incident light in certain cyanobacteria. Phycocyanin synthesis is enhanced when cells are grown in red light, and phycoerythrin synthesis is enhanced when cells are grown in green light. Chromatic adaptation (change in the accessory pigment composition in response to alterations in the energy distribution in the available light) is a convenient way for the cell to vary its biliprotein composition in order to absorb most effectively the wavelengths of light available. The response was first described by Gaidukov (1902), who showed that when *Oscillatoria* was illuminated with a spectrum of light, the filaments soon became colored in a fashion complementary to the part of the spectrum in which they were growing. Although these observations may have been partially due to light intensity changes, detailed studies in *Tolypothrix* begun in 1959 (Hattori and Fujita, 1959) showed that differential stimulation of either phycocyanin or phycoerythrin synthesis could be induced within a few minutes using appropriate wavelengths of light. Recent information on chromatic adaptation can be found in reviews by Bogorod (1975) and Glazer (1982).

Not all blue-greens, however, adapt chromatically. Jones and Myers (1965) showed that the chlorophyll content of *Anacystis nidulans* grown in red light was about one-quarter of that of cells grown using tungsten bulbs. The phycocyanin and carotenoid contents showed little change. The hypothesis to explain this phenomenon was that since PS II controlled PS I activity and in red light PS II was light-limited, the chlorophyll of PS I absorbed more light than could be used by PS I. Myers and Kratz (1955) first established the inverse relationship of pigment content to the light intensity at which *A. nidulans* was grown, but the ratio of chlorophyll to phycocyanin remained constant at varying intensities of white light. Thylakoid content also varied inversely with light intensity (Allen, 1968b), and estimations of relative amounts of thylakoids in cells showed that the amount of thylakoid membrane was directly proportional to the chlorophyll content.

Temperature Effects

Higher-than-ambient temperatures have been used to enrich for cyanobacteria in the laboratory (Fogg, 1956; Allen and Stanier, 1968). Although few strains are truly thermophilic (capable of growth at temperatures above 50°C) (Fogg, 1956), many strains grow well at 35°C (Allen and Stanier, 1968), a temperature well above that tolerated by eukaryotic algae such as *Chlorella* (Fogg, 1956). This selection, using natural material, greatly favors enrichment and development of the cyanobacterial population initially present and allows observation and analysis of cyanobacteria from material in which they could not initially be detected microscopically (Allen and Stanier, 1968). In laboratory studies, as temperature was increased from 25 to 35°C, the amount of chlorophyll and phycocyanin per milligram of protein and of thylakoid membrane increased in *Anacystis nidulans* (Allen, 1968b; Myers and Kratz, 1955). The ratio of chlorophyll to phycocyanin did not change. Cyanobacteria can grow at constant temperatures as high as 73–74°C (Brock, 1967); Brock has suggested that the photosynthetic apparatus itself is the most temperature-sensitive component of the cell.

Other Nutritional Adaptations

Boresch (1910) had referred to a "yellowing" of older cultures of *Oscillatoria* as a result of nitrogen deficiency, a phenomenon he termed nitrogen chlorosis. Nitrogen deficiency caused a selective decrease in phycocyanin in *A. nidulans* (Allen and Smith, 1969), but the pigment reappeared 8 hr after addition of nitrate to the culture; chlorophyll and carotenoid contents remained constant throughout the period of loss and resynthesis of phycocyanin. Lau et al. (1977) later showed that the loss of phycocyanin absorbance was accompanied by the coordinate loss of the two phycocyanin apoprotein subunits and the specific repression of phycocyanin apoprotein synthesis.

Yamanaka and Glazer (1980) have recently shown decrease in size of *Synechococcus* phycobilisomes following a preferential loss of phycocyanin and one linker polypeptide in the first stage of nitrogen starvation; the smaller particle showed unimpaired energy transfer. Prolonged nitrogen starvation then caused degradation of two more linker polypeptides, detachment of the particles from thylakoids, and then complete breakdown of particles. Eley (1971) studied the effect of CO_2 on pigment concentration in the same organism using two CO_2 concentrations (1% and 0.03%). Decrease in pigment concentration (determined to be due almost entirely to changes in phycocyanin) was directly related to CO_2 concentration; Eley hypothesized that CO_2 control of phycocyanin synthesis allowed maximum light capture and maintained a balance of quantum input to the two photoreactions. Miller and Holt (1977) showed that after 120 hr of CO_2 deprivation of *Synechococcus,* all chlorophyll a and C-phycocyanin were lost, while the amount of carotenoid increased 1.8 fold. Electron microscopy revealed cells devoid of thylakoid membrane; addition of CO_2 to bleached cultures resulted in a rapid resynthesis of both chlorophyll a and phycocyanin as well as resynthesis (or reassembly) of thylakoid membrane. As in nitrogen-starved cells (Allen and Smith, 1969), growth did not resume after nutrient provision until the normal C-phycocyanin level was reached. Action spectra of *Aphanocapsa* depleted of phycobiliproteins by nitrogen chlorosis showed that the substantial capacity to produce oxygen in the light was due to chlorophyll a as the principal light-harvesting pigment

(Lemasson et al., 1973). When the cellular biliprotein level becomes too low to maintain photosynthetic activity, the light-harvesting role of these pigments is assumed by chlorophyll a. This implies that the way energy is transferred to PS II can undergo major physiologically induced changes in cyanobacteria.

FACULTATIVE ANOXYGENIC PHOTOSYNTHESIS

Water has usually been considered the ultimate hydrogen or electron donor in cyanobacterial photosynthesis, as in oxygenic eukaryotes. However, many blue-greens are also capable of a photoreduction resembling that of anoxygenic bacterial photosynthesis, by utilizing hydrogen as electron donor in a hydrogenase reaction induced by anaerobic incubation (Frenkel et al., 1950). Nakamura (1938) grew *Oscillatoria* in 1 mM H_2S and did not detect O_2 evolution; sulfur deposition in cells suggested CO_2 assimilation using H_2S as electron donor.

Cohen et al. (1975) substantiated and extended this observaton by demonstrating that *Oscillatoria limnetica* can carry out facultative anoxygenic CO_2 photoassimilation with sulfide as an electron donor in a PS I-driven process similar to that carried out by anoxygenic photosynthetic bacteria. Assimilation of CO_2 is accompanied by a stoichiometric oxidation of H_2S to elemental sulfur, which is deposited outside the filament. This cyanobacterium, isolated from the sulfide-rich layers of Solar Lake in Israel, can carry out DCMU-insensitive photoassimilation in high sulfide concentrations (3 mM), while 0.1–0.2 mM sulfide or DCMU inhibits its typical oxygenic photosynthesis (Cohen et al., 1975). Light and sulfide are both absolute requirements for this anoxygenic photosynthesis, which is independent of PS II. The immediate effect of sulfide is to act as a selective inhibitor of PS II, converting oxygenic photosynthesis into a purple- or green-bacterial type photosynthesis. A delayed effect of sulfide is to induce a sulfide-oxidizing enzyme system allowing the resumption of photosynthetic CO_2 assimilation. Oren et al. (1977) showed that neither enhancement nor PS II activation in the presence of far-red light was observed in the anoxygenic process. Anoxygenic photosynthesis is widespread among cyanobacteria, and thorough reviews on the impact of this physiological character on cyanobacterial ecology are available (Padan, 1979; Padan and Cohen, 1982).

In a number of aquatic systems, photoanaerobic and photoaerobic conditions alternate with time and distance from stream origins. Preference of many cyanobacteria for microaerophilic conditions with low redox potential may give them a selective advantage in these aquatic systems. Cyanobacteria able to carry out facultative anoxygenic photosynthesis therefore occupy a more intermediate ecological and physiological position than do strictly aerobic or strictly anaerobic phototrophs (Padan, 1979). When growing anaerobically, nonheterocystous cyanobacteria can express the genetic capacity to fix dinitrogen (Rippka and Waterbury, 1977); the facultative anoxygenic photosynthetic cyanobacteria may therefore play a role as dinitrogen-fixers in microenvironments that fluctuate between aerobic and anaerobic states.

CONCLUSIONS

There are distinct experimental advantages in studying the process of photosynthesis in cyanobacteria. As with other prokaryotes, axenic growth in a precisely controlled environment allows easy manipulation of their physiology. Heterotrophic capabilities of some species should allow growth of cells with specific genetic deletions in their photosynthetic apparatus, and recent advances in recombinant DNA and molecular cloning (Sherman and Guilkema, 1982; Kuhlenmeier et al., 1981) should allow studies of exact functional characteristics of membrane components. The simpler and more loosely organized photosynthetic thylakoid system should facilitate these studies, by allowing manipulation and reconstitution of the membranes.

The ubiquitous cyanobacteria with capabilities both of oxygenic and anoxygenic photosynthesis are adaptable to many environments, and their ability to fix dinitrogen either aerobically or anaerobically also extends their potential niches in dimensions not possible for other organisms. The use of phycobiliproteins as accessory pigments extends the useful range of efficient energy transfer and allows utilization of light that other organisms have not absorbed. The existence of phenomena such as chromatic adaptation allows a regulation of the photosynthetic process, based on changing environmental conditions, which has ecological significance.

The discovery of a widespread deep-living population of unicellular cyanobacteria at sea (Waterbury et al., 1979) has recently raised questions about the extent of their contribution to the total phototrophic biomass. Johnson and Sieburth (1979) estimated their biomass as 20% of the total microbial plankton. Yopp et al. (1978) found cyanobacteria in a hypersaline environment where the salt concentration was saturating and the lake floor covered with salt crystals.

Growth in these many different habitats, their apparent antiquity, and their influence on the early environment of earth, make the cyanobacteria among the most successful and most ecologically significant microbes.

REFERENCES

Allen, M. B., 1952, The Cultivation of Myxophyceae, *Arch. Mikrobiol.* **17:**34–53.

Allen, M. M., 1968a, Simple conditions for growth of unicellular blue-green algae on plates, *J. Phycol.* **4:**1–4.

Allen, M. M., 1968b, Photosynthetic membrane system in *Anacystis nidulans*, *J. Bacteriol.* **96:**836–841.

Allen, M. M., and Smith, A. M., 1969, Nitrogen chlorosis in blue-green algae, *Archiv. Mikrobiol.* **69:**114–120.

Allen, M. M., and Stanier, R. Y., 1968, Selective isolation of blue-green algae from water and soil, *J. Gen. Microbiol.* **51**:203–209.

Amesz, J., 1964, Spectrophotometric evidence for the participation of a quinone in photosynthesis of intact blue-green algae, *Biochim. Biophys. Acta* **79**:257–265.

Amesz, J., and Duysens, I. N. M., 1962, Action spectrum, kinetics and quantum requirement of phosphopyridene nucleotide reduction and cytochrome oxidation in the blue-green alga *Anacystis nidulans, Biochim. Biophys. Acta* **64**:261–278.

Arnold, W., and Oppenheimer, J. R., 1950, Internal conversion in the photosynthetic mechanism of blue-green algae, *J. Gen. Physiol.* **33**:423–435.

Arnon, D. I., 1963, Photosynthetic electron transport and phosphorylation in chloroplasts, in: *Photosynthetic Mechanisms of Green Plants* (B. Kok and A. T. Jagendorf, eds.), Natl. Acad. Sci.—Natl. Res. Council Publ. 1145, Washington, D.C., pp. 195–212.

Arnon, D. I., 1965, Ferredoxin and photosynthesis, *Science* **149**:1460–1470.

Astier, C., Vernotte, C., Der Vartanian, M., and Joset-Espardellier, F., 1979, Isolation and characterization of three DCMU-resistant mutants of the blue-green algae *Aphanocapsa* 6714, *Plant Cell Physiol.* **20**:1501–1510.

Binder, A., Tel-Or, E., and Avron, M., 1976, Photosynthetic activities of membrane preparations of the blue-green alga *Phormidium luridum, Eur. J. Biochem.* **67**:187–196.

Blinks, L. R., 1960, Action spectra of chromatic transients and the Emerson effect in marine algae, *Proc. Natl. Acad. Sci. U. S. A.* **46**:327–333.

Bogarod, L., 1975, Phycobiliproteins and complementary chromatic adaptation, *Ann. Rev. Plant Physiol.* **26**:369–401.

Bonen, L., and Doolittle, W. F., 1976, Partial sequences of l6 SrRNA and the phylogeny of blue-green algae and chloroplasts, *Nature (London)* **261**:669–673.

Boresch, K., 1910, Zur Physiologie der Blaualgenfarbstoffe, *Lotos (Prag.)* **58**:344–345.

Brock, T. D., 1967, Life at high temperatures, *Science* **158**:1012–1019.

Buchanan, R. E., and Gibbons, N. E. (eds.), 1974, *Bergey's Manual of Determinative Bacteriology*, Williams & Wilkins, Baltimore.

Calvin, M., and Lynch, V., 1952, Grana-like structures of *Synechococcus cedorum, Nature* **169**:455–456.

Cameron, R. E., and Fuller, W. H., 1960, Nitrogen fixation by some soil algae in Arizona soil, *Soil Sci. Soc. Am. Proc.* **24**:353–356.

Castenholz, R. W., 1969, Thermophilic blue-green algae and the thermal environment, *Bateriol. Rev.* **33**:476–504.

Cohen, Y., Padan, E., and Shilo, M., 1975, Facultative anoxygenic photosynthesis in the cyanobacterium *Oscillatoria limnetica, J. Bacteriol.* **123**:855–861.

Cohen-Bazire, G., and Bryant, D. A., 1982, Phycobilisomes: Composition and structure, in: *The Biology of Cyanobacteria* (N. G. Carr and B. A. Whitton, eds.), University of California Press, pp. 143–190.

Cohn, F., 1853, Untersuchungen über die Entwickelungsgeschickte mikroskopischer Algen and Pilze, *Nov. Act. Acad. Leop. Carol.* **24**:103–256.

Cohn, F., 1875, Untersuchungen über Bakterien II, *Beitr. Biol. Pflanz.* **1** (3):141–207.

Copeland, H. F., 1938, The kingdoms of organisms, *Quart. Rev. Biol.* **13**:383–420.

Cox, R. M., Fay, P., and Fogg, G. E., 1964, Nitrogen fixation and photosynthesis in a subcellular fraction of the blue-green alga *Anabaena cylindrica, Biochim. Biophys. Acta* **88**:208–210.

Doolittle, W. F., 1979, The cyanobacterial genome, its expression, and the control of that expression, *Adv. Microb. Physiol.* **20**:1–102.

Doolittle, W. F., 1982, Molecular evolution, in: *The Biology of Cyanobacteria,* (N. G. Carr and B. A. Whitton, eds.), University of California Press, pp. 307–332.

Duysens, L. N. M., 1951, Transfer of light energy within pigment systems present in photosynthesizing cells, *Nature* **68**:548–550.

Duysens, L. N. M., and Amesz, J. 1962, Function and identification of two photosystems in photosynthesis, *Biochim. Biophys. Acta* **64**:243–260.

Edelman, M., Swinton, D., Schiff, J. A., Epstein, H. T., and Zeldin, B., 1967, Deoxyribonucleic acid of the blue-green algae (Cyanophyta), *Bacteriol. Rev.* **37**:315–331.

Eley, J. H., 1971, Effect of carbon dioxide concentration on pigmentation in the blue-green alga *Anacystis nidulans, Pl. Cell Physiol. (Tokyo)* **12:**311–316.

Emerson, R., 1958, The quantum yield of photosynthesis, *Ann. Rev. Pl. Physiol.* **9:**1–24.

Emerson, R., and Lewis, C. M., 1941a, Carbon dioxide exchange and the measurement of the quantum yield of photosynthesis, *Am. Journ. Bot.* **28:**789–804.

Emerson, R., and Lewis, C. M.,1941b, The photosynthetic efficiency of phycocyanin in Chroococcus and the problem of carotenoid participation in photosynthesis, *J. Gen. Physiol.* **25:**579–595.

Emerson, R., and Lewis, C. M., 1943, The dependence of the quantum yield of *Chlorella* photosynthesis on wavelength of light, *Am. J. Bot.* **30:**165–178.

Engelmann, T. W., 1883, Farbe and Assimilation, *Bot. Z.* **41:**1–13, 17–29.

England, R. R., and Evans, E. H., 1981, A rapid method for extraction of oxygenevolving photosystem 2 preparations from cyanobacteria, *FEBS Lett.* **134:**175–177.

Fogg, G. E., 1956, The comparative physiology and biochemistry of the blue-green alga, *Bacteriol. Rev.* **20:**148–165.

Fogg, G. E., and Horne, A. J., 1970, The physiology of antarctic freshwater algae, in: *Antarctic Ecology,* vol. 2 (M. W. Holdgate, ed.), Academic Press, pp. 632–638.

Frenkel, A., Gaffron, H., and Battley, E. H., 1950, Photosynthesis and photoreduction by the blue-green alga, *Synechococcus elongatus* Näg., *Biol. Bull.* **99:**157–162.

Friedmann, I., Lipkin, Y., and Ocampo-Paus, R., 1967, Desert algae of the Negev (Israel), *Phycologia* **6:**185–200.

Fujita, Y., and Myers, J., 1965, Light induced redox reactions of cytochrome c by cell-free preparations of *Anabaena cylindrica, Archs. Biochem. Biophys.* **111:**619–625.

Gaidukov, N., 1902, Über der Einfluss farbigen Lichtes auf die Färbung lebender Oscillarien., *Abh. dt. Akad. Wiss. Berl.* **5:**1–36.

Gantt, E., 1980, Structure and function of phycobilisomes: light harvesting pigment complexes in red and blue-green algae, *Int. Rev. Cytol.* **66:**45–80.

Gantt, E., and Conti, S. F., 1966, Phycobiliprotein localization in algae, *Brookhaven Symp. Biol.* **19:**393–405.

Gantt, E., and Lipschultz, C. A., 1972, Phycobilisomes of *Porphyridium cruentum, J. Cell Biol.* **54:**313–324.

Gantt, E., and Lipschulz, C. A.,1973, Energy transfer in phycobilisomes from phycoerythrin to allophycocyanin, *Biochim. Biophys. Acta* **292:**858–861.

Gerloff, G. C., Fitzgerald, G. P., and Skoog, F., 1950, The isolation, purification and nutrient solution requirements of blue-green algae, in: *The Culturing of Algae* (J. Brunel, G. W. Prescott, and L. H. Tiffany, eds.), Charles F. Kettering Foundation, Yellow Springs, pp. 27–44.

Glazer, A. N., 1982, Phycobilisomes: structure and dynamics, *Annu. Rev. Microbiol.* **36:**173–198.

Glazer, A. N., 1983, Comparative biochemistry of photosynthetic light-harvesting systems, *Annu Rev. Biochem.* **52:**125–157.

Goedheer, J. C., 1968, On the low temperature fluorescence spectra of blue-green algae and red algae, *Biochim. Biophys. Acta* **153:**903–906.

Guilkema, J. A., and Sherman, L. A., 1980, Metronidazole and the isolation of temperature sensitive photosynthetic mutants in cyanobacteria, *J. Bioenerg. Biomemb.* **12:**277–295.

Guilkema, J. A., and Sherman, L. A., 1982, Protein composition and architecture of the photosynthetic membranes from the cyanobacterium *Anacystis nidulans* R2, *Biochim. Biophys. Acta* **681:**440–450.

Hattori, A., and Fujita, Y., 1959, Formation of phycobilin pigments in a blue-green alga, *Tolypothrix tenius,* as induced by illumination with coloured lights, *J. Biochem. (Tokyo)* **46:**521–524.

Haxo, F., and Blinks, L. R., 1950, Photosynthetic action spectra of marine algae, *J. Gen. Phys.* **33:**389–422.

Herdman, M., Janvier, M., Waterbury, J. B., Rippka, R., and Stanier, R. Y., 1979a, Deoxyribonucleic acid base composition of cyanobacteria, *J. Gen. Microbiol.* **111:**63–71.

Herdman, M., Janvier, M., Rippka, R., and Stanier, R. Y., 1979b, Genome size of cyanobacteria, *J. Gen. Microbiol.* **111:**73–85.

Ho, K. K., and Krogmann, D. W., 1982, PHotosynthesis, in: *The Biology of Cyanobacteria,* (N. G. Carr and B. A. Whitton, eds.), University of California Press, Berkeley, pp. 191–214.
Holton, R. W., and Myers, J., 1963, Cytochromes of a blue-green alga: extraction of a c-type with a strongly negative redox potential, *Science* **142:**234–235.
Ihlenfeldt, M. J. A., and Gibson, J., 1975, CO_2 fixation and its regulation in *Anacystis nidulans, Arch. Microbiol.* **102:**13–21.
Johnson, P. W., and Sieburth, J. Mc.N., 1979, Chroococcoid cyanobacteria in the sea: a ubiquitous and diverse phototrophic biomass, *Limnol. Oceanogr.* **24:**928–935.
Jones, L. W., and Myers, J., 1964, Enhancement in the blue-green alga, *Anacystis nidulans. Plant Physiol.* **39:**938–946.
Jones, L. W., and Myers, J., 1965, Pigment variations in *Anacystis nidulans* induced by light of selected wavelengths, *J. Phycol.* **1:**6–13.
Katoh, S., 1969, Studies on the algal cytochrome of c-type, *J. Biochem. (Tokyo)* **46:**629–632.
Knoff, D. B., Light-induced oxidation-reduction reactions in a cell-free preparation from the blue-green alga *Nostoc muscorum:* The role of cytochrome f_1 cytochrome b_{558}, C550 and P700 in noncyclic electron transport, *Biochim. Biophys. Acta* **325:**284–296.
Kok, B., 1956, On the reversible absorption change at 705 mμ in photosynthetic organisms, *Biochim. Biophys. Acta* **22:**399–401.
Kok, B., and Hoch, G., 1961, Spectral changes in photosynthesis, in: *Light and Life* (W. D. McElroy and B. Glass, eds.), Johns Hopkins University Press, Baltimore, pp. 397–416.
Krogmann, D. W., 1973, Photosynthetic reactions and components of thylakoids, in: *The Biology of Blue-green Algae* (N. G. Carr and B. A. Whitton, eds.), Blackwell Scientific Publications, Oxford, pp. 80–98.
Krogmann, D. W., 1977, Blue-green algae, in: *Encyclopedia of Plant Physiology* (A. Trebst and M. Avron, eds.), Vol. 5, Springer-Verlag, Berlin, pp. 625–636.
Kuhlenmeier, C. J., Borrias, W. E., Van den Hondel, C. A. M. J., Jr., and Van Arkel, G. A., 1981, Vectors for cloning in cyanobacteria: construction and characterization of two recombinant plasmids capable of transformation to *Escherichia coli* and *Anacystis nidulans* R2, *Mol. Gen. Genet.* **184:**249–254.
Lau, R. H., MacKenzie, M. M., and Doolittle, W. F., 1977, Phycocyanin synthesis and degradation in *Anacystis nidulans, J. Bacteriol.* **132:**771–778.
Lemasson, C., Tandeau de Marsac, N., Cohen-Bazire, G., 1973, Role of allophycocyanin as a light-harvesting pigment in cyanobacteria. *Proc. Natl. Acad. Sci.,* **70:**3130–3133.
Lewin, R. A., 1981, The prochlorophytes, in: *The Prokaryotes* (M. P. Starr, H. Stolp, H. G. Truper, A. Balows, H. G. Schlegel, eds.), Berlin, Springer-Verlag, pp. 257–266.
Lewin, R. A., and Cheng, L., 1975, Associations of microscoscopic algae with didemnid ascidians, *Phycologia,* **14:**149–152.
Lightbody, J. J., and Krogmann, D. W., 1967, Isolation and properties of plastocyanin from *Anabaena variabilis, Biochim. Biophys. Acta* **191:**508–515.
Linko, P., Holm-Hansen, O., Bassham, J. A., and Calvin, M. J., 1957, Formation of radioactive citrulline during photosynthetic $C^{14}O_2$-fixation by blue-green algae, *J. Exp. Botany* **8:**147–156.
Mereschkowsky, C., 1905, Ueber Natur und Ursprung der Chromatophoren in Pflangenreiche, *Biol. Zentralbl.* **25:**593–604.
Miller, L. S., and Holt, S. C., 1977, Effect of carbon dioxide on pigment and membrane content in *Synechococcus lividus, Arch. Microbiol.* **115:**185–198.
Myers, J., and Kratz, W. A., 1955, Relation between pigment content and photosynthesis characteristics in a blue-green alga, *J. Gen. Physiol.* **39:**11–22.
Myers, J., Graham, J. R., and Wang, R. T., 1980, Spontaneous pigment mutants of *Anacystis nidulans* selected by growth under far-red light, *Arch. Microbiol.* **124:**143–148.
Nakamura, H., 1938, Über die Kohlenäsureassimilation bei niederen algen in Anwesenheit des Schwefelwasserstaffes, *Acta Phytochim. (Tokyo)* **10:**271–281.
Niklowitz, W., and Drews, G., 1956, Beiträge zur Cytologie der Blaualgen. I. Mitt.: Untersuchungen zur substruktur vom *Phormidium uncinatum* Gomont.; *Archiv. Mikrobiol.* **24:**134–146.

Norris, L., Norris, R. E., and Calvin, M., 1955, A survey of the rates and products of short-term photosynthesis in plants of nine phyla, *J. Exp. Botany* **6:**64–74.

Ogawa, T., Vernon, L. P., and Mollenhauer, H. H., 1969, Properties and structure of fractions prepared from *Anabaena variabilis* by the action of Triton X-100, *Biochim. Biophys. Acta* **172:**216–229.

Oren, A., Padan, E., and Avron, M., 1977, Quantum yields for oxygenic and anoxygenic photosynthesis in the cyanobacterium *Oscillatoria limnetica, Proc. Natl. Acad. Sci. USA* **74:**2152–2156.

Padan, E., 1979, Impact of facultatively anaerobic photoautotrophic metabolism on ecology of cyanobacteria (blue-green algae), *Adv. Microb. Ecol.* **3:**1–48.

Padan, E., and Cohen, Y., 1982, Anoxygenic photosynthesis, in: *The Biology of Cyanobacteria,* (N. G. Carr and B. A. Whitton, eds.), University of California Press, Berkeley, pp. 215–235.

Pelroy, R. A., and Bassham, J. A., 1972, Photosynthesis and dark carbon metabolism in unicellular blue-green algae, *Arch. Mikrobiol.* **86:**25–38.

Petrack, B., and Lipmann, F., 1961, *A Symposium on Light and Life* (W. D. McElroy and B. Glass, eds.), Johns Hopkins Press, Baltimore, pp. 621–630.

Pringsheim, E. G., 1949, The relationship between bacteria and myxophyceae, *Bacteriol. Rev.* **13:**47–98.

Rippka, R., 1972, Photoheterotrophy and chemoheterotrophy among unicellular blue-green algae, *Arch. Mikrobiol.* **87:**93–98.

Rippka, R., Waterbury, J. B., and Cohen-Bazire, G., 1974, A cyanobacterium which lacks thylakoids, *Arch. Microbiol.* **100:**419–436.

Rippka, R., and Waterbury, J. B., 1977, The synthesis of nitrogenase by nonheterocystous cyanobacteria, *FEMS Microbiol. Lett.* **2:**83–86.

Rippka, R., Derulles, J., Waterbury, J. B., Herdman, M., and Stanier, R. Y., 1979, Generic asssignments, strain histories and properties of pure cultures of cyanobacteria, *J. Gen. Microbiol.* **111:**1–61.

Ris, H., and Singh, R. N., 1961, Electron microscope studies on blue-green algae, *J. Biophys. Biochem. Cytol.* **9:**63–80.

Sachs, J., 1874, *Lehrbuch der Botanik,* 4th ed., 928 pp., W. Engelmann, Leipzig.

Sadewasser, D. A., and Sherman, L. A., 1981, Internal and external membrane proteins of the cyanobacterium, *Synechococcus cedrorum, Biochim. Biophys. Acta* **640:**326–340.

Sagan, L., 1967, On the origin of mitosing cells, *J. Theoret. Biol.* **14:**225–274.

Schopf, J. W., 1974, Paleobiology of the Precambrian: The age of blue-green algae, in: *Evolutionary Biology* (T. Dobhansky, M. K. Hecht, and W. C. Steere, eds.), Vol. 7, Plenum Press, N.Y., pp. 1–43.

Schopf, J. W., and Walter, M. R., 1982, Origin and early evolution of cyanobacteria: The geological evidence, in: *The Biology of Cyanobacteria* (N. G. Carr and B. A. Whitton, eds.), University of California Press, Berkeley, pp. 543–564.

Seewaldt, E., and Stackebrandt, E., 1982, Partial sequence of 16 S ribosomal RNA and the phylogeny of *Prochloron, Nature* **295:**618–620.

Shatkin, A., 1960, A chlorophyll-containing cell fraction from the blue-green alga, *Anabaena variabilis, J. Biophys. Biochem. Cytol.* **7:**583–584.

Sherman, L. A., and Cunningham, J., 1977, Isolation and characterization of temperature-sensitive, high fluorescence mutations of the blue-green alga, *Synechococcus cedrorum, Plant Sci. Lett.* **8:**319–326.

Sherman, L. A., and Guilkema, J., 1982, Photosynthesis and cloning in cyanobacteria, in: *Genetic Engineering of Microorganisms for Chemicals* (A. Hollaender et al., eds.) Plenum Press, New York, pp. 103–131.

Shields, L. M., 1957, Algal and lichen floras in relation to nitrogen content of certain volcanic and arid range soils, *Ecology* **38:**661–663.

Stanier, R. Y., 1977, The position of cyanobacteria in the world of phototrophs, *Carlsberg Res. Commun.* **42:**77–98.

Stanier, R. Y., and Cohen-Bazire, G., 1977, Phototrophic prokaryotes: The cyanobacteria, *Annu. Rev. Microbiol.* **31:**225–274.

Stanier, R. Y., and van Niel, C. B., 1941, The main outlines of bacterial classification, *J. Bacteriol.* **42:**437–466.

Stanier, R. Y., and van Niel, C. B., 1962, The concept of a bacterium, *Archiv. Mikrobiol.* **42:**17–35.

Stanier, R. Y., Kunisawa, R., Mandel, M., and Cohen-Bazire, G., 1971, Purification and properties of unicellular blue-green algae (Order Chroococcales) *Bacteriol. Rev.* **35:**171–205.

Stanier, R. Y ., Sistrom, W. R., Hansen, T. A., Whitton, B. A., Castenholtz, R. W., Pfennig, N., Gorlenko, V. N., Kondratieva, E. N., Eimhjellen, K. E., Whittenbury, R., Gherna, R. L., and Trüper, H. G., 1978, Proposal to place the nomenclature of the cyanobacteria (blue-green algae) under the rules of the International Code of Nomenclature of Bacteria, *Int. J. Syst. Bact.* **28:**335–336.

Stevens, C. L. R., and Myers, J., 1976, Characterization of pigment mutants in a blue-green alga *Anacystis nidulans, J. Phycol.* **12:**99–105.

Stewart, A. C., and Bendall, D. S., 1981, Properties of oxygen-evolving photosystem II particles from *Phormidium laminosum,* a thermophilic blue-green alga, *Biochem. J.* **194:**877–887.

Stizenberger, E., 1860, *Dr. Ludwig Rabenhorst's Algen Sachsens resp. Mitteleuropa's Systematisca geordnet (mit Zugrund elegung lines neuen systems),* Heinrich, Dresden, 41 pp.

Susor, W. A., and Krogmann, D. W., 1964, Hill activity in cell-free preparations of a blue-green alga, *Biochim. Biophys. Acta* **88:**11–19.

Susor, W. A., and Krogmann, D. W., 1967, Triphosphopyridine nucleotide photoreduction with cell-free preparations of *Anabaena variabilis, Biochim. Biophys. Acta* **120:**65–72.

Tandeau de Marsac, N., and Cohen-Bazire, G., 1977, Molecular composition of cyanobacterial phycobilisomes, *Proc. Natl. Acad. Sci. USA* **74:**1635–1639.

Taylor, M. M., and Storck, R., 1964, Uniqueness of bacterial ribosomes, *Proc. Natl. Acad. Sci. USA* **52:**958–965.

Thomas, J. B., and DeRover, W., 1955, On phycocyanin participation in the hill reaction in the blue-green alga *Synechococcus cedrorum, Biochim. Biophs. Acta* **16:**391–395.

Waterbury, J. B., Watson, S. W., Guillard, R. R., and Brand, L. E., 1979, Widespread occurrence of a unicellular marine, planktonic cyanobacterium, *Nature* **277:**293–294.

Whitton, B. A., and Potts, M., 1982, Marine littoral, in: *The Biology of Cyanobacteria* (N. G. Carr and B. A. Whitton, eds.), University of California Press, Berkeley, pp. 515–542.

Withers, N. W., Alberte, R. S., Lewin, R. A., Thorber, J. P., Britton, G., and Goodwin, T. W., 1978, Photosynthetic unit size, carotenoids, and chlorophyll-protein composition of *Prochloron* sp., a prokaryotic green alga, *Proc. Natl. Acad. Sci. USA* **75:**2301–2305.

Wyatt, J. J., and Silvey, J. K. G., 1969, Nitrogen fixation by *Gleocapsa, Science* **165:**908–909.

Yamanaka, G., and Glazer, A. N., 1980, Dynamic aspects of phycobilisome structure. Phycobilisome turnover during nitrogen starvation in *Synechococcus* sp., *Arch. Microbiol.* **124:**39–47.

Yopp, J. H., Tindall, D. R., Miller, D. H., and Schmid, W. E., 1978, Isolation, purification and evidence for a halophytic nature of the blue-green alga *Aphanatheca halophytica* Fremy (Chroococales) *Phycologia* **17:**172–178.

6

CONSUMPTION OF ATMOSPHERIC NITROGEN

David R. Benson

INTRODUCTION

Almost a century ago, Hellriegel and Wilfarth demonstrated that legumes fixed N_2 with the aid of microorganisms living in root nodules (Hellriegel, 1886a; Hellriegel and Wilfarth, 1888). They finally explained an observation, made by Boussingault a half-century earlier, that legumes gathered more nitrogen than other plants. Since then, microbiologists, ecologists, chemists, biochemists, agronomists, botanists, geneticists, and, more recently, molecular geneticists have investigated the problems of biological N_2 fixation. Our perception of this important process has evolved from the empirical observations of ancient farmers on the beneficial effects that legumes have on the soil to the point today where many of the genes involved in N_2 fixation have been isolated and their nucleotide sequences determined. The meaning of nitrogen fixation has changed from the early 19th century to the present. Originally the term implied the fixing of any nitrogen into organic matter, whether combined nitrogen or N_2. Today, the term nitrogen fixation is most commonly used when referring to the conversion of N_2 to the oxidation state of ammonia—a process carried out biologically only by prokaryotes.

This chapter outlines the major historical developments that have led to our current understanding of the distribution and importance of N_2 fixation in the environment. Given the enormous volume of literature that has accumulated over the last century, a complete historical review would exhaust both the reader and the writer. The reader therefore is directed to several excellent reviews and monographs for a more complete treatment of individual topics (Aulie, 1970; Allen and Allen, 1958; Burris, 1974, 1979; Fred et al., 1932; Jensen, 1940, 1965; Mishustin and Shil'nikova, 1971; Postgate, 1982;

David R. Benson • Microbiology Section, University of Connecticut, Storrs, Connecticut 06268.

Stewart, 1966; Wilson, 1940, 1957, 1958, 1969). This chapter presents a broad view of the subject and reviews the sometimes stormy controversies that have surfaced since N_2 fixation first was entertained as a possibility early in the nineteenth century.

Our present state of knowledge can be viewed as part of a continuum that began millenia ago with the use of legumes in stable farming communities. In relatively recent years, the continuum has included Boussingault's discovery of nitrogen accumulation by legumes, Hellriegel and Wilfarth's experiments on the root nodule symbiosis, the discovery of free-living N_2-fixing bacteria, and the ultimate unraveling of the biochemistry and genetics of N_2 fixation. Major advances in each of these areas have usually followed the description of new techniques. The lineage of techniques includes the Dumas technique for elemental analysis described in 1834, the titration technique used by Peligot in 1847, the Kjeldahl technique described in 1883, the $^{15}N_2$ isotopic enrichment technique developed in 1940, and the acetylene reduction technique described in 1966. Many of these techniques form natural milestones in the historical development of studies on N_2 fixation, and this chapter is subdivided accordingly.

EARLY OBSERVATIONS

The Ancient Roots

The process of N_2 fixation was exploited several millenia before it was scientifically understood. Although written records started after legumes were already in use, archaeological evidence from Neolithic and Bronze Age farming communities in the Near East and part of Europe has demonstrated that legumes were founder crops of Old World agriculture (Zohary and Hopf, 1973). Remains of pea (*Pisum sativum*) and lentil (*Lens culinaris*) are usually found in association with barley and wheat in settlements dating back to the seventh millenium B.C. (Zohary and Hopf, 1973). Fred et al. (1932) and Wilson (1940) mention several examples, gleaned from the ancient literature, of crop management practices centered around the manuring properties of legumes or around the beneficial effects of letting fields lie fallow for a period of time. Theophrastus, a Greek philosopher who lived around 285 B.C., wrote:

> Beans . . . are in other ways not a burdensome crop to the ground, they even seem to manure it, because the plant is of loose growth and rots easily; Wherefore the people of Macedonia and Thessaly turn over the ground when it is in flower. (Theophrastus, 1916)

Subsequent Roman writers, such as Varro and Columella, left remarkable records of contemporary agricultural practices that described green manuring with alfalfa and lupines (Fred et al., 1932; Wilson, 1940). Such practices persisted until the decline of the Roman Empire, and their record was carried into the Dark Ages through the faithful recopying of ancient manuscripts by medieval monks (Wilson, 1940).

That legumes had a reinvigorating effect on the soil was certainly noticed by civilizations other than those in the Mediterranean Basin. Few written records describing the agricultural practices of such civilizations remain, but archaeological evidence does suggest an early appreciation for legumes by many cultures. Inhabitants of Mesoamerica were cultivating beans (*Phaseolus*) around the fourth millenium B.C. (Smartt, 1969), and groundnuts were probably in use when the first Spanish explorers arrived in the New World (Krapovickas, 1969). An interesting parallelism is apparent between Old World and New World agriculture. Lentils and peas supplied the high protein crop and wheats and barley supplied the starch to early agricultural communities in the Near East and Europe; beans and maize, respectively, fulfilled the same roles in the New World.

The general observation that some plants reinvigorated the soil was not limited to legumes. For example, the water fern *Azolla,* in symbiosis with a diazotrophic (N_2-fixing) cyanobacterium, *Anabaena azollae,* was used centuries ago for manuring rice fields in Vietnam; the practice was so successful that villagers offered sacrifices in the autumn to the "goddess of *Azolla*" (Mishustin and Shil'nikova, 1971). In other areas, certain nonleguminous plants such as the alder (*Alnus glutinosa*) were noted as having a beneficial effect on associated plants. William Browne of Tavistock penned these words in 1613:

> The alder, whose fat shadow nourisheth—Each plant set neere to him long flourisheth. (quoted in Goldman, 1961)

However pleasant these lyrical sentiments may sound, the actual basis of such an observation remained unknown until more than 250 years after the words were written.

Beginnings of Scientific Agriculture

Discovery of Nitrogen

The age of scientific agriculture emerged in the 18th and 19th centuries against a background of empirical observation. Before nitrogen had been discovered as an element, the first methodological advance came when Johan Rudolph Glauber learned how to detect ordinary saltpeter (KNO_3) in plants, in animals, and in the soil. He made the connection that ". . . Salt-petre was of necessity in the Herbs and Grass, afore the Beasts feeding on them . . ." (Glauber, 1658, quoted in Aulie, 1970). The Renaissance appreciation of saltpeter grew to the point where several writers were convinced, like the Englishman John Evelyn, that "were saltpeter . . . to be obtained in plenty we should need but few other composts to meliorate our ground" (Evelyn, 1676, quoted in Aulie, 1970). The connection between plants, animals and the atmosphere was obscure to the early naturalists, since little in the way of experimental evidence was available, but the cyclic flow of matter was sensed by a few prescient individuals. For example, in 1717, the French chemist Louis Lemery described the slow production of nitre (KNO_3) in the superficial layers of the soil. He ascribed its presence to the decay of animals and plants and was one of the earliest workers to demonstrate the process of nitrification (Aulie, 1970). He drew a connection between soil fertility and the processes of decay, although the role played by microorganisms could not be known.

The elemental nature of nitrogen was not established until English chemists isolated "mephitic air" and the French chemists, chief among them Lavoisier, who later was beheaded in the French Revolution, recognized the elemental nature of the gas and in 1790 named it "azote" (without life) and "gaz nitrogene" (Aulie, 1970). Berthollet (1785) subsequently discovered the element in plants and animals, although his analytical techniques were not sensitive enough to detect nitrogen in all plants tested. By the beginning of the 19th century, investigators were in a position to examine the interrelationship between nitrogen in plants, animals, and the soil.

In the mid-17th century, clover was introduced into English agriculture by Sir Richard Weston as part of a crop rotation (Aulie, 1970). The practice of using legumes in crop rotation subsequently spread throughout much of England and France particularly through the influence of Jethro Tull, who in 1829 wrote an influential book on agricultural practices entitled *The Horse-Hoeing Husbandry*. Although the book dealt with increasing soil fertility by repeatedly pulverizing the earth, it also included descriptions of benefits obtained by planting legumes in rotation with other crops (Aulie, 1970). Agricultural improvements, including legume rotations, that were introduced from the mid-17th century through the 18th century led to an estimated 40–50% increase in farm productivity (Aulie, 1970). This agricultural revolution set the stage for Boussingault's later experiments, conducted between 1838 and 1841, wherein the unusual property of legumes in accumulating nitrogen would be discovered.

In other fields of plant science the roles played by CO_2 and O_2 in plant growth were being worked out in the last twenty years of the 18th century with contributions from Priestley, Ingenhousz, Senebier, and de Saussure (Nash, 1959). They showed that the atmosphere was the major source of plant carbon and that plants also produced O_2. The concept of aerial nutrition suggested the possibility that plants might obtain other nutrients, as well, from the air. Since N_2 comprised the bulk of the atmosphere, it was logical to look to the air as a nitrogen source. In fact, Priestley and Ingenhousz performed experiments that seemed to show a decrease in N_2 in closed chambers during plant growth, but their results were not confirmed by the more careful experiments of de Saussure (Wilson, 1940). Davy (1836) also suggested that ". . . when glutinous and albuminous substances exist in plants the azote they contain may be suspected to be derived from the atmosphere." Since reliable techniques for measuring the nitrogen content of organic matter were not available, the question of whether plants could obtain nitrogen from the air could not be effectively addressed until satisfactory methods of analysis were developed.

Early Controversies Regarding the Sources of Plant Nitrogen

In 1957, P. W. Wilson delivered a delightful lecture entitled "In Defense of Imperfection" to the Society of American Bacteriologists. In it he described the work of some early investigators who had demonstrated N_2 fixation in legumes and in the soil prior to Hellriegel and Wilfarth, but who had discarded the notion of N_2 fixation in favor of other, more facile, explanations. The French chemist, Jean-Baptiste Boussingault, was given as the classic example of an individual who ". . . listened to an authority that he thought superior to his own and then made the fatal mistake of running another experiment" (Wilson, 1957). A splendid account of Boussingault's life and work has been provided by Aulie (1970).

Boussingault was a self-taught scientist and son of a shop-keeper who nurtured his appetite for science by auditing lectures at the educational and scientific institutions of Paris. In his experiments from 1836–1838, carried out at Bechelbronn, Alsace, he ranked plants according to nitrogen content. He took advantage of the copper-oxide technique of elemental analysis, developed by Dumas in 1834, which was the first useful method for measuring total nitrogen in plant or vegetable matter. Boussingault's initial experiments of 1836 showed that a legume, white bean, had a much greater nitrogen content than hay or corn; 25 parts of bean, by weight, and 63 parts of corn equaled the nitrogen content of 100 parts of hay. With these deceptively simple observations, Boussingault showed, for the first time, that the superior nitrogen content of legumes accounted for their usefulness as a manuring crop.

After these initial results, Boussingault performed a series of experiments that stimulated much controversy in the mid-19th century. His crop rotation experiments, probably begun around 1834 or 1835, had indicated that crops in rotation with legumes actually accumulated more nitrogen than could be accounted for from other sources. In 1838, after recognizing that some plants depleted the soil of nutrients but, paradoxically, some actually enriched the soil, he declared:

> . . . there are certain undeniable facts of agriculture that lead me to think that under some circumstances plants find in the atmosphere a part of the nitrogen which contributes to their organization . . . (Boussingault, 1838a).

To test this idea, Boussingault designed a pot experiment in which he sought to eliminate any possible external source of nitrogen. His growth substrate was crushed brick that had been heated to redness to remove organic impurities, and he grew the experimental plants under bell jars in a greenhouse. Air entering the bell jars was filtered in order to eliminate any particulate contaminants. With the Dumas technique, he measured the amounts of car-

TABLE I
Boussingault's First Greenhouse Experiments on Nitrogen Fixation[a]

Plant	Initial nitrogen content (mg)	Final nitrogen content (mg)	Gain in nitrogen (mg)	Substrate
First series				
Clover	72	79	+7	Ignited sand
Clover	114	156	+42	Ignited sand
Clover[b]	61	69	+8	Ignited sand
Wheat	43	40	−3	Ignited sand
Wheat	57	60	+3	Ignited sand
Second series				
Peas	46	101	+55	Burned clay
Clover[c]	33	56	+23	Ignited sand
Oats[b]	59	53	−6	Water

[a] Adapted from Wilson (1940).
[b] Grown in closed containers through which washed air was passed.
[c] Plants grown in field and transplanted to greenhouse.

bon, hydrogen, oxygen, and nitrogen present in seed samples at the beginning of the experiment and in the crop at the end of the experiment in an attempt to detect any accumulation. The results, reproduced in Table I, show that only legumes had increased significantly in nitrogen content. Apparently, in spite of precautions, rhizobial contamination had occurred and allowed nodulation of the peas and clover; Boussingault, of course, was unaware of the role played by microorganisms in any of these early experiments and noted that he could not explain how this nitrogen increase had occurred. He ventured that nitrogen ". . . can enter the plant by aerated rain water, which is absorbed by the roots. . . . it is possible, as some physicians think, that it exists in the air as an infinitely small percentage of ammoniacal vapor" (Boussingault, 1838c). Taken together, the results of Boussingault's crop rotation and pot experiments were so dramatic that he announced in 1838 that ". . . the nitrogen of the atmosphere can be assimilated during the life of a plant . . ." (Boussingault, 1838b). In spite of this pronouncement, he failed to recognize the connection between root nodules on legumes and enhanced plant growth. In later years, he drastically altered his ideas on plant nitrogen assimilation, for he and Dumas published a discourse in 1840 in which they stated that:

> Every plant fixes nitrogen during its life, whether it obtains this element from the atmosphere, or from manures added to the soil. In either case, it seems probable that the azote only enters the plant, is only consumed, under the form of ammonia or nitric acid. (Dumas and Boussingault, 1844).

By 1860, Boussingault was even more uncertain about the source of plant nitrogen:

> Azote may enter the living frame of plants directly, or, as M. Piobert has maintained, in the state of solution in the water, always aerated, which is taken up by their roots. The observations of vegetable physiologists are not generaly (loc cit.) favorable to this view. It is farther possible that the element in question may be derived from ammoniacal vapors, which, according to some philosophers, exist in infinitely small proportion in our atmosphere. These vapors, dissolved by rains and dews, would readily make their way into plants, and might there undergo elaboration.
>
> It is long since Saussure alluded to the probable influence of ammoniacal vapors upon vegetation. Prof. Liebig has more recently maintained the same opinion . . . (Boussingault, 1860).

To this he added electricity as a possible source of nitrogen compounds in the soil. Thus, although he was correct in his original view that some plants were different in their ability to accumulate nitrogen, it appears that he gave only passing consideration to N_2 fixation to explain his results. This was due, in part, to the appeal of simpler explanations being espoused by some of the more insistent scientists at the time.

To Baron Justus von Liebig, the most famous German chemist of the time and gadfly to the early microbiologists, the source of plant nitrogen was eminently clear. In a report prepared for the British Association for the Advancement of Science, Liebig enunciated his famous "mineral hypothesis" of plant nutrition (Liebig, 1840). He argued that plants obtained nitrogen in the form of ammonia that was washed from the air by rainwater. The ammonia originated from decaying animal and vegetable matter and could be trapped in mineral form in the soil or could be absorbed directly by the plant through the leaves. In retrospect, his idea was not particularly revolutionary since De Saussure and Boussingault already had speculated in that direction (Boussingault, 1860). However, in view of Liebig's reputation as a chemist and with the lack of any experimental evidence to the contrary, the ammonia hypothesis was congenially accepted by many scientists at the time.

One of the outcomes of Liebig's hypothesis was that ". . . a farmer need trouble himself as little about a compensating supply of nitrogen, as of carbon. Both are, in fact, originally constituents of the air . . ." (Liebig, 1863). Thus it was the removal of other inorganic constituents by a plant that depleted soil fertility rather than the removal of nitrogen. As to Boussingault's crop rotation experiments with legumes, Liebig pounced on an analytical error in Boussingault's technique and recalculated his data to show that manure nitrogen added at the beginning of the experiment was sufficient to account for any increases in the nitrogen content of the crops (Wilson, 1957).

The mineral hypothesis, supported with few data and expounded by an individual who had little practical farming experience, touched off a rather lively debate between Liebig and the painfully precise investigators at Rothamsted, England. The latter included a former student from Liebig's own laboratory, Joseph Henry Gilbert, and the founder and benefactor of the Rothamsted Experiment Station, Sir John Bennett Lawes. With a large set of crop rotation experiments they confirmed that fields in rotation with legumes accumulated nitrogen, and showed that parts of the "mineral theory" were not very tenable (Lawes and Gilbert, 1851). Liebig (1855) answered:

> The results of Mr. Lawes have no value for his next-door neighbor, nay, they have no value for himself. . . . None of these facts are new, and what is new in the opinion of Mr. Lawes is erroneous.
>
> The experiments of Messrs. Lawes and Gilbert are very far, indeed, from proving the conclusions which they wish to draw; they establish rather the fact that these gentlemen have not the slightest notion of what is meant by argument or proof.

The debate between the two laboratories raged for several years. It slowly became obvious that the workers at Rothamsted had based their opinions on reproducible experimental evidence whereas Liebig had relied on the work of others and upon his own prejudices to form an opinion. By 1855, enough evidence had been accumulated to overthrow the mineral theory of nitrogen nutrition.

The final demise of that part of the mineral theory dealing with the source of plant nitrogen came when Boussingault (1853a, 1853b, 1854) quantitated the ammonia in 9 rivers, 20 wells and springs, 130 rain samples from several locations in France, and even samples of fog, dew, and snow. He was able to do this by using a technique, introduced by Peligot in 1847, for measuring quantities of combined nitrogen in solution by titration against a known acid; it was sufficiently sensitive to detect 30 to 40 μg of ammonia per liter after sample concentration. Boussingault's results clearly demonstrated that only traces of ammonia could be found in rainwater, even when it was collected in the countryside where, according to Liebig, the decay of vegetable matter should be the greatest.

Boussingault's ammonia determination studies were confirmed and extended by Lawes and Gilbert (1863) and James Thomas Way (1856). They concluded that there was insufficient nitric acid and ammonia in rainwater to account for the amount of nitrogen found in crop plants. Although these experiments eventually eliminated Liebig's ideas about ammonia in the air, the question still remained: From where did nitrogen for plant growth originate?

During the 1850s, N_2 fixation by plants was being reexamined by the now established Boussingault and a young French chemist, Georges Ville. Boussingault had already abandoned his opinion that plants could fix N_2 and was convinced that the majority of plant nitrogen came from the soil. Ville's experiments employed techniques similar to those developed by Boussingault, including growing the plants in calcined soil in an enclosed system. He consistently achieved large increases in nitrogen content, not only in legumes, but also in wheat, rye, and cress, among other plants (Aulie, 1970).

To support his view, Boussingault performed two additional experiments on N_2 fixation by plants (Aulie, 1970). In both cases he failed to obtain any increase in nitrogen, even for the legumes, probably because his careful techniques eliminated rhizobial contamination.

As was common at the time, a commission was appointed to settle the dispute between Ville and Boussingault by watching Ville perform his experiments. Lawes et al. (1861) have provided us with a characteristically restrained account of some of the experimental errors committed during the test of Ville's methods:

> Unfortunately, an element of uncertainty attached to the evidence afforded by these experiments made under the superintendence of the Commission, which is very much to be regretted. A quantity of distilled water taken from the same bulk as that used for watering the experimental plants was saved for analysis. The examination of this water devolved on M. Cloez; who, unfortunately, was called away for some days, during the evaporation of the water with oxalic acid, with a view to the after-determination of any ammonia it might contain. M. Peligot determined the ammonia in the acid residue of the evaporation of this water, as well as in that of the water removed from the cases, after it had served in the experiments. The result was, that there was indicated such an excess of ammonia in the water before being used, over that in the residual water after removal from the larger case, as more than covered the increase in Nitrogen of the plants over that in the seeds sown. M. Cloez found, however, that, in his absence, the evaporation of the water had been conducted by the side of ammoniacal emanations from other processes. But when new portions of the original water were evaporated with proper precautions, less ammonia was indicated in it than in the water at the close of the experiment; and then, also, a gain of Nitrogen by the plants in the larger apparatus was indicated.

In spite of the inconclusive nature of the experiments, the Commission inexplicably favored Ville's conclusions over Boussingault's findings (Ville, 1855).

Lawes, Gilbert, and Evan Pugh designed experiments to examine the problem for themselves at Rothamsted (1861). Like Boussingault, they essentially sterilized all their growth substrates, thereby eliminating any possible rhizobial contamination, and grew their plants in closed chambers. They came to the same conclusions as had Boussingault. Even when they used Ville's own apparatus, they could not repeat his results and concluded that plants do not obtain nitrogen from the air (Lawes et al., 1861). The matter rested there for the next 25 years, with many investigators turning their attention to the soil to search for sources of nitrogen. It is unfortunate that men like Boussingault and Lawes, Gilbert, and Pugh did not recognize the relationship between the presence of leguminous root nodules and nitrogen accumulation; nor did they entertain the possibility that calcining the soil may have removed microorganisms beneficial to their test plants. The rudimentary knowledge about microorganisms at the time was the primary cause for this failure.

Discovery of Nitrogen Fixation

In the course of experiments on nitrification conducted around 1857, Boussingault discovered that the total amount of nitrogen in fallow soil actually increased over time (Boussingault, 1886–1891). He thus provided a scientific basis for an ancient practice that was used in Biblical times (Leviticus 45:4,5). In addition, he came close to a microbiological explanation for this phenomenon by recognizing that:

> "Without doubt these *mycodermes* have only an ephemeral existence, and ultimately, they leave behind in the soil what they will have taken; their detritus will result in giving rise to ammonia and nitric acid. . . .
>
> That is, I believe, the reason why, in supposing that the nitrogen of the air enters into nitrification, one does not find all of it in the nitric acid; what is missing has entered temporarily into the constitution of living *mycodermes* or their remains" (Boussingault, 1886–1891).

Unfortunately, Boussingault lacked experimental support for his contention that the nitrogen of the air entered temporarily into microorganisms. His renewed emphasis on microorganisms was due in part to the work of Louis Pasteur, who was busily demonstrating the role of microorganisms in a variety of fermentations (Pasteur, 1860, 1861, 1862). In Boussingault's only attempt to prove this view of N_2 fixation, he unluckily chose milk as a substrate for demonstrating nitrogen increases. Since he was dealing with a lactic acid fermentation, he was able only to demonstrate a net loss of ammonia that was due presumably to protein and amino acid decomposition (Boussingault, 1861). He proceeded no further with that line of research and instead became preoccupied with the process of nitrification. Although many observations pointed to the reality of N_2 fixation, a fundamental appreciation of the distribution and importance of microorganisms in the environment was lacking. In the ensuing years, the stage would be more appropriately set through the efforts of Pasteur, Cohn, Koch, Schloesing, Muntz, Warington, and other pioneers of microbiology.

Indications that microorganisms could fix N_2 appeared again in the 1860s. In a little noticed report in 1862, M. Jodin reported N_2 fixation by *végétaux phanerogames* and *mucédinées* present in mixed cultures. His conclusions were based on the disappearance of N_2 from closed chambers containing an infusion. His preliminary report ended with the promise that a future communication would describe additional experiments. Apparently, that communication was presented at a later meeting; it was sent to be examined by another Commission composed of five members, three of whom had participated in the Ville Commission. Later in 1862, the memoir was withdrawn: "M. Jodin demande l'autorisation de reprendre un Memoire presente a l'Academie le 20 octobre dernier sur le role physiologique de l'azote chez les mucédinées et les ferments" (*C. R. Acad. Sci. (Paris)* **55:**801).

Nitrogen fixation in soils was reported by Berthelot in 1888 and 1890.

However, he was indecisive about whether the cause was physical or biological. His results have been criticized because the amount of nitrogen accumulated in the soil samples seems far beyond what one would expect if free-living N_2-fixing bacteria were involved (Jensen, 1965). Furthermore, his contemporaries could not confirm his findings (Schloesing, 1888). Berthelot did, however, demonstrate an appreciation for the importance of microorganisms in the soil and probably made the possibility that N_2 fixation occurred more acceptable.

The idea that legumes were somehow different from other plants in their nitrogen-gathering properties persisted through the experiments of Lawes, Gilbert and Pugh (1861), Atwater (1885, 1886), and Schultz-Lupitz (1881). W. O. Atwater and his associates at Wesleyan University in Middletown, Connecticut, illustrate the typical attitude of the time. Their experiments in the field and in the greenhouse demonstrated clearly that pea plants accumulated nitrogen. In a report presented to the British Association for the Advancement of Science, Atwater (1885) suggested that N_2 fixation was occurring, but concluded that such a hypothesis was ". . . contrary to the general belief and to the results of the best investigations of the subject." He could not decide whether the fixation could be attributed to biological or perhaps even electrical sources.

The problem was finally solved by the classical experiments of Hellriegel and Wilfarth that began in the spring of 1886, and were reported by Hellriegel to the *Versammlung Deutscher Naturforscher und Aertze* in Berlin in September 1886 (Hellriegel, 1886a). In preliminary experiments on nitrogen nutrition in plants, Hellriegel and Wilfarth noted that legumes with root nodules showed excellent growth when combined nitrogen sources were withheld. Other workers had shown previously that the root nodules contained particles variously described as vibriolike (Lachmann, 1858), bacterialike (Woronin, 1866, 1867), or fungal (Frank, 1879, Ward, 1887). Recognizing that a microorganism was involved in the formation of root nodules and being aware of contemporary notions concerning bacterial distribution in the air and soil, Hellriegel and Wilfarth hypothesized that microorganisms were contaminating their legume plants, growing within them, and providing nitrogen for their growth. The unpredictable nature of nodulation in their open pots led them to think that air-borne contamination was occurring.

In their definitive experiment, Hellriegel and Wilfarth set up three series of plants. In the first series, they sterilized the seeds, sand, pots, and nutrient solution and took precautions to avoid air-borne contamination of the sand. The second series included identically treated plants except that unsterilized garden soil was added to the sand. In the third series, plants were placed in unsterilized soil so that infection would be left to chance. Plants grown in sterilized soil showed growth characteristics like those of nitrogen-starved cereals. The plants grown in sand inoculated with garden soil and those grown in the unsterilized soil that were well-developed were nodulated; those that developed poorly in the unsterilized soil were not nodulated. The classical paper describing these experiments was published by Hellriegel and Wilfarth in 1888, although earlier communications contain their basic hypothesis (Hellriegel, 1886a, 1886b).

The controversy over the nature of the root nodule organism was finally ended when Martinus Beijerinck (1888) at the Technical University of Delft isolated a bacterium that he named *Bacillus radicicola* (renamed *Rhizobium leguminosarum* by Frank in 1890). His isolation, together with Prazmowski's (1890) and Laurent's (1890) demonstrations that pure cultures of the bacterium elicited nodule formation, established the symbiotic nature of the association. In the closing years of the 19th century, several workers confirmed the results of Hellriegel and Wilfarth (Lawes and Gilbert, 1889, 1891; Atwater and Woods, 1890; Schloesing and Laurent, 1890; Alpe and Menozzi, 1892). Problems associated with nitrogen input to the soil passed from the hands of agricultural chemists, who had dominated the field beginning with Boussingault in 1836, to the early soil microbiologists of the time including such luminaries as Beijerinck, Frank, Winogradsky, Nobbe, and Hiltner. These individuals made fundamental discoveries in the field of N_2 fixation while concentrating on the role of microorganisms in the soil in general. Out of their work proceeded our basic understanding of how nitrogen transformations fit together to form our present concept of a nitrogen cycle. The role that bacteria play in the cycle was uncovered in a relatively short time span with the discovery of bacterial nitrification by Schloesing and Muntz (1877) and Warington (1878–1891), and the process of denitrification described by Gayon and Dupetit (1886).

Once N_2 fixation had been discovered, it was easier to understand the results of Boussingault (1886–1891), who had found that soil devoid of plants could actually increase in nitrogen content. The credit for proving that some free-living bacteria fix N_2 is difficult to assign. Sergei Winogradsky usually is accorded priority because he isolated *Clostridium pastorianum* (now *Clostridium pasteurianum*) in pure culture and demonstrated its ability to accumulate nitrogen (Winogradsky, 1893). Berthelot (1888) previously had demonstrated that soils could increase in nitrogen content, but his vacillation over the cause of the increase and the questions over his analytical technique leave the significance of his findings open to question. For these reasons, most subsequent workers accepted the work of Winogradsky as more conclusive.

Following the demonstration of free-living N_2-fixing bacteria in the soil, new discoveries were slow to emerge. Beijerinck (1910) isolated an organism from soil that he named *Azotobacter chroococcum,* and one from canal water in Delft that he named *Azotobacter agile,* thereby demonstrating that certain aerobic bacteria were capable of N_2 fixation. A facultatively anaerobic bacterium, named *Bacillus asterosporus* (later *Bacillus polymyxa*) was also claimed to be capable of fixing N_2 (Bredemann, 1908), as were several other organisms. However, severe methodological deficiencies shed considerable doubt on the significance of new discoveries.

ADVANCES PRIOR TO 1940

Methodological Problems

In the years following the successes of Hellriegel and Wilfarth, Winogradsky, and Beijerinck, relatively few advances were made in understanding biological N_2 fixation as an important ecological process until the 1940s. The major stumbling block to rapid advance, aside from political and economic upheaval, was the absence of a reliable and sensitive assay for N_2 fixation. Many investigators were forced to use criteria such as visible growth occurring in "nitrogen-free" medium, small increases in nitrogen content that were on the borderline of detection by the Kjeldahl method, or small increases in nitrogen content of a medium already containing large amounts of combined nitrogen in the form of plant or yeast extracts.

The Kjeldahl technique, developed originally in 1883, was at that time incapable of detecting nitrates and several other nitrogen-containing compounds. Although numerous modifications made the technique more reliable for certain classes of compounds, it was clear that the procedure, in general, gave lower values for total nitrogen content than did the more difficult Dumas method (Wilson, 1940). For this reason, the reliability of many experiments using the Kjeldahl procedures was in serious doubt (Wilson, 1940). These problems account, in part, for the observations in the early 1900's of N_2 fixation by free-living rhizobia (reviewed by Fred et al., 1932), actinomycetes and fungi (Emerson, 1917; Carter and Greaves, 1928; Greaves, 1929; Greaves and Greaves, 1932), and other organisms. Taken together, these reports contributed to a general confusion over the distribution and significance of N_2-fixing organisms in the environment. The early literature on this subject and the problems encountered in detecting N_2 fixation have been reviewed by Fred et al. (1932), Wilson (1940, 1958), and others with first-hand knowledge of the methodological pitfalls.

The problems encountered in determining whether an organism could fix N_2 are understandable considering how little was known about the process. Much of the basic biochemistry that is taken for granted today was only matter for speculation in the early years. Since isotopes generally were unavailable prior to 1940, and since biochemical techniques were still in their infancy, workers had to rely on indirect evidence for intermediates. The form to which N_2 was fixed was unknown; oxidation, reduction, direct combination with an organic molecule, or even direct combination with molecular oxygen all were debated as possible biochemical mechanisms of fixation (Fred et al., 1932; Wilson, 1940). A controversy arose over whether the key anabolic intermediate of fixation was hydroxylamine (Blom, 1931; Virtanen and Laine, 1938) or, as we know today, ammonia (Kostytschew and Rsyskaltschuk, 1925; Winogradsky, 1930, 1932). In 1940, Wilson examined the sparse evidence for and against each intermediate and concluded, with considerable qualification, that the most likely candidate was hydroxylamine. Later studies by Burris and

Wilson (1946) provided strong evidence that ammonium was the "key intermediate". The problem was not completely resolved until 1951 when Zelitch et al. used $^{15}N_2$ to show that *Clostridium pasteurianum* would excrete labeled ammonium under conditions where α-ketoglutarate, the acceptor of ammonium during assimilation, was limiting.

In spite of inadequate analytical techniques and a poor understanding of the biochemistry, significant discoveries were made prior to 1940. Burk (1934) and Wilson (1940) used the developing theories of enzyme kinetics to determine the apparent K_m of the nitrogenase systems from *Azotobacter* and red clover, respectively. Burk and colleagues introduced microrespirometer methods for studying the physical constants of nitrogenase (Wilson, 1958). During their work on red clover nodules, Wilson and Umbreit (1937) accidentally discovered that H_2 competitively inhibited N_2 fixation by nodules. That discovery initiated the long-standing belief that H_2 metabolism was associated with N_2 fixation.

Development of an Ecological Understanding

Free-Living Microorganisms

With the discovery of *Azotobacter* and *Clostridium,* several workers set out to determine the distribution and importance of diazotrophs in soil samples collected from around the world. Soil microbiologists were particularly interested in determining the distribution of diazotrophs as influenced by soil type, soil acidity, soil temperature, elemental content and geographical origin (Jensen, 1940; Waksman, 1932). Not surprisingly, N_2-fixing organisms were isolated from most of the soil samples tested. Bacteriologists were thus provided with one of the first examples of the world-wide distribution of a particular physiologic type of organism. All the work, however, failed to uncover many new diazotrophs, probably because the methods employed selected for either clostridia or azotobacters.

Cyanobacteria were among the few organisms that were accepted as being diazotrophic in the early 1900's. Beginning with Frank (1889), who worked with blue-green algae in the soil, and Prantl (1889), several workers suspected that cyanobacteria were capable of fixing N_2 (Schloesing and Laurent, 1892; Kossowitsch, 1894; Beijerinck, 1901). However, since these early reports were based on work with nonaxenic cultures, they were criticized on the basis that nonphotosynthetic bacteria could be responsible for the observed fixation (Waksman, 1932). The situation was complicated when other workers reported the absence of fixation in various pure cultures of cyanobacteria (Pringsheim, 1913). Drewes (1928), using pure cultures, was the first to demonstrate convincingly that heterocystous cyanobacteria accumulated nitrogen;

other workers later confirmed and extended his findings (Allison and Morris, 1930; Allison et al., 1937; De, 1939; Winter, 1935).

The discovery of diazotrophy in cyanobacteria opened up the possibility that significant amounts of N_2 were fixed by cyanobacteria in aquatic and soil environments. Some individuals speculated that cyanobacteria were the major source of combined nitrogen in freshwater habitats (Allison et al., 1937), especially in rice paddies (De, 1939; Singh, 1942).

In the early 1900s, many workers were estimating the N_2-fixing power of the soil; a sample of their results can be found in reviews by Jensen (1940) and Waksman (1932). Estimates of nitrogen input into the soil due to non-symbiotic fixation varied from 50 kg nitrogen per hectare (Remy, 1909) to insignificant quantities (Winogradsky, 1932; Lochhead and Thexton, 1936). By 1940, it was generally acknowledged that nonsymbiotic fixation contributed only small quantities of combined nitrogen to the soil under normal conditions. The amount of N_2 fixed was sufficient to keep the soil at an equilibrium level.

The search for N_2-fixing microorganisms in the soil and the concern over the amount of N_2 fixed were prompted by the hope of exploiting diazotrophs for improving the nitrogen content of agricultural soils, much as legumes already were employed. Caron, in 1895, was the first to claim that increased growth could be obtained by inoculating plants with diazotrophs. He inoculated nonlegumes with organisms isolated from soil and composts and obtained dramatic increases in yields. In the closing years of the 19th century a powdered preparation of *Bacillus ellenbachensis,* called "alinit" was marketed for use as a soil inoculum. Although this first attempt at marketing free-living diazotrophs failed—the organism was later shown unable to fix N_2—it stimulated interest within the agricultural community (Waksman, 1932). The discovery of *Azotobacter* in 1901 provided new impetus to trials on soil inoculation. Many workers achieved striking increases from inoculation, while others could find no benefits (Allison, 1947). The only place where soil inoculation was eventually put into practice on a large scale was in the Soviet Union. Soviet workers were almost unanimous in considering inoculation effective in improving the nitrogen status of crops (Allison, 1947). An *Azotobacter chroococcum*–peat–soil–calcium carbonate mixture known as "Azotogen" was used extensively in Russia beginning around 1932. Allison (1947) quotes an estimate that 5 million acres of crops were treated with Azotogen in 1942. However, soil inoculation was considered of dubious value elsewhere in the world; the positive effects were suggested to be due to the production of plant growth accelerators by bacteria rather than due to N_2 fixation (Allison, 1947). Later experiments generally failed to confirm the Russian results (Allison et al., 1947; Lochhead, 1952; Mishustin and Shil'nikova, 1971; Starkey, 1938).

By 1940, only members of the genera *Azotobacter, Clostridium,* and some cyanobacteria, were generally accepted as free-living diazotrophs. Claims of N_2 fixation in other organisms were usually based on marginal evidence.

Symbioses

Advances in the Rhizobium–Legume Symbiosis. By 1932, when Fred, Baldwin, and McCoy published their classic monograph, *Root Nodule Bacteria and Leguminous Plants,* many questions had already been answered about the legume symbiosis: The cytology of the infection process and nodule development had been described; initial work had been done on the physiology of the rhizobia with regard to the differences between fast and slow-growers; cross-inoculation groups were known to exist; and rhizobia had been demonstrated to be free-living microorganisms in the soil (Fred et al., 1932).

The question of whether free-living *Rhizobium* could fix N_2 was addressed by several early workers. Many of them, starting with Beijerinck (1888), were of the opinion that rhizobia could not fix N_2 outside the nodule. The tide of opinion swung from the negative findings of the late 19th century to the positive results of an embarrassingly large number of workers in the early 20th century (see Fred et al., 1932, for a listing). Around 1930, several workers analyzed a large number of cultures, over 1500 combined, to convincingly show that free-living rhizobia did not accumulate nitrogen when in culture (Allison, 1929; Lohnis, 1930; Pohlman, 1931). These studies apparently laid the controversy to rest as most subsequent workers accepted the findings (Fred et al., 1932; Waksman, 1932). Almost a half century would elapse before the opinion was again modified (see below).

Further developments concerning the leguminous symbiosis prior to 1940 came when Kubo (1939) demonstrated that the red pigment in the root nodules was quite similar to hemoglobin. This was confirmed later by Virtanen (1945), who showed its presence to be associated with nitrogenase activity and who, together with Laine, suggested the name "leghemoglobin" (Virtanen and Laine, 1946). Kubo's original idea that leghemoglobin functions in respiratory activity still has some validity. Other suggestions that the protein binds N_2 (Bergersen and Wilson, 1959) or that it reduces N_2 to hydroxylamine (Virtanen and Laine, 1945) have proven incorrect. Cappelletti (1923a,b; 1924) reported finding hemagglutinins in root nodules of various legumes that were specific for their rhizobial symbionts. His findings were largely overlooked or discarded by other workers of the time (Fred et al., 1932), but are interesting to reconsider in light of the current controversy over the lectin-mediated host specificity hypothesis (Bohlool and Schmidt, 1975).

Soon after rhizobia were isolated from leguminous root nodules, the next obvious step was to enhance nodulation by direct inoculation. The first study on direct inoculation was published by Nobbe and Hiltner in 1896. Their initial success led them to patent preparations of nodule bacteria for inoculating crops. They eventually marketed 17 types of rhizobia under the name "Nitragin." Although the efficacy of such commercial inoculants was the subject of considerable debate, other entrepreneurs soon launched their products onto the market. In the battle for name recognition, the rhizobial preparations were colorfully labeled Azotogen, Stimugerm, Shur-Inoc, Nod-O-Gen, Farmogen, and Nodule-Bacter, among others. So many products appeared on the market that, by the early 1920s, several states in the U.S. had passed laws with the intention of protecting farmers from bogus rhizobial preparations sold as seed inoculants (Fred et al., 1932).

Other Symbioses. Actinorhizal symbioses (thus named by Torrey and Tjepkema, 1979) between trees and shrubs belonging to nonleguminous plant genera and actinomycetes from the genus *Frankia* were considered to be diazotrophic by many early workers. Nodulated plants were known to grow more vigorously than nonnodulated plants (Hiltner, 1896; Nobbe et al., 1892), and combined nitrogen was known to suppress nodule growth (Hiltner, 1896). The first measurement that demonstrated an enriched nitrogen content of nodules over the remainder of the plant was made in 1910 on *Shepherdia* (Buffalo berry) nodules (Warren, 1910). Increased nitrogen content in nodulated versus nonnodulated *Casuarina* plants was shown in the early 1930s (Aldrich-Blake, 1932; Mowry, 1933).

The main problem in dealing with the actinorhizal symbiosis was the difficulty of obtaining the endophyte in pure culture. At various times, the organism was identified as *Rhizobium* (Bottomley, 1911), *Azotobacter* (Ziegenspeck, 1929), a bacterium (Chaudari, 1931; Mowry, 1933), an actinomycete (Hiltner, 1898; Youngken, 1919), a fungus (Woronin, 1866), or as a bacterium lysed by a bacteriophage (Borm, 1931). The identity of the actinorhizal endophyte was not resolved until recently by Callaham et al. (1978). By 1932, there remained some confusion over whether the nonleguminous root nodules were in fact involved in N_2 fixation (Waksman, 1932). In spite of this, nonleguminous N_2-fixing plants were used to some extent in Northern Europe and Japan (Kohnke, 1941), Germany (Bond, 1974) and in Korea and Formosa (Wilson, 1920) to fertilize agricultural and forest soils.

Other habitats proposed as sites of symbiotic N_2 fixation included the intestinal tract of termites (Cleveland, 1925), the rhizosphere of plants (Krasil'nikov, 1958), cyanobacterial-plant associations such as the *Azolla–Anabaena* symbiosis (Bortels, 1940; Oes, 1913), and lichen cyanobacterial-fungal associations (Ward, 1895). Winter (1935) isolated *Nostoc punctiforme* from the root nodules of cycads and from the angiosperm *Gunnera;* he demonstrated nitrogen increases in pure cultures and considered these unusual associations as N_2-fixing symbioses. Given the problems in proving that an organism could fix N_2, many of these early claims were based on marginal evidence. However, it is a tribute to the instincts of the early workers that many of these habitats subsequently have been confirmed to be suitable for N_2 fixation.

Claims for N_2 fixation in some other habitats and associations have proven to be incorrect. Fixation was reported to occur in the leaf nodules of plants from the families Myrsinaceae and Rubiaceae (Miehe, 1917; von Faber, 1912), the leaf glands of the West African *Diascorea macroura* (family Diascoreaceae) (Orr, 1923), the mycorrhizal nodules of *Podocarpus* (Nobbe and Hiltner, 1899; Spratt, 1912), mycorrhizal associations of *Phoma* sp. on Ericaceous plants (Rayner, 1915; 1922; Ternetz, 1907; Jones and Smith, 1929), and mycorrhizae in orchids (Wolff, 1927). All of these were shown subsequently not to fix N_2, although in some cases more than a half-century would elapse before that conclusion could be drawn.

ADVENT OF ISOTOPIC TECHNIQUES

Burris and Miller (1941) were the first to demonstrate N_2 fixation directly by using $^{15}N_2$ as an isotopic tracer. They used mass spectrometry to detect ^{15}N enrichment in cells of *Azotobacter.* In rapid succession, legume root nodules (Burris et al., 1942), *Clostridium* (Burris et al., 1943), and *Nostoc* (Burris et al., 1943) were all confirmed as N_2-fixing agents. Early attempts to extend the list of diazotrophs were unsuccessful, although the isotopic technique was approximately 100-fold more sensitive than previously used variations on the Kjeldahl technique (Wilson, 1958).

Discovery of New N_2-Fixing Organisms

After an extended hiatus caused by the Second World War, the first breakthrough emerged from work on photosynthetic bacteria. Kamen and Gest (1949) noted that N_2 inhibited photoproduction of H_2 from *Rhodospirillum rubrum.* Hydrogen gas metabolism had been associated with N_2 fixation since 1937 when Wilson and Umbreit reported that H_2 inhibited fixation in red clover nodules. Since N_2 inhibition of H_2 evolution was the counterpart to the H_2 inhibition of N_2 fixation, Kamen and Gest postulated that *R. rubrum* was likely to fix N_2. This hypothesis was validated by ^{15}N incorporation experiments. Lindstrom and co-workers quickly applied the $^{15}N_2$ technique to other representatives of the known photosynthetic bacterial groups including *Rhodopseudomonas, Chromatium, Chlorobium,* and *Rhodomicrobium,* and showed that the ability was common among the photosynthetic bacteria (Lindstrom et al., 1949, 1950, 1951).

Since the O_2-lability of nitrogenase was not recognized until the 1960's, demonstrating fixation by facultative anaerobes proved to be somewhat more difficult. Reports in the literature claiming fixation by *Aerobacter aerogenes* had first appeared in 1928 (Skinner, 1928). Later workers found only borderline activity when measuring nitrogen increases (Bhat and Palacios, 1949), or even when using the $^{15}N_2$ technique (Hamilton et al., 1953). The definitive demonstration of fixation occurred when Hamilton and Wilson (1955) recognized that anaerobic conditions and a neutral pH were required for consistent results.

Expansion of Ecological Understanding

The pace of N_2 fixation research was slow from 1940 to the 1960s. Progress was inhibited by political upheaval in World War II and its aftermath, the scarcity of institutions equipped for ^{15}N analysis, and the refractory nature of many of the problems in N_2 fixation. Relatively few studies on direct incorporation of isotopic N_2 into soil systems were conducted. Those that were done involved incubating soil samples for weeks before the amount of $^{15}N_2$ incorporated was determined (Delwiche and Wijler, 1956; Brouzes et al., 1969). Studies of N_2-fixing bacteria in unusual symbioses or habitats continued on a limited scale. For example, Bond and Scott (1955) demonstrated fixation in lichens and liverworts that was attributable to the cyanobacterial symbionts. Virtanen et al. (1954) and Bond (1955) showed conclusively that the root nodules of actinorhizal plants incorporated ^{15}N. A controversy over the N_2-fixing ability of mycorrhizal associations was not so easily resolved. Bond and Scott (1955) found no evidence for incorporation of $^{15}N_2$ into mycorrhizal roots of *Calluna vulgaris* and *Pinus sylvestris.* However, later workers found slight but positive fixation in mycorrhizal roots from *Podocarpus* (Bergersen and Costin, 1964) and other conifers (Richards and Voigt, 1964; Morrison and English, 1967). This conflict was not resolved until after the acetylene reduction technique was introduced in 1966 (see below).

With the unambiguous demonstration of N_2 fixation in several cyanobacterial genera, particularly those in the *Nostocaceae,* a new appreciation for the role of these organisms in various environments began to emerge. In particular, cyanobacteria and cyanobacterial symbioses were recognized as important nitrogen sources in marine environments (Dugdale et al., 1959, 1961; Goering et al., 1966; Stewart, 1965), in lakes and deserts (Tchan and Beadle, 1955), in volcanic areas (Fogg, 1947; Shields et al., 1957), and in the rather bleak expanses of Antarctica (Brightman, 1959). Work with cyanobacteria in rice fields showed that remarkable increases in nitrogen could be obtained largely through the activity of cyanobacteria (Fogg, 1947; Allen and Arnon, 1955; Singh, 1942, 1961; Stewart, 1966).

Even prior to 1960, when research on the biochemistry of N_2 fixation began to accelerate, some environmental limitations on the process had been described by bacterial physiologists and biochemists. For example, Bortels had shown in 1930 that *Azotobacter chroococcum* required molybdenum for N_2 fixation. The requirement was noted later in the azotobacters and clostridia (Wilson, 1958), the beijerinckias (Becking, 1962), *Derxia* (Jensen et al., 1960), the cyanobacteria (Fogg and Wolfe, 1954), *Klebsiella pneumoniae* (Pengra and Wilson, 1959), and *Bacillus polymyxa* (Grau and Wilson, 1962), among others. That nitrogenase was a molybdenum-containing enzyme remained unproven until 1966 (Mortenson, 1966).

Oxygen also was recognized as having an unusual, but inconsistent, effect on N_2 fixation. Aerobes, like *Azotobacter,* were shown to use carbohydrate less efficiently at elevated O_2 concentrations (Mishustin and Shil'nikova, 1971). It is now well established that this observation bears on the mechanism that *Azotobacter* uses to maintain a low internal pO_2. Increased respiration serves to protect nitrogenase from inactivation by O_2 (Phillips and Johnson, 1961; Yates and Jones, 1974). In the facultative anaerobe *Klebsiella pneumoniae,* O_2

abolished N_2 fixation at concentrations as low as 0.05 atm (Pengra and Wilson, 1958). In photoautotrophs such as *Rhodopseudomonas*, fixation occurred only when the organism grew phototrophically under anaerobic conditions (Wilson, 1958). On the other hand, cyanobacteria evolved O_2 and thus apparently did not require anaerobiosis for N_2 fixation. Evidence that many cyanobacteria compartmentalize nitrogenase in heterocysts, which do not evolve O_2, was not available until the late 1960s (Fay et al., 1968; Stewart et al., 1969). Similarly, nonheterocystous cyanobacteria like *Gloeocapsa* (now *Gloeothece*) or *Plectonema* were shown to fix N_2 in the late 1960s and early 1970s (Wyatt and Silvey, 1969; Stewart and Lex, 1970). The anomalous effects of O_2 no doubt delayed the consistent demonstration of N_2-fixing activity in cell-free extracts.

Development of a Biochemical Understanding

Although isotopic enrichment provided a sensitive assay for nitrogenase, attempts to detect N_2 fixation in cell-free preparations usually failed or were only marginally positive (Burris, 1966). The reason for early frustrations became clear when reproducible methods for obtaining cell-free nitrogenase were developed. Carnahan et al. (1960) solved the problem by preparing extracts of *Clostridium pasteurianum* anaerobically without freezing them. Their original preparation depended on pyruvate oxidation in the phosphoroclastic reaction to supply reducing power to the enzyme. It was not until 1962, when McNary and Burris reported that glucose-hexokinase inhibited N_2 fixation in crude extracts, that the requirement for ATP was suggested; the ATP requirement was confirmed and extended later by Hardy and D'Eustachio (1964) and Mortenson (1964). Subsequent work led to the introduction of dithionite as a source of reducing power (Bulen et al., 1965), resolution of the nitrogenase complex into two proteins (Mortenson, 1966), and identification of one as a molybdenum-containing protein and of both as containing iron (Mortenson, 1966). Cell-free preparations of nitrogenase were shown to possess ATP-dependent hydrogen evolution, and for a time this was used as a convenient assay for nitrogenase (Bulen et al., 1965).

Understanding the biochemistry of N_2 fixation turned out to be essential for understanding the ecology of the process. Enzyme activity depended on anaerobic conditions, a source of low potential electrons, energy in the form of ATP, and trace metals including molybdenum and iron. Knowing these requirements, a more rational approach for locating and identifying organisms that could fix N_2 could be developed. More rapid progress lacked only a sensitive technique for detecting N_2 fixation under field conditions in mixed populations. Such a technique also was needed for examining "borderline" diazotrophs having low nitrogenase activity. The major advance came when acetylene was recognized as an analogue of dinitrogen.

ADVENT OF THE ACETYLENE REDUCTION TECHNIQUE

The events leading to the development of acetylene reduction as a specific and sensitive assay for nitrogenase have been presented in detail by Burris (1975). The technique was developed as an outgrowth of studies on inhibition of nitrogenase by various gases, notably nitrous oxide (Mozen and Burris, 1954; Lockshin and Burris, 1965). Independently, Schollhorn and Burris (1966) and Dilworth (1966) made the critical observation that acetylene gas inhibited N_2 reduction; ethylene was the product of the reaction. Koch and Evans (1966) demonstrated the usefulness of the technique by measuring nitrogenase in soybean root nodules; they were followed quickly by Sloger and Silver (1967), Stewart et al. (1967), and Hardy and Knight (1967). Since then, the simplicity and the great sensitivity of the assay has led to an explosive increase in the literature on all aspects of N_2 fixation, particularly with respect to the ecology of diazotrophs.

Further Advances in Ecological Understanding

Acetylene reduction developed into the first useful technique for field studies of N_2 fixation. Older estimates of the significance and scope of N_2 fixation in the environment were severely limited by the methods available. As a consequence, comparatively little ecological information was obtained in the period from 1940 to the mid-1960s. Once the acetylene reduction technique had been described in detail (Hardy et al., 1968), field work became both feasible and fashionable. Although much of the work was more voluminous than enlightening, one positive outcome was the finding that N_2-fixing activity was surprisingly widespread in the environment. In fact, work no longer centered on whether diazotrophs could be found in a given habitat; rather, it became possible to assess the significance of N_2 fixation in any habitat.

The acetylene reduction technique simplified the identification of many new diazotrophs. Before $^{15}N_2$ was introduced, Waksman (1932) listed a handful of prokaryotes that were generally regarded as capable of fixing N_2. He included organisms that today would be placed in the genera *Clostridium, Azotobacter, Klebsiella,* and *Bacillus*. Symbiotic bacteria included *Bacterium radicicola* (*Rhizobium* sp.) and the inhabitants of nonlegume root and leaf nodules. By 1957, Wilson (1958) counted 14 bacterial genera, the inhabitant of actinorhizal root nodules, and a soil yeast (*Rhodotorula sp.*). Stewart listed 29 genera in 1967, with around half of them accounted for by cyanobacteria. By 1982, that number had risen to 58 genera (Postgate, 1982), largely as a result of the acetylene reduction technique, and new genera are added each year (Table II). Currently, the ability to fix N_2 is thought to be limited to prokaryotes. However, as Table II shows, the ability is distributed among most of the major physiologic groups of prokaryotes. In keeping with their diverse physiology, diazotrophs have been found in diverse habitats, including various terrestrial

TABLE II
Bacterial Genera Containing N_2-Fixing Species[a]

Physiologic group	Family or group	Genus
Aerobes	*Azotobacteraceae*	*Azotobacter, Azomonas, Azotococcus, Beijerinckia, Derxia, Xanthobacter*
	Frankiaceae	*Frankia*
	Spirillaceae	*Azospirillum, Aquaspirillum, Campylobacter*
	Corynebacteriaceae	*Arthrobacter*
	Methylomonadaceae	*Methylosinus, Methylocystis, Methylococcus, Methylomonas, Methylobacter*
	Thiobacteriaceae	*Thiobacillus*
	Rhizobiaceae	*Rhizobium*
	Uncertain	*Alcaligenes*
Facultative anaerobes	*Enterobacteriaceae*	*Klebsiella, Enterobacter, Erwinia, Citrobacter, Escherichia*
	Bacillaceae	*Bacillus*
Anaerobes	*Bacillaceae*	*Clostridium, Desulfotomaculatum*
	Uncertain	*Desulfovibrio*
Photosynthetic bacteria	*Rhodospirillaceae*	*Rhodospirillum, Rhodopseudomonas, Rhodomicrobium*
	Chromatiaceae	*Chromatium, Thiocystis, Thiocapsa, Amoebobacter, Ectothiorhodospira*
	Chlorobiaceae	*Chlorobium, Pelodictyon*
Cyanobacteria	Heterocystous	*Anabaena, Cylindrosperma, Nostoc, Scytonema, Calothrix, Chlorogloeopsis, Fischerella*
	Non-heterocystous	*Spirulina, Oscillatoria, Pseudoanabaena, Lyngbia, Plectonema, Phormidium*
	Unicellular	*Synechococcus, Dermocarpa, Xenococcus, Myxosarcina, Chroococcidiopsis, Pleurocapsa, Gloeothece*

[a]Modified from Postgate (1982).

and aquatic environments and in an increasing number of symbiotic associations.

Terrestrial Habitats

Older studies of free-living N_2-fixing organisms were limited by the sensitivity of the methods employed for detecting the process. The acetylene reduction technique made it possible to detect N_2 fixation by free-living bacteria in the soil with a relatively short incubation time. Many types of soils have been tested for asymbiotic N_2-fixing activity, including silt loam (Hardy et al., 1971), dune sand (Waughman et al., 1981), beach sand (Henriksson,

1971), prairie grassland (Paul et al., 1971), paddy soil (Rinaudo et al., 1971), forest soils (Waughman et al., 1981), arctic soils (Stutz and Bliss, 1975), desert soil (Rychert and Skujins, 1974), Hawaiian pastures (Koch and Oya, 1974), and peat (Waughman and Bellamy, 1972). The values reported are highly variable, but the overall impression from a host of studies supports the contention of earlier workers that N_2-fixing bacteria can add substantial nitrogen to soil systems. This is especially true in climax vegetational communities where slow but constant inputs of nitrogen are required. On the other hand, the contribution of free-living diazotrophs to the fertility of agricultural soils is considered to be marginal because of the large requirement for nitrogen in continually cropped fields. The hopes of early workers who sought to fertilize fields with N_2-fixing bacteria have not yet been realized.

Cyanobacteria apparently are responsible for most of the N_2 fixation in the soil (Postgate, 1982). They are important in "extreme" environments, such as arctic, subarctic, and desert regions, and usually they are the pioneer organisms in arid ground and recently formed lava soil (Fay, 1981). In tropical soils, cyanobacteria may fix economically significant quantities of N_2 (Stewart et al., 1978). In most agricultural situations, however, cyanobacteria have been found to contribute little to the soil nitrogen economy. The exception is in rice paddies, where inputs as high as 20–30 kg N/ha/annum have been reported (Venkatamaran, 1981).

A problem in estimating N_2 fixation rates with the acetylene reduction assay is the variability encountered in the results of field tests. Knowles (1977) suggested that short but significant "flushes" of N_2 fixation occur in soils because of seasonal or episodic increases in energy supplies. Many studies have demonstrated that acetylene reduction can be stimulated by amending soils with an oxidizable substrate (Knowles, 1977). Since heterotrophic diazotrophs rarely find themselves in a natural situation where oxidizable substrates are not limiting, this may help explain their relatively small contribution to the nitrogen economy of most soil systems.

Despite the vast literature compiled over the last century on the distribution of N_2-fixing bacteria in the soil, few estimates are available on the amount of N_2 that may be fixed over decades or centuries. A series of studies that began at Rothamsted Experiment Station between 1843 and 1856 have spanned the years. Field plots were initially set up by Lawes and Gilbert to compare the effects of various inorganic compounds and manure on crop yield (Day et al., 1975). Between 1885 and 1967 an average increase in the nitrogen balance of 34 kg N/ha/annum was found in plots planted with grain and straw. In an unharvested area where mixed woodland developed (Broadbalk Wilderness) an estimated 65 kg N/ha/annum was accumulated. A field that was annually "stubbed" to remove woody species received 55 kg N/ha/annum from N_2 fixation and other sources. The origin of nitrogen in the various field plots was not clear. Around 70–75% of the nitrogen was estimated to have been fixed by nonsymbiotic soil bacteria (Day et al., 1975). Day et al.

(1975) and Witty et al. (1977) suggested important roles for rhizosphere, cyanobacterial, and free-living N_2 fixation. Seasonal or episodic inputs of nitrogen probably occurred when moisture and other environmental conditions were optimal for N_2 fixation. Rainfall and animal activities probably also contributed to the input of nitrogen. The overall impression left by the Rothamsted experiment on Broadbalk Field is that free-living diazotrophs can contribute substantial nitrogen to a variety of environments provided that their basic requirements of sufficient moisture and nutrition are met. The Broadbalk experiment also illustrates how little the basic questions about N_2 fixation have changed since the time of Boussingault.

Rhizosphere Associations

Hiltner first described the accumulation of microorganisms in the region of the root in contact with the soil in 1905. He proposed the term "rhizosphere" to describe the environment (Waksman, 1932). That N_2-fixing bacteria selectively inhabited the plant rhizosphere was suggested early in this century and the older literature is replete with descriptions of N_2-fixing bacteria dwelling in the rhizosphere (Krasil'nikov, 1958). However, because methods were not available for detecting low levels of N_2 fixation, the significance of rhizosphere diazotrophs was unknown. The acetylene reduction assay gave new life to rhizosphere studies, and, in a few instances, considerable fixation has been reported.

In the rhizosphere of temperate plants, the number and activity of N_2-fixing bacteria has been considered to be low (Jensen, 1981). Most of the N_2-fixing organisms isolated from such environments are enteric bacteria or bacilli. In general, the associations seem to be rather nonspecific for the plants involved (Nelson et al., 1976).

The situation may differ for tropical and subtropical plants. Prior to 1968, several reports appeared on N_2 fixation in the rhizosphere of tropical grasses (Dobereiner, 1961; Jaiyebo and Moore, 1963; Moore, 1963; Parker, 1957). They were not taken seriously until Dobereiner et al. (1972) found considerable acetylene-reducing activity in the roots of *Paspalum notatum* (Bahia grass). The activity was attributed to *Azotobacter paspali,* which was isolated from the rhizosphere. In other tropical grasses, *Spirillum lipoferum* (now *Azospirillum lipoferum* and *Azospirillum brasilense*) was found to be a common rhizosphere-inhabiting diazotroph (Dobereiner and Day, 1976). These discoveries received great attention because of the possibility that the organisms form a so-called associative symbiosis (a term used by Burns and Hardy, 1975) or rhizocoenose (a term preferred by Postgate, 1982) in the rhizosphere of grasses, notably that of maize (Dobereiner and Day, 1976). Although the initial excitement was tempered later when more careful experiments were conducted (Dobereiner, 1977; Burris, 1977; Tjepkema and van Berkum, 1977), the appeal of the associations remains because their conceptually simple nature makes them potentially applicable to agricultural systems.

The methods used in rhizosphere studies are controversial and have sometimes involved prolonged incubation of samples under rather artificial conditions. Consequently, the actual significance of rhizosphere N_2 fixation has yet to be determined. Van Berkum and Bohlool (1980) critically reviewed the information on rhizosphere associations and concluded that rhizosphere diazotrophs can fix significant amounts of N_2. The quantities fixed are important mainly to the long-term enrichment of the soil. Recent work has shown that in certain circumstances marginal increases in crop yields can be obtained by direct inoculation (Cohen et al., 1980; Okon et al., 1983). The application of modern genetic engineering techniques to such rhizosphere-dwelling microorganisms could eventually lead to the development of more useful strains.

Symbioses Revisited

On a global basis, symbiotic N_2-fixing systems contribute the majority of nitrogen both to agricultural soils and to the biosphere (Postgate, 1982). Because of their importance to agriculture, leguminous symbioses traditionally have attracted the greatest attention.

Since acetylene reduction was described in 1966, more information has emerged regarding the biology of *Rhizobium* than had been discovered in the nearly 80 years from 1888 when Beijerinck first isolated *Bacillus radicicola.* In 1975, several groups reported that certain slow-growing rhizobia could fix N_2 in pure culture under the appropriate conditions (Keister, 1975; Kurz and LaRue, 1975; McComb et al., 1975; Pagan et al., 1975; Tjepkema and Evans, 1975). These findings reversed the consensus achieved when Fred, Baldwin, and McCoy published their classic monograph in 1932. Another dimension was added to the ecology of rhizobia when some strains were shown to grow autotrophically at the expense of H_2 (Hanus et al., 1979; Lepo et al., 1980). The hydrogenase involved also acted as an energy conserving system by oxidizing the H_2 evolved by nitrogenase (Emerich et al., 1979). Furthermore, the presence of hydrogenase seems to make the symbiosis more efficient in field tests (Eisbrenner and Evans, 1983).

In a radical departure from traditional assumptions, certain "cowpea" rhizobia have been found to nodulate nonleguminous plants from the genus *Parasponia* in the family Ulmaceae (Trinick, 1973). This is the only known diazotrophic association between *Rhizobium* and a nonlegume to date. It has provided agriculturalists the hope of extending the host range of rhizobia in the future. Interestingly, although nodulation of members of the Ulmaceae was reported as long ago as 1909 (Akkermans, 1979), not until 1973 did microbiologists rediscover the symbiosis.

The development of a simple assay for nitrogenase attracted talented investigators from all areas of biology to the problems of N_2 fixation in the early 1970's. As a result, spectacular advances have been made in understanding the genetics, biochemistry, and molecular biology of N_2 fixation in general and the rhizobial symbiosis in particular. For example, Higashi (1967) was the first to suggest plasmid involvement in rhizobial host specificity. He mixed cultures of *R. trifolii* (clover-specific rhizobia) and *R. phaseoli* (bean-specific rhizobia), and then killed the *R. trifolii* cells by treating with a specific phage. Cells surviving this treatment infected clover, leading to the conclusion that conjugal transfer of host specificity genes had occurred. This first bit of evidence has been criticized because the possibility exists that not all of the clover rhizobia were killed by phage treatment. However, in the same paper, Higashi also reported the loss of infectiveness after treatment of rhizobia with acridine dyes, a common method for curing bacteria of plasmids. Later, physical methods revealed that fast-growing rhizobia possessed high molecular weight (160–>500 kb) pSym megaplasmids (Nuti et al., 1977). Located on the pSym plasmids are many of the genes responsible for determining host range and function of the symbiosis (Denarie et al., 1981). Sequence homology exists between pSym plasmids from different rhizobial strains and between pSym plasmids and the large Ti (tumor-inciting) plasmids of *Agrobacterium* (Prakash et al., 1981). In fact, transfer of the pSym plasmid from *Rhizobium* to *Agrobacterium* allows the *Agrobacterium* to ineffectively nodulate the appropriate legume host (Hooykaas et al., 1981, 1982; Truchet et al., 1984). These observations have stimulated work on the molecular genetics of the rhizobial symbiosis that will continue for many years.

Many recent developments in studies of the interactions between legumes and rhizobia have as their origin discoveries made during the first half of this century. For example, phytohemagglutinins (lectins) have been known to exist in legumes since Landsteiner and Raubitschek reported their presence in *Phaseolus* seed extracts in 1908. More recently, lectins were implicated in the recognition and root hair-binding stages of rhizobial infection by the work of Bohlool and Schmidt (1974) and Hamblin and Kent (1973). Although the hypothesis remains controversial (Graham, 1981), it is an attractive idea that offers a starting point for studies on the molecular interactions between plant and microorganism.

Leghemoglobin was identified as a plant hemoglobin in 1939 by Kubo; considerable work has been done on the biochemistry of the protein since that time. Recent work suggests that the globin chains of leghemoglobin are synthesized by the plant, while the heme portion is synthesized by the bacterium (Godfrey et al., 1975; Sutton et al., 1981). This finding raises interesting questions about the evolution of the legume–*Rhizobium* symbiosis. The advances made in understanding the molecular biology, genetics and biochemistry of the legume symbiosis all contribute to our knowledge of the ecology of rhizobial N_2 fixation. The reader is directed to several recent reviews that deal with those aspects more fully (Postgate, 1982; Graham, 1981; Sutton et al., 1981; Imsande, 1981; Kondorosi and Johnston, 1981; Denarie et al., 1981; Kuykendall, 1981; Verma and Long, 1983).

Aside from the *Rhizobium* associations, other symbioses have been recognized since the 19th century. Ecologically, the most important of these is the actinorhizal symbiosis between an actinomycete of the genus *Frankia* and plants found in eight families and seven orders of dicotyledonous angiosperms (Akkermans and van Dijk, 1981). Many studies have demonstrated the important successional role played by these plants in nitrogen-impoverished environments. Their favored habitats include sand dunes, swamps and wetlands, range land, chapparal, borders of salt marshes, strip mined areas, and abandoned fields and farmlands (Benson, 1978). Nitrogen inputs by this association rival those reported for leguminous plants. Estimates of N_2 fixed range from a low of 0.06 kg N/ha/annum by *Purshia tridentata* in range land (Dalton and Zobel, 1977), to a high of over 200 kg N/ha/annum by *Alnus* sp. in the Netherlands (Akkermans, 1971) and the Pacific Northwest (Zavitkovski and Newton, 1968). The ecology of actinorhizal associations has been extensively reviewed (Akkermans and van Dijk, 1981; Becking, 1977; Silvester, 1977) and the consensus is that the actinorhizal association is one of the most important N_2-fixing symbioses in nonagricultural habitats.

In contrast to the information available about the plants, relatively little information was available about *Frankia* prior to 1978. No confirmed isolates had been obtained despite many attempts since the turn of the century. In 1978, Callaham et al. finally reported the isolation of a slow-growing actinomycete from *Comptonia peregrina* root nodules. The organism closely resembles an isolate obtained by Pommer from *Alnus glutinosa* root nodules in Germany in 1959. Information on the biology of the microorganism is still rudimentary, but some studies have been done on the diversity and taxonomy of nodule inhabitants (Benson and Hanna, 1983; Normand and Lalonde, 1982; Lechevalier and Lechevalier, 1979), on the infection process (Callaham et al., 1979), and on the biochemistry (Benson et al., 1979; Benson et al., 1980).

Prior to the introduction of acetylene reduction, many unusual mycorrhizal, bacterial-leaf, and bacterial-animal associations had been reported to fix N_2. N_2 fixation was difficult to detect in these symbioses because the nitrogenase activity was low or episodic. Since the development of acetylene reduction, "loose" symbiotic associations between plants and prokaryotes and betweeen animals and prokaryotes have quite commonly been found in habitats where the carbon/nitrogen ratio is high.

Reports have appeared of N_2-fixing organisms in the phyllosphere of trees (Ruinen, 1975), although the significance of such fixation is unknown and the findings are difficult to reproduce even when the same plant species are used (Becking, 1975). The prokaryotes involved include cyanobacteria, both free-living and in symbiosis with lichens and mosses, and heterotrophic bacteria (Ruinen, 1975). Nitrogen-fixing bacteria have also been found on the surfaces of sea grasses (Goering and Parker, 1972) and in the rhizosphere of marine and freshwater angiosperms (Knowles, 1977).

Since the early 1900s, symbioses of diazotrophs with animals have been suspected to occur. Cleveland (1925) suggested that termites might obtain part of their nitrogen from the atmosphere and Peklo (1946) claimed that several insects fixed N_2 in association with *Azotobacter*. Peklo further considered the fixation to be similar in magnitude to the legume-*Rhizobium* association, but work since then has not borne out this contention. However, using acetylene reduction, N_2 fixation has indeed been found in termites (Breznak et al., 1973; Benemann, 1973), in wood-eating cockroaches (Breznak, 1975), and in other insects. Many insects tested have proven negative (Waughman et al., 1981), but it is clear that some insects whose diets are high in carbohydrate and low in nitrogen may benefit from an N_2-fixing association with microorganisms in the gut. Conceivably, symbiotic N_2 fixation in insects may result in considerable nitrogen input into the environment largely because of the sheer abundance of insects (Waughman et al., 1981).

In other animals, N_2 fixation has been detected in the guts of sea urchins (Guerinot et al., 1977), in sea squirts from the Red Sea (Wilkinson and Fay, 1979), in shipworms (Waterbury et al., 1983), in the gastroenteric cavity of terrestrial snails and earthworms (Citernesi et al., 1977), in guinea pigs (Bergersen and Hipsley, 1970), and in the rumens of steers (Hardy et al., 1968), sheep, goats and reindeer (Granhall and Ciszuk, 1971; Elleway et al., 1971). In an unusual case involving humans, Ooman (1972) suggested that diazotrophs contributed fixed N_2 to New Guinea natives subsisting on a high carbohydrate-low protein diet of yams. Bergersen and Hipsley (1970) identified several diazotrophic enteric bacteria in the feces of such individuals. In general, because the rates of acetylene reduction are usually quite low, N_2-fixing bacteria probably do not contribute significant amounts of nitrogen to higher animals. Similarly, the amount of fixed N_2 added to the environment by these higher animal symbioses is probably insignificant compared to the enormous quantity fixed by leguminous and nonleguminous root nodule symbioses (Waughman et al., 1981).

In the early 1970s, the acetylene reduction technique was instrumental in demonstrating that a number of symbioses were not diazotrophic. For example, the leaf nodule associations found in members of the *Myrsinaceae* and *Rubiaceae,* and the leaf gland association in the *Dioscoraceae* have all been demonstrated to be inactive in fixing N_2; the bacteria probably contribute to plant growth by hormonal interactions (Becking, 1974; Silver, 1977). The controversy over N_2 fixation in mycorrhizal fungi was particularly difficult to settle. Old claims of N_2 fixation by mycorrhizal root nodules in the Podocarpaceae (Nobbe and Hiltner, 1899) were supported by ^{15}N studies performed by Bergersen and Costin (1964), and Becking (1965), among others. Silvester and Bennett (1973) resolved the issue by using the acetylene reduction assay to show that *Podocarpus* nodules did not fix N_2. Other mycorrhizal associations also were suggested to be diazotrophic. They included fungal associations with various conifers, ericaceous plants, orchids, and members of the Gramineae. The evidence for and against fixation in mycorrhizal roots was reviewed by Becking (1974); his major conclusion was that mycorrhizae do not fix N_2. Since many studies on N_2 fixation in mycorrhizal roots were performed on field material, rhizosphere diazotrophs may have been responsible for any observed fixation. As in the leaf nodule associations, activities other than N_2

fixation probably account for enhanced growth of mycorrhizal plants. In particular, mycorrhizal fungi mobilize phosphate and other immobile ions that can be taken up by plant roots (Ruehle and Marx, 1979).

Aquatic Environments

Any discussion of N_2 fixation in fresh water or marine environments necessarily focuses on phototrophic organisms. Both the photosynthetic bacteria and the cyanobacteria have been implicated in the nitrogen economy of aquatic habitats since the reports by Drewes (1928) on N_2 fixation in *Nostoc* and the discovery by Kamen and Gest (1949) of N_2 fixation in *Rhodospirillum rubrum*. In general, because they are widely distributed, cyanobacteria contribute substantially more nitrogen to the environment than the other photosynthetic bacteria. The latter fix N_2 when growing phototrophically or chemotrophically under anaerobic conditions (Madigan et al., 1984). Information on N_2 fixation in aquatic habitats is not extensive, probably because of complications inherent in measuring the process in excessively dynamic systems.

For many years, planktonic cyanobacteria were assumed to contribute nitrogen to lakes and ponds. However, it was not until 1959 that Dugdale et al. (1959) used $^{15}N_2$ as a tracer to demonstrate fixation in lake waters in Alaska and Pennsylvania. Since then, fixation has been shown to coincide with blooms of heterocystous cyanobacteria, particularly species of *Anabaena, Gloeotrichia,* and *Aphanizomenon,* in lakes in Antarctica (Horne, 1972), the U.S. (Horne and Goldman, 1972; Stewart et al., 1967; Stewart et al., 1971), Great Britain (Horne and Fogg, 1970), and Sweden (Granhall and Lundgren, 1971), among others. In tropical and subtropical areas, where conditions are suitable for year-round growth of cyanobacteria, remarkably high rates of fixation have been reported—up to 44 kg N/ha/annum in Lake George, Uganda (Horne and Viner, 1971). This value probably represents the upper limit for nitrogen input by free-living cyanobacteria (Mague, 1977).

Pearsall (1932) and Hutchinson (1944) long ago suggested that the N:P ratio of a lake is an important determinant in the proliferation of cyanobacteria in eutrophic lakes. Their hypothesis has been strengthened by more recent work that has taken advantage of the acetylene reduction technique (Smith, 1983). In general, fixation in eutrophic lakes has been shown to exceed fixation in oligotrophic lakes by a factor of 10^4, and diazotrophy in such environments has been considered to exacerbate the eutrophication process (Mague, 1977).

Nitrogen fixation in rice paddies has received considerable attention over the years. The importance of N_2 fixation in this habitat is understandable, since over a billion people depend on rice for their existence. Suggestions concerning the importance of cyanobacteria in paddy waters have persisted since De (1936) attributed nitrogen fertility of wet-land rice to the presence of blue-green algae (cyanobacteria). Numerous studies have demonstrated that cyanobacteria, either free-living or in symbiosis with the water fern *Azolla,* fix a considerable amount of N_2 in the paddy environment (Watanabe and

Brotonegoro, 1981; Singh, 1961; Venkataraman, 1975; Stewart et al., 1978). During blooms of *Azolla,* as much as 600 kg N/ha/annum may be fixed, a record for any diazotrophic association (Postgate, 1982). More recently, other photosynthetic bacteria on anaerobic mud surfaces and heterotrophic bacteria in the flooded soil and associated with rice roots have also been suggested to contribute nitrogen to paddy environments (Fay, 1981; Watanabe and Brotonegoro, 1981).

Diazotrophy in the marine environment remains somewhat difficult to assess. Heterocystous cyanobacteria are generally not major components of plankton in the open ocean (Fay, 1981). However, N_2 fixation is associated with blooms of the nonheterocystous cyanobacterium *Oscillatoria* (*Trichodesmium*) (Dugdale et al., 1961; Carpenter and McCarthy, 1975). The magnitude of such blooms can be immense; they have been recorded to cover areas as large as 40,000–50,000 km^2 (Mague, 1977). Recent information suggests that oceanic fixation may have been grossly underestimated in the past because of sampling techniques (Martinez et al., 1983).

Significant fixation has sometimes been found in more limited marine environments. For example, species of the planktonic diatoms *Rhizoselenia* and *Chaetoceros* house a symbiotic heterocystous cyanobacterium, *Richelia intracellularis,* that fixes N_2 (Mague et al., 1974). Other *Rhizoselenia* sp. have associated heterotrophic bacteria that have recently been shown to reduce acetylene at high rates (Martinez et al., 1983). Cyanobacteria also may contribute to the nitrogen economy of coral reef communities (Burris, 1976; Capone et al., 1977). Photosynthetic bacteria and particularly cyanobacteria add to the productivity of littoral and supralittoral zones, salt marshes, and sand-dune slack areas (Mague, 1977). Given the immensity of the oceans, the magnitude and importance of marine N_2 fixation is extremely difficult to estimate, and many more studies will have to be performed before accurate conclusions can be drawn.

CURRENT PERSPECTIVES AND FUTURE PROSPECTS

Looking back on the historical development of N_2 fixation, it is clear that many of our current concepts about the process evolved largely from the speculations of the early soil microbiologists. They were addressing questions posed by 19th century agricultural chemists such as Boussingault and Liebig who, in their turn, were trying to build a scientific foundation under thousands of years of empirical observations. As Burris (1974) stated when referring to the monograph published by Fred, Baldwin, and McCoy in 1932:

> Anyone who feels that our knowledge of biological N_2 fixation has all been acquired in the last decade would be disabused of the idea by reading this 1932 monograph.

This admonition holds true for much of the information on the practical applications of diazotrophs in agriculture, particularly with regard to legumes.

However, problems concerning the ecological importance and distribution of diazotrophs could not be approached effectively until the last two decades, although many of the basic directions of research into the ecology of diazotrophy had already been pointed out by early workers.

While our ecological understanding of N_2 fixation has increased dramatically over the past 20 years, there are still many issues that remain unresolved. For example, the question of whether rhizosphere inhabitants such as *Azospirillum* can in fact contribute substantial nitrogen to plants in rhizosphere associations needs further study with improved methodology. Further research on rhizosphere microorganisms is needed, particularly with regard to identifying and genetically manipulating competitive strains. Nitrogen fixation in marine and freshwater environments needs to be reassessed in light of a recent report suggesting that earlier methods were inadequate (Martinez et al., 1983). The controversy over lectin involvement in *Rhizobium* recognition by the host plant is a matter for debate, as is the question of whether nitrogenase and hydrogenase contribute to the survival of rhizobia in the soil. Now that *Frankia* has been isolated from actinorhizal root nodules, an entirely new field of symbiotic N_2 fixation has been introduced complete with all of the questions that have hitherto been directed at the leguminous symbiosis. The tremendous diversity of cyanobacterial diazotrophic associations also begs for exploration. Studies of these organisms may prove the most fruitful with regard to long-term application to agricultural problems. Discovery of a cellulolytic diazotroph inhabiting specific glands in wood-eating shipworms (Waterbury et al., 1983) highlights the possibility that many more diazotrophic symbioses remain to be found.

The main thrust of research on N_2 fixation has taken a more molecular direction in recent years. Our understanding of the ecology of diazotrophs will no doubt benefit from what is learned. For example, the biochemical homology of the nitrogenase proteins from distantly related prokaryotes (Emerich and Burris, 1978) and the sequence homology of the *nif* genes (Ruvkun and Ausubel, 1980) raise questions about why the homology was preserved and how the genetic information was transferred to such a wide variety of microorganisms. These are questions that must be addressed with some knowledge of the interactions of diazotrophs with the environment and with other microorganisms. Studies on the ecology of N_2 fixation have an interesting future, provided that new techniques and perspectives are applied to the field. Since rapid advance in the understanding of N_2 fixation has usually been catalyzed by methodological improvements in the past, a new era of understanding may be upon us in the near future.

REFERENCES

Akkermans, A. D. L., 1971, Nitrogen fixation and nodulation of *Alnus* and *Hippophae* under natural conditions, Ph.D. Thesis, Leiden.

Akkermans, A. D. L., 1979, Symbiotic nitrogen fixers available for use in temperate forestry, in: *Symbiotic Nitrogen Fixation in the Management of Temperate Forests* (J. C. Gordon, C. T. Wheeler, and D. A. Perry, eds.), Forest Research Laboratory, Corvallis, Oregon, pp. 23–35.

Akkermans, A. D. L., and van Dijk, C., 1981, Non-leguminous root-nodule symbioses with actinomycetes and *Rhizobium,* in: *Nitrogen Fixation, Vol. 1, Ecology* (W. J. Broughton, ed.), Clarendon Press, Oxford, pp. 57–103.

Aldrich-Blake, R. N., 1932, On the fixation of atmospheric nitrogen by bacteria living symbiotically in root nodules of *Casuarina equisetifolia, Oxford Forestry Mem.* **14:**20 pp.

Allen, E. K., and Allen, O. N., 1958, Biological aspects of symbiotic nitrogen fixation, in: *Handbuch der Pflanzenphysiologie, Vol. 8, Nitrogen Metabolism* (W. Ruhland, ed.), Springer-Verlag, Berlin, pp. 48–118.

Allen, M., and Arnon, D., 1955, Studies on nitrogen-fixing blue-green algae. I. growth and nitrogen fixation by *Anabaena cylindrica, Plant Physiol.* **30:**366–372.

Allison, F. E., 1929, Can nodule bacteria of leguminous plants fix atmospheric nitrogen in the absence of the host? *J. Agr. Res. (U.S.)* **39:**893–924.

Allison, F. E., 1947, *Azotobacter* inoculation of crops. I. Historical, *Soil Sci.* **64:**413–429.

Allison, F. E., and Morris, H. J., 1930, Nitrogen fixation by blue-green algae, *Science* **71:**221–223.

Allison, F. E., Caddy, V. L., Pinck, L. A., and Armiger, W. H., 1947, *Azotobacter* inoculation of crops. 2. Effect on crops under greenhouse conditions, *Soil Sci.* **64:**489–497.

Allison, F. E., Hoover, S. R., and Morris, H. J., 1937, Physiological studies with the nitrogen-fixing algae *Nostoc muscorum, Bot. Gaz.* **82:**433–463.

Alpe, V., and Menozzi, A., 1892, Studi e richerche sulla questione dell' assimilazione dell' azoto per parte delle piante, *Bol. Not. Agrarie (Rome)* **14:**747–779.

Atwater, W. O., 1885, On the acquisition of atmospheric nitrogen by plants, *Am. Chem. J.* **6:**365–388.

Atwater, W. O., 1886, On the liberation of nitrogen from its compounds and the acquisition of atmospheric nitrogen by plants, *Am. Chem. J.* **8:**398–420.

Atwater, W. O., and Woods, C. D., 1890, The acquisition of atmospheric nitrogen by plants, *Conn. Agr. Expt. Sta. Ann. Rpt. (Storrs),* **2:**11–51.

Aulie, R. P., 1970, Boussingault and the nitrogen cycle, *Proc. American Phil. Soc.* **114:**435–479.

Becking, J. H., 1962, Species differences in molybdenum and vanadium requirements and combined nitrogen utilization by *Azotobacteriaceae, Plant Soil* **16:**171–201.

Becking, J. H., 1965, Nitrogen fixation and mycorrhiza in podocarpus root nodules, *Plant Soil* **23:**213–226.

Becking, J. H., 1974, Putative nitrogen fixation in other symbioses, in: *The Biology of Nitrogen Fixation* (A. Quispel, ed.), American Elsevier, New York, pp. 583–613.

Becking, J. H., 1975, Nitrogen fixation in some natural ecosystems in Indonesia, in: *Symbiotic Nitrogen Fixation in Plants* (P. S. Nutman, ed.), Cambridge University Press, New York, pp. 539–550.

Becking, J. H., 1977, Dinitrogen-fixing associations in higher plants other than legumes, in: *A Treatise on Dinitrogen Fixation, Section III—Biology* (R. W. F. Hardy and W. S. Silver, eds.), John Wiley and Sons, New York, pp. 185–275.

Beijerinck, M. W., 188, Die Bacterien der Papilionaceenknollchen, *Bot. Ztg.* **46:**726–735, 741–750, 757–771, 781–790, 797–804.

Beijerinck, M. W., 1901, Uber Oligonitrophile Mikroben, *Zentralbl. Bakteriol.* **7:**561–582.

Benemann, J. R., 1973, Nitrogen fixation in termites, *Science* **181:**164–165.

Benson, D. R., 1978, Root nodules of *Myrica pensylvanica* (bayberry): structure, ultrastructure, and preparation of nitrogen-fixing homogenates, Ph.D. Thesis, Rutgers University.

Benson, D. R., and Eveleigh, D. E., 1979, Nitrogen-fixing homogenates from *Myrica pennsylvanica* (bayberry) root nodules, *Soil Biol. Biochem.* **11:**331–334.

Benson, D. R., Arp, D. J., and Burris, R. H., 1980, Cell-free nitrogenase and hydrogenase from actinorhizal root nodules, *Science* **205:**688–689.

Benson, D. R., and Hanna, D., 1983, *Frankia* diversity in an alder stand as estimated by sodium dodecyl sulfate—polyacrylamide gel electrophoresis of whole-cell proteins, *Can. J. Bot.* **61:**2919–2923.

Bergersen, F. J., and Costin, A. B., 1964, Root nodules on *Podocarpus lawrencii* and their ecological significance, *Aust. J. Biol. Sci.* **17:**44–48.

Bergersen, F. J., and Hipsley, E. H., 1970, The presence of N_2-fixing bacteria in the intestines of man and animals, *J. Gen. Microbiol.* **60:**61–65.

Bergersen, F. J., and Wilson, P. J., 1959, Spectrophotometric studies of the effects of nitrogen on soybean nodule extracts, *Proc. Natl. Acad. Sci. U.S.A.* **45:**1641–1646.

Berkum, P. van, and Bohlool, B. B., 1980, Evaluation of nitrogen fixation by bacteria in association with roots of tropical grasses, *Microbiol. Rev.* **44:**491–517.

Berthelot, M., 1888, Fixation de l'azote atmosphérique sur la terre végétale, *Ann. Chim. Phys.* **13:**1–119.

Berthelot, M., 1890, Recherches nouvelles sur la fixation de l'azote par la terre végétale et les plantes et sur l'influence de l'électricité sur ce phénomène, *Ann. Chim. Phys.* **29** (Ser. 6):434–492.

Berthollet, C. L., 1788, Analyse de l'alcali volatil, *Histoire Mem. Acad. Roy. des Sciences (Paris)*, pp. 316–326.

Bhat, J., and Palacios, G., 1949, Influence of *Aerobacter aerogenes* in the nitrogen status of the soil, *J. Univ. Bombay* **71B:**84–87; *Chem. Abst.* **44:**779d.

Blom, J., 1931, Ein Versuch, die chemischen Vorgange bei der Assimilation des molekularen Stickstoffs durch Microorganism zu erklaren, *Zentralbl. Bakt.* **84:**60–86.

Bohlool, B. B., and Schmidt, E. L., 1974, Lectins: a possible basis for specificity in the *Rhizobium*-legume root nodule symbiosis, *Science* **185:**269–271.

Bond, G., 1955, An isotopic study of the fixation of nitrogen associated with nodulated plants of *Alnus, Myrica* and *Hippophae, J. Exp. Bot.* **6:**303–311.

Bond, G., 1974, Root nodule symbioses with actinomycete-like organisms, in: *The Biology of Nitrogen Fixation* (A. Quispel, ed.), American Elsevier, New York, pp. 342–378.

Bond, G., and Scott, G. D., 1955, An examination of some symbiotic systems for fixation of nitrogen, *Ann. Bot.* **19:**67–77.

Borm, L., 1931, Die wurzelknollchen von *Hippophae rhamnoides* und *Alnus glutinosa, Bot. Arch.* **31:**441–488.

Bortels, H., 1930, Molybdan als Katalysator bei der biologischen Stickstoffbindung, *Arch. Mikrobiol.* **1:**333–342.

Bortels, H., 1940, Uber die Bedeutung des Molybdans fur stickstoffbindende *Nostocaceen, Arch. Microbiol.* **11:**155–186.

Bottomley, W. B., 1911, The structure and physiological significance of the root-nodules of *Myrica gale, Proc. Roy. Soc. (London)* **84B:**215–216.

Boussingault, J. B., 1836, Recherches sur la quantité d'azote contenue dans les fourrages et leur équivalens, *Ann. Chim. Phys.* **63** (Ser. 2):225–244.

Boussingault, J. B., 1838a, Recherches chimiques sur la végétation. entreprises dans le but d'examiner si les plantes prennent de l'azote de l'atmosphère, *Ann. Chim. Phys.* **67** (Ser. 2): 5–54.

Boussingault, J. B., 1838b, Recherches chimiques sur la végétation. Troisieme Memoire. De la discussion de la valeur relative des assolements par l'analyse élémentaire, *C. R. Acad. Sci. (Paris)* **7:**1149–1155.

Boussingault, J. B., 1838c, Recherches chimiques sur la végétation entreprises dans le but d'examiner si les plants prennent de l'azote de l'atmosphère, *Ann. Chim. Phys.* **69:**353–367.

Boussingault, J. B., 1841, De la discussion de la valeur relative des assolements, par les résultats de l'anayse élémentaire, *Ann. Chim. Phys.* **1** (Ser. 3):208–246.

Boussingault, J. B., 1853a, Mémoire sur le dosage de l'ammoniaque contenue dans les eaux, *Ann. Chim. Phys.* **39** (Ser. 3):257–291.

Boussingault, J. B., 1853b, Sur la quantité d'ammoniaque contenue dans l'eau de pluie recueillie loin des villes, *C. R. Acad. Sci. (Paris)* **37:**207, 208, 798–806.

Boussingault, J. B., 1854, Sur la quantité d'ammoniaque dans la pluie, la rosée et le brouillard recueillie loin des villes, *Ann. Chim. Phys.* **40** (Ser. 3):257–291.

Boussingault, J. B., 1860, *Rural Economy, in its Relations with Chemistry, Physics, and Meteorology or, Chemistry Applied to Agriculture,* C. M. Saxton, Barker and Co., New York.

Boussingault, J. B., 1861, Observations relatives au développement des mycodermes, *Ann. Chim. Phys.* **61** (Ser. 3):363–367.

Boussingault, J. B., 1886–1891, *Agronomie, Chimie Agricole et Physiologie,* Gauthier, Paris.

Bredemann, G., 1908, Untersuchungen uber die Variation und das Stickstoffbindungsvermogen des *Bacillus asterosporus* A. M. *Zentrabl. Bakt. (II)* **22:**44–89.

Breznak, J. A., 1975, Symbiotic relationships between termites and their intestinal microbiota, in: *Symbiosis,* (D. H. Jennings and D. L. Lee, eds.), Society of Experimental Biology Series No. 29, Cambridge University Press, New York, pp. 559–580.

Breznak, J. A., Brill, W. J., Mertins, J. W., and Coppel, H. C., 1973, Nitrogen fixation in termites, *Nature (London)* **244:**577–580.

Brightman, F. H., 1959, Neglected plants—lichens, *New Biol.* **29:**75–94.

Brouzes, R., Lasik, J., and Knowles, R., 1969, The effect of organic amendment, water content, and oxygen on the incorporation of $^{15}N_2$ by some agricultural and forest soils, *Can. J. Microbiol.* **15:**899–905.

Bulen, W. A., Burns, R. C., and LeComte, J. R., 1965, Nitrogen fixation: hydrosulfite as electron donor with cell-free preparation of *Azotobacter vinelandii* and *Rhodospirillum rubrum, Proc. Natl. Acad. Sci. USA* **53:**532–539.

Bulen, W. A., and LeComte, J. R., 1966, The nitrogenase system from *Azotobacter:* two enzyme requirements for N_2 reduction, ATP-dependent H_2 evolution and ATP hydrolysis, *Proc. Nat. Acad. Sci. USA* **56:**979–986.

Burk, D., 1934, Azotase and nitrogenase in *Azotobacter, Ergeb. Enzymforsch.* **3:**23–56.

Burns, R. C., and Hardy, R. W. F., 1975, *Nitrogen Fixation in Bacteria and Higher Plants,* Springer Verlag, New York.

Burris, R. H., 1966, Biological nitrogen fixation, *Ann. Rev. Plant Physiol.* **17:**155–184.

Burris, R. H., 1974, Biological nitrogen fixation, 1924–1974, *Plant Physiol.* **54:**443–449.

Burris, R. H., 1975, The acetylene-reduction technique, in: *Nitrogen fixation by Free-Living Microorganisms* (W. D. P. Stewart, ed.), Cambridge University Press, New York, pp. 249–257.

Burris, R. H., 1976, Nitrogen fixation by blue-green algae of Lizard Island area of the Great Barrier Reef, *Aust. J. Plant Physiol.* **3:**41–51.

Burris, R. H., 1977, A synthesis paper on non-leguminous N_2-fixing systems, in: *Recent Developments in Nitrogen Fixation* (W. E. Newton, J. R. Postgate, and C. Rodriguez-Barrueco, eds.), Academic Press, New York, pp. 487–511.

Burris, R. H., 1979, The early biochemistry, in: *A Treatise on Dinitrogen Fixation, Sections I and II: Inorganic and Physical Chemistry and Biochemistry* (R. W. F. Hardy, F. Bottomely, and R. C. Burns, eds.), John Wiley and Sons, pp. 383–398.

Burris, R. H., and Miller, C. E., 1941, Application of N^{15} to the study of biological nitrogen fixation, *Science* **93:**114–115.

Burris, R. H., and Wilson, P. W., 1946, Ammonia as an intermediate in nitrogen fixation by *Azotobacter, J. Bacteriol.* **52:**505–512.

Burris, R. H., Eppling, F. J., Wahlin, H. B., and Wilson, P. W., 1942, Studies of biological nitrogen fixation with isotopic nitrogen, *Proc. Soil Sci. Soc. Am.* **7:**258–262.

Burris, R. H., Eppling, T. S., Wahlin, H. B., and Wilson, P. W., 1943, Detection of nitrogen fixation with isotopic nitrogen, *J. Biol. Chem.* **148:**349–357.

Callaham, D., DelTredici, P., and Torrey, J. G., 1978, Isolation and cultivation in vitro of the actinomycete causing root nodulation in *Comptonia, Science* **199:**899–902.

Callaham, D., Newcomb, W., Torrey, J. G., and Peterson, R. L., 1979, Root hair infection in actinomycete-induced root nodule initiation in *Casuarina, Myrica,* and *Comptonia, Bot. Gaz.* **140** (Suppl.):S1–S9.

Capone, D. G., Taylor, D. L., and Taylor, B. F., 1977, Nitrogen fixation (acetylene reduction) associated with macroalgae in a coral-reef community in the Bahamas, *Mar. Biol.* **40:**29–32.

Cappelletti, C., 1923a, Reazioni immunitarie nei tubercoli radicali delle Leguminose, *Atti Soc. Medico-Chirurgica Padova, pp.* 5–7.

Capelletti, C., 1923b, Reazioni immunitarie nei tubercoli radicali delle Leguminose, *Gior. Biol. e Med. Sper.* **1:**Fasc. 2, 1–16.

Capelletti, C., 1924, Reazioni immunitarie nei tubercoli radicali di Leguminose, *Ann. Bot. (Rome)* **16:**1–16.

Carnahan, J. E., Mortenson, L. E., Hower, H. F., and Castle, J. E., 1960, Nitrogen fixation in cell-free extracts of *Clostridium pasteurianum, Biochim. Biophys. Acta* **44:**520–535.

Caron, A., 1895, Landwirtschaftlich—bakteriologische Probleme, *Landw. Vers. Sta.* **45:**401–418.

Carpenter, E. J., and McCarthy, J. J., 1975, Nitrogen fixation and uptake of combined nitrogenous nutrients by *Oscillatoria (Trichodesmium) thiebautii* in the Western Sargasso Sea, *Limnol. Oceanogr.* **20:**389–401.

Carter, E. G., and Greaves, J. D., 1928, The nitrogen-fixing organisms of an arid soil, *Soil Sci.* **26:**179–192.

Chaudari, H., 1931, Recherches sur la bacterie des nodosites radiculaires du *Casuarina equisetifolia* (Fort.), *Bull. Soc. Bot. France* **78:**447–452.

Citernesi, U., Neglia, R., Seritti, A., Lepidi, A. A., Filippi, C., Bagnoli, G., Nuti, M. P., and Galluzzi, R., 1977, Nitrogen fixation in the gastro-enteric cavity of soil animals, *Soil Biol. Biochem.* **9:***71–72*.

Cleveland, L. R., 1925, The ability of termites to live perhaps indefinitely on a diet of pure cellulose, *Biol. Bull. Marine Biol. Lab. (Woods Hole)* **48:**289–293.

Cohen, E., Okon, Y., Kigel, J., Nur, I., and Henis, Y., 1980, Increases in dry weight and total nitrogen content in *Zea mays* and *Setaria italica* associated with nitrogen fixing *Azospirillum* spp., *Plant Physiol.* **66:**246–249.

Dalton, D. A., and Zobel, D. B., 1977, Ecological aspects of nitrogen fixation by *Purshia tridentata, Plant Soil* **48:**57–80.

Davy, H., 1836, *Elements of Agricultural Chemistry,* 5th Ed., Longmans, London.

Day, J. M., Harris, D., Dart, P. J., and van Berkum, P., 1975, The Broadbalk experiment. An investigation of nitrogen gains from nonsymbiotic nitrogen fixation, in: *Nitrogen Fixation by Free-living Micro-organisms* (W. D. P. Stewart, ed.), Cambridge University Press, New York, pp. 71–84.

De, P. K., 1936, The problem of the nitrogen supply of rice. I. Fixation of nitrogen in the rice soils under water-logged conditions, *Indian J. Agr. Sci.* **6:**1237–1245.

De, P. K., 1939, The role of blue-green algae in nitrogen fixation in rice-fields, *Proc. Roy. Soc. (London) Ser. B.* **127:**121–139.

Delwiche, C. C., and Wijler, J., 1956, Non-symbiotic nitrogen fixation in soil, *Plant Soil* **7:**113–129.

Dénarié, J., Boistard, P., Casse-Delbart, F., Atherly, A. G., Berry, J. O., and Russell, P., 1981, Indigenous plasmids of *Rhizobium,* in: *Biology of the Rhizobiaceae* (K. L. Giles and A. B. Atherly, eds.), Academic Press, New York, pp. 225–246.

Dilworth, M. J., 1966, Acetylene reduction by nitrogen-fixing preparations from *Clostridium pasteurianum, Biochim. Biophys. Acta* **127:**285–294.

Dobereiner, J., 1961, Nitrogen fixing bacteria of the genus *Beijerinckia Derx* in the rhizosphere of sugar cane, *Plant Soil* **14:**211–217.

Dobereiner, J., 1977, Physiological aspects of N_2-fixation in grass–bacteria associations, in: *Recent Developments in Nitrogen Fixation* (W. Newton and J. R. Postgate, eds.), Academic Press, New York, pp. 513–522.

Dobereiner, J., and Day, J. M., 1976, Associative symbioses in tropical grasses: Characterization of microorganisms and dinitrogen-fixing sites, in: *Proceedings of the 1st International Symposium on Nitrogen Fixation, Vol. 2* (W. E. Newton and C. J. Nyman, eds.), Washington State University Press, Pullman, pp. 518–538.

Dobereiner, J., Day, J. M., and Dart, P. J., 1972, Nitrogenase activity and oxygen sensitivity of the *Paspalum notatum–Azotobacter paspali* association, *J. Gen. Microbiol.* **71:**103–116.

Drewes, K., 1928, Uber die Assimilation des Luftstickstoffs durch Blaualgen, *Zentralbl. Bakteriol. (II)* **76:**88–101.

Dugdale, R. C., Dugdale, V., Neese, J. C., and Goering, J., 1959, Nitrogen fixation in lakes, *Science* **130:**859–860.

Dugdale, R. C., Menzel, D. W., and Ryther, J. G., 1961, Nitrogen fixation in the Sargasso Sea, *Deep Sea Res.* **7:**297–302.

Dumas, J. B., 1834, De l'analyse élémentaire des substances organiques, *J. Phar. Sci. Acc.* **20:**129–156.

Dumas, J. B., and Boussingault, J. B., 1844, *The Chemical and Physiological Balance of Organic Nature,* Saxton and Miles, New York.

Eisbrenner, G., and Evans, H. J., 1983, Aspects of hydrogen metabolism in nitrogen-fixing legumes and other plant-microbe associations, *Annu. Rev. Plant Physiol.* **34:**105–136.

Elleway, R. F., Sabine, J. R., and Nicholas, D. J. D., 1971, Acetylene reduction by rumen microflora, *Arch. Mikrobiol.* **76:**277–291.

Emerich, D. W., and Burris, R. H., 1978, Complementary functioning of the component proteins of nitrogenase from several bacteria, *J. Bacteriol.* **134:**936–943.

Emerich, D. W., Ruis-Argueso, T., Ching, T. M., and Evans, H. J., 1979, Hydrogen-dependent nitrogenase activity and ATP formation in *Rhizobium japonicum* bacteroids, *J. Bacteriol.* **137:**153–160.

Emerson, R., 1917, Are all the soil bacteria and streptothrices that develop on dextrose agar azofiers? *Soil Sci.* **3:**417–422.

Faber, F. C. von, 1912, Das erbliche Zusammenleben von Bakterien und tropischen Pflanzen, *Jahrb. Wiss. Bot.* **51:**283–375.

Fay, P., 1981, Photosynthetic micro-organisms, in: *Nitrogen Fixation Volume I, Ecology* (W. J. Broughton, ed.), Clarendon Press, Oxford, pp. 1–29.

Fay, P., Stewart, W. D. P., Walsby, A. E., and Fogg, G. E., 1968, Is the heterocyst the site of nitrogen fixation in blue-green algae? *Nature* **220:**810–812.

Fogg, G. E., 1947, Nitrogen fixation by blue-green algae, *Endeavor* **6:**172–175.

Fogg, G. E., and Wolfe, M., 1954, The nitrogen metabolism of blue-green algae (Myxophyceae), *Symp. Soc. Gen. Microbiol.* **4:**99–125.

Frank, A. B., 1879, Ueber die Parasiten in den Wurzelanschwellungen der Papilionaceen, *Bot. Ztg.* **37:**377–388, 393–400.

Frank, B., 1889, Uber den gegenwartigen Stand unserer Kenntnis der Assimilation elementaren Stickstoffs durch die Pflanzen, *Ber. Dtsch. bot. Ces.* **7:**34–42.

Frank, B., 1890, Ueber die Pilzsymbiose der Leguminosen, *Landw. Jahrb.* **19:**523–640.

Fred, E. B., Baldwin, I. L., and McCoy, E., 1932, *Root Nodule Bacteria and Leguminous Plants,* University of Wisconsin Press, Madison.

Gayon, U., and Dupetit, G., 1886, Recherches sur la réduction des nitrates par les infiniments petits, *Soc. Sci. Phys. Nat. Bordeaux* (Ser. 3) **2:**201–207.

Godfrey, C. A., Coventry, D. R., and Dilworth, M. J., 1975, Some aspects of leghemoglobin biosynthesis, in: *Nitrogen Fixation by Free-Living Microorganisms,* (W. D. P. Stewart, ed.), Cambridge University Press, New York, pp. 311–332.

Goering, J. J., Dugdale, R. C., and Menzel, D. W., 1966, Estimates of *in situ* rates of nitrogen uptake by *Trichodesmium* sp. in the tropical Atlantic Ocean, *Limnol. Oceanogr.* **11:**614.

Goering, J. J., and Parker, P. L., 1972, Nitrogen fixation by epiphytes on sea grasses, *Limnol. Oceanogr.* **15:**320–323.

Goldman, C. R., 1961, The contribution of alder trees (*Alnus tenuifolia*) to the primary productivity of Castle Lake, Calif., *Ecology* **42:**282–288.

Graham, T. L., 1981, Recognition in *Rhizobium*-legume symbioses, in: *Biology of the Rhizobiaceae* (K. L. Giles and A. G. Atherly, eds.), Academic Press, New York, pp. 127–148.

Granhall, U., and Ciszuk, P., 1971, Nitrogen fixation in rumen contents indicated by the acetylene reduction test, *J. Gen. Microbiol.* **65:**91–93.

Granhall, U., and Lundgren, A., 1971, Nitrogen fixation in Lake Erken, *Limnol. Oceanogr.* **16:**711–719.

Grau, F. H., and Wilson, P. W., 1962, Physiology of nitrogen fixation by *Bacillus polymyxa, J. Bacteriol.* **85:**446–450.

Greaves, J. D., 1929, The microflora of leached alkali soils. ii. *Soil Sci.* **29:**79–83.

Greaves, J. E., and Greaves, J. D., 1932, Nitrogen fixation of leached alkali soils, *Soil Sci.* **34:**375–382.

Guerinot, M. L., Fong, W., and Patriquin, D. G., 1977, Nitrogen fixation (acetylene reduction) associated with sea urchins (*Strongylocentrotus droebachiensis*) feeding on seaweeds and eelgrass, *J. Fish. Res. Board Canada* **34:**416–420.

Hamblin, J., and Kent, S. P., 1973, Possible role of phytohaemagglutinin in *Phaseolus vulgaris* L., *Nature New Biol.* **245:**28–30.
Hamilton, P. B., and Wilson, P. W., 1955, Nitrogen fixation by *Aerobacter aerogenes. Ann. Acad. Sci. Fennicae, Ser. A.* **2:**139–150.
Hamilton, P. B., Magee, W. E., and Mortenson, L. E., 1953, Nitrogen fixation by *Aerobacter aerogenes* and cell-free extracts of the *Azotobacter vinelandii, Bacteriol. Proc.* p. 82.
Hanus, F. J., Maier, R. J., and Evans, H. J., 1979, Autotrophic growth of H_2-uptake positive strains of *R. japonicum* in an atmosphere supplied with hydrogen gas, *Proc. Natl. Acad. Sci. USA* **76:**1788–1792.
Hardy, R. W. F., Burns, R. C., Hebert, R. R., Holsten, R. D. , and Jackson, E. K., 1971, Biological nitrogen fixation: a key to world protein, *Plant Soil, Special Vol.*, pp. 561–590.
Hardy, R. W. F., and D'Eustachio, A. J., 1964, The dual role of pyruvate and the energy requirement in nitrogen fixation, *Biochem. Biophys. Res. Commun.* **15:**314–318.
Hardy, R. W. F., Holsten, R. D., Jackson, E. K., and Burns, R. C., 1968, The acetylene-ethylene assay for N_2 fixation: laboratory and field evaluation, *Plant Physiol.* **43:**1185–1207.
Hardy, R. W. F., and Knight Jr., E., 1967, ATP-dependent reduction of azide and HCN by N_2-fixing enzymes of *Azotobacter vinelandii* and *Clostridium pasteurianum, Biochim. Biophys. Acta* **139:**69–90.
Hellriegel, H., 1886a, *Welche Stickstoffquellen stehen der Pflanze zu Gebote? Tageblatt der 50 Versammlung Deutscher Naturforscher und Aerzte* in Berlin, 18–24 Sept., p. 290.
Hellriegel, H., 1886b, *Welche Stickstoffquellen stehen der Pflanze zu Gebote? Ztschr. Ver. Rubenzucker–Industrie Deutschen Reichs* **36:**863–877.
Hellriegel, H., and Wilfarth, H., 1888, Untersuchungen uber die Stickstoffnahrung der Gramineen und Leguminosen, *Beilageheft zu der Ztschr. Ver. Rubenzucker-Industrie Deutschen Reich.*, 234 pp.
Henriksson, E., 1971, Algal nitrogen fixation in temperate regions, *Plant Soil, Special Vol.* pp. 415–419.
Higashi, S., 1967, Transfer of clover infectivity of *R. trifolii* to *R. phaseoli* as mediated by an episomic factor, *J. Gen. Appl. Microbiol.* **13:**391–403.
Hiltner, 1896, Ueber die Bedeutung der Wurzelknollchen von *Alnus glutinosa* fur die Stickstoffernahrung dieser Pflanze, *Landw. Vers. Sta.* **46:**153–161.
Hiltner, 1898, Ueber Entstehung und physiologische Bedeutung der Wurzelknollchen, *Forstl. Naturw. Ztschr.* **7:**415–423.
Hiltner, L., 1905, Ueber neuere Erfahrungen und Probleme auf dem Gebiete der Boden bakteriologie und unter besonderer Berucksichtigung der Grundungung und Brache, *Zentralbl. Bakteriol.* **14:**46–48.
Hiltner, L., 1920, *Zentralbl. Bakteriol. (II)* **58:**351.
Hooykaas, P. J. J., van Brussel, A. A. N., den Dulk-Ras, H., van Slogteren, G. M. S., and Schilperoort, R. A., 1981, Sym plasmid of *Rhizobium trifolii* expressed in different rhizobial species and *Agrobacterium tumefaciens, Nature* **291:**351–353.
Horne, A. J., 1972, The ecology of nitrogen fixation on Signy Island, South Orkney Islands, *Br. Antarct. Surv. Bull.* **27:**1.
Horne, A. J., and Fogg, G. E., 1970, *Proc. Roy. Soc. London, Ser. B* **175:**351.
Horne, A. J., and C. R. Goldman, 1972, Nitrogen fixation in Clear Lake, California. I. Seasonal variations and the role of heterocysts, *Limnol. Oceanogr.* **17:**678.
Horne, A. J., and Viner, A. B., 1971, Nitrogen fixation and its significance in tropical Lake George, Uganda, *Nature (London)* **232:**417.
Hutchinson, G. E., 1944, Limnological studies in Connecticut. VII. A critical examination of the supposed relationship between phytoplankton periodicity and chemical changes in lake waters, *Ecology* **25:**2–26.
Imsande, J., 1981, Exchange of metabolites and energy between legume and *Rhizobium,* in: *Biology of the Rhizobiaceae* (K. L. Giles and A. G. Atherly, eds.), Academic Press, New York, pp. 179–190.
Jaiyebo, E. O., and Moore, A. W., 1963, Soil nitrogen accretion in a tropical rain forest, *Nature (London)* **197:**317–318.

Jensen, H. L., 1940, Contributions to the nitrogen economy of Australian wheat soils, with particular reference to New South Wales, *Proc. Linn. Soc. N. S. Wales.* **65:**1–122.

Jensen, H. L., 1965, Nonsymbiotic nitrogen fixation, in: *Soil Nitrogen,* (W. V. Bartholomew and F. E. Clark, eds.), Amer. Soc. of Agronomy, Inc., Madison, Wisconsin, pp. 436–480.

Jensen, H. L., 1981, Heterotrophic micro-organisms, in: *Nitrogen Fixation, Volume I, Ecology* (W. J. Broughton, ed.), Clarendon Press, Oxford, pp. 30–56.

Jensen, H. L., Peterson, E. J., De, P. K., and Bhattachrya, R., 1960, A new nitrogen-fixing bacterium: *Derxia gummosa* nov. gen. nov. spec., *Arch. Mikrobiol.* **36:**182–195.

Jodin, M., 1862, Du rôle physiologique de l'azote (etc.), *C. R. Acad. Sci. (Paris)* **55:**612–615.

Jones, W. N., and Smith, M. L., 1929, On the fixation of atmospheric nitrogen by *Phoma radicis callunae,* including a new method for investigating nitrogen-fixation in microorganisms, *Br. J. Exp. Biol.* **6:**167–189.

Kamen, M. D., and Gest, H., 1949, Evidence for a nitrogenase system in the photosynthetic bacterium *Rhodospirillum rubrum, Science* **109:**560.

Keister, D. L., 1975, Acetylene reduction by pure cultures of rhizobia, *J. Bacteriol.* **123:**1265–1268.

Kjeldahl, J., 1883, Neue Methode zur Bestimmung des Stickstoffs in organischen Korpern, *Z. anal. Chem.* **22:**366–382.

Knowles, R., 1977, The significance of asymbiotic dinitrogen fixation by bacteria, in: *A Treatise on Dinitrogen Fixation, Section IV: Agronomy and Ecology* (R. W. F. Hardy and A. H. Gibson, eds.), John Wiley and Sons, New York, pp. 33–84.

Koch, B., and Evans, H. J., 1966, Reduction of acetylene to ethylene by soybean root nodules, *Plant Physiol.* **41:**1748–1750.

Koch, B. L., and Oya, J., 1974, Nonsymbiotic nitrogen fixation in some Hawaiian pasture soils, *Soil Biol. Biochem.* **6:**363.

Kohnke, H., 1941, The black alder as a pioneer plant on sand dunes and eroded land, *J. Forestry* **39:**333–334.

Kondorosi, A., and Johnston, A. W. B., 1981, The genetics of *Rhizobium,* in: *Biology of the Rhizobiaceae* (K. L. Giles and A. G. Atherly, eds.), Academic Press, New York, pp. 191–224.

Kostytschew, S., and Ryskaltschuk, A., 1925, Les produits de la fixation de l'azote atmospherique par l'*Azotobacter agile, C. R. Acad. Sci. (Paris)* **180:**2070–2072.

Kossowitsch, P., 1894, Durch welche Organe nehman die Leguminosen den freien stickstoff auf? *Bot. Z.* **53:**199–202.

Krapovickas, A., 1969, The origin, variability and spread of the groundnut (*Arachis hypogaea*), in: *The Domestication and Exploitation of Plants and Animals* (P. J. Ucko and G. W. Dimbleby, eds.), Gerald Duckworth and Co., London, pp. 427–442.

Krasil'nikov, N. A., 1958, *Soil Microorganisms and Higher Plants,* Office of Technical Services, U. S. Dept. Commerce, Washington.

Kubo, H., 1939, Uber hamoprotein aus den Wurzelknollchen von Leguminosen, *Acta Phytochim. (Japan)* **11:**195–200.

Kurz, W. G. W., and LaRue, T. A., 1975, Nitrogenase activity in rhizobia in absence of plant host, *Nature (London)* **256:**407–409.

Kuykendall, L. D., 1981, Mutants of *Rhizobium* that are altered in legume interaction and nitrogen fixation, in: *Biology of the Rhizobiaceae* (K. L. Giles and A. G. Atherly, eds.), Academic Press, New York, pp. 299–310.

Lachmann, J., 1858, Ueber Knollchen der Leguminosen, *Landw. Mitt. Atschr. K. Lehranst. Vers. Sta. Poppelsdorf (Bonn),* p. 37.

Landsteiner, K., and Raubitschek, H., 1908, Beobachtungen über Hamolyse und Hamagglutination, *Zentralbl. Bakteriol.* **45:**660.

Laurent, E., 1890, Sur le microbe des nodosites des Légumineuses, *C. R. Acad. Sci. (Paris)* **111:**754–756.

Lawes, B., and Gilbert, J., 1851, Agricultural chemistry, especially in relation to the mineral theory of Baron Liebig, in: *Rothamsted Memoirs on Agricultural Chemistry and Physiology, Volume I,* W. Clowes and Sons, Ltd., London (1893), pp. 1–4.

Lawes, B., and Gilbert, J., 1863, On the amounts of, and methods of estimating ammonia and nitric acid in rain water, in: *Rothamsted Memoirs on Agricultural Chemistry and Physiology, Volume I,* W. Clowes and Sons, London (1893), 1–15.

Lawes, J. B., and Gilbert, J. H., 1889, On the present position of the question of the sources of the nitrogen of vegetation, with some new results, and preliminary notice of new lines of investigation, *Trans. Roy. Soc. (London)* **180B:**1–107.

Lawes, J. B., and Gilbert, J. H., 1891, The sources of the nitrogen of our leguminous crops, *J. Roy. Agr. Soc. England* **2**(Ser. 3):657–702.

Lawes, B., Gilbert, J., and Pugh, E., 1861, On the sources of nitrogen of vegetation; with special reference to the question whether plants assimilate free or uncombined nitrogen. *Trans. Roy. Soc. (London)* **151B:**431–577.

Lechevalier, M., and Lechevalier, H., 1979, The taxonomic position of the actinomycetic endophytes, in: *Symbiotic Nitrogen Fixation in the Management of Temperate Forests,* (J. D. Gordon, C. T. Wheeler, and D. A. Perry, eds.), Oregon State University, Corvallis, pp. 111–123.

Lepo, J. E., Hanus, F. J., and Evans, H. J., 1980, Further studies on the chemoautotrophic growth of hydrogen uptake positive strains of *R. japonicum, J. Bacteriol.* **141:**664–670.

Liebig, J., 1840, *Organic Chemistry in its Applications to Agriculture and Physiology,* Taylor, London.

Liebig, J., 1855, *Principles of Agricultural Chemistry with Special Reference to the Late Researches Made in England,* Wiley, New York.

Liebig, J., 1863, *The Natural Laws of Husbandry,* D. Appleton and Co., New York.

Lindstrom, E. S., Burris, R. H., and Wilson, P. W., 1949, Nitrogen fixation by photosynthetic bacteria, *J. Bacteriol.* **58:**313–316.

Lindstrom, E. S., Lewis, S. M., and Pinsky, M. I., 1951, Nitrogen fixation and hydrogenase in various bacterial species, *J. Bacteriol.* **61:**481–487.

Lindstrom, E. S., Tove, R. R., and P. W. Wilson, 1950, Nitrogen fixation by the green and purple sulfur bacteria, *Science* **112:**197–198.

Lochhead, A. G., 1952, Soil microbiology, *Annu. Rev. Microbiol.* **6:**185–206.

Lochhead, A. G., and Thexton, R. H., 1936, A four-year quantitative study of nitrogen-fixing bacteria in soils of different fertilizer treatments, *Can. J. Res.* **14C:**166–177.

Lockshin, A., and Burris, R. H., 1965, Inhibitors of nitrogen fixation in extracts from *Clostridium pasteurianum, Biochim. Biophys. Acta* **111:**1–10.

Lohnis, M. P., 1930, Can *Bacterium radicicola* assimilate nitrogen in the absence of the host plant? *Soil Sci.* **29:**37–57.

Madigan, M., Cox, S. S., and Stegeman, R. A., 1984, Nitrogen fixation and nitrogenase activities in members of the family *Rhodospirillaceae, J. Bacteriol.* **157:**73–78.

Mague, T. H., 1977, Ecological aspects of dinitrogen fixation by blue-green algae, in: *A Treatise on Dinitrogen Fixation, Section IV: Agronomy and Ecology,* John Wiley and Sons, New York, pp. 85–140.

Mague, T. H., Weare, N. W., and Holm-Hansen, O., 1974, Nitrogen fixation in the North Pacific Ocean, *Mar. Biol.* **24:**109–119.

Martinez, L., Silver, M. W., King, J. M., and Alldredge, A. L., 1983, Nitrogen fixation by floating diatom mats: a source of new nitrogen to oligotrophic ocean waters, *Science* **221:** 152–154.

McComb, J. A., Elliot, J., and Dilworth, M. J., 1975, Acetylene reduction by *Rhizobium* in pure culture, *Nature (London)* **256:**409–410.

McNary, J. E., and Burris, R. H., 1962, Energy requirements for nitrogen fixation by cell-free preparations from *Clostridium pasteurianum, Biochim. Biophys. Acta* **111:**1–10.

Miehe, H., 1917, Weitere Untersuchungen uber die Bakteriensymbiose bei *Ardisia crispa.* II. Die Pflanzen ohne Bakterien, *Jahrb. Wiss. Bot.* **58:**29–65.

Mishustin, E. N., and Shil'nikova, V. K., 1971, *Biological Fixation of Atmospheric Nitrogen,* Pennsylvania State University Press, University Park.

Moore, A. V., 1963, Nitrogen fixation in a latosolic soil under grass, *Plant Soil* **19:**127–138.

Morrison, T. M., and English, D. A., 1967, The significance of mycorrhizal nodules of *Agathis australis, New Phytol.* **66:**245–250.

Mortenson, L. E., 1964, Ferredoxin and ATP, requirements for nitrogen fixation in cell-free extracts of *Clostridium pasteurianum, Proc. Natl. Acad. Sci. USA* **52:**272–279.

Mortenson, L. E., 1966, Components of cell-free extracts of *Clostridium pasteurianum* required for ATP-dependent H_2 evolution from dithionite and for N_2 fixation, *Biochim. Biophys. Acta* **127:**18–25.

Mowry, H., 1933, Symbiotic nitrogen fixation in the genus *Casuarina, Soil Sci.* **36:**409–421.

Mozen, M. M., and Burris, R. H., 1954, The incorporation of ^{15}N-labelled nitrous oxide by nitrogen-fixing agents. *Biochim. Biophys. Acta* **14:**577–578.

Nash, L. K., 1959, *Plants and the Atmosphere,* Harvard University Press, Cambridge.

Nelson, A. D., Barber, L. E., Tjepkema, J., Russell, S. A., Powelson, R., Evans, H. J., and Seidler, R. J., 1976, Nitrogen fixation associated with grasses in Oregon, *Can. J. Microbiol.* **22:**523–530.

Nobbe, F., and Hiltner, L., 1896, Bodenimpfung fur Anbau von Leguminosen, *Sachs. Landw. Ztschr.* **44:**90–92.

Nobbe, F., and Hiltner, L., 1899, Die endotrophe Mycorhiza von *Podocarpus* und ihre physiologische Bedeutung, *Landw. Vers. Sta.* **51:**241–245.

Nobbe, F., Schmid, E., Hiltner, L., and Hotter, E., 1892, Ueber die physiologische Bedeutung der Wurzelknollchen von *Elaeagnus angustifolius, Landw. Vers. Sta.* **41:**138–140.

Normand, P., and Lalonde, M., 1982, Evaluation of *Frankia* strains isolated from provenances of two *Alnus* species, *Can. J. Microbiol.* **28:**1133–1142.

Nuti, M. P., Ledeboer, A. M., Lipidi, A. A., and Schilperoort, R. A., 1977, Large plasmids in different *Rhizobium* species, *J. Gen. Microbiol.* **100:**241–256.

Oes, A., 1913, Uber die Assimilation des freien Stickstoffs durch *Azolla, Z. Botan.* **5:**145–163.

Okon, Y., Heytler, P. G., and Hardy, R. W. F., 1983, N_2 fixation by *Azospirillum brasilense* and its incorporation into host *Setaria italica, Appl. Env. Microbiol.* **46:**694–697.

Ooman, H. A. P. C., 1972, Distribution of nitrogen and composition of nitrogen compounds in food, urine and faeces in habitual consumers of sweet potato and taro, *Nutr. Metabol.* **14:**65–82.

Orr, M. Y., 1923, The leaf glands of *Dioscorea macroura* Harms., *Notes from the Roy. Bot. Gard., Edinburgh* **14:**57–72.

Pagan, J. D., Child, J. J., Scowcroft, W. R., and Gibson, A. H., 1975, Nitrogen fixation by *Rhizobium* cultured on a defined medium, *Nature (London)* **256:**406–407.

Parker, C. A., 1957, Non-symbiotic nitrogen fixing bacteria in soil. III. Total nitrogen changes in a field soil, *J. Soil Sci.* **8:**48–59.

Pasteur, L., 1860, De l'origine des ferments. Nouvelles experiences relatives aux générations dites spontanées, *C. R. Acad. Sci. (Paris)* **50:**849–854.

Pasteur, L., 1861, Mémoire sur les corpuscles organisés qui existent dans l'atmosphère. Examen de la doctrine des générations spontanées, *Ann. Sci. Nat.* **16** (Ser. 4):5–98.

Pasteur, L., 1862, Etudes sur les mycodermes. Rôles de ces plantes dans la fermentation acétique, *C. R. Acad. Sci. (Paris)* **44:**265–270.

Paul, E. A., Myers, R. J. K., and Rice, W. A., 1971, Nitrogen fixation in grassland and associated cultivated ecosystems, *Plant Soil Special Vol.,* pp. 495–507.

Pearsall, W. H., 1932, Phytoplankton in the English lakes. II. The composition of the phytoplankton in relation to dissolved substances, *J. Ecol.* **20:**241–262.

Peklo, J., 1946, Symbiosis of *Azotobacter* with insects, *Nature (London)* **158:**795–796.

Peligot, E. M., 1847, Sur un procédé propre à déterminer d'une manière sensible rapide la quantité d'azote contenue dans les substances organique, *J. Pharm. Chim.* **11** (Ser. 3):334–337.

Pengra, R. M., and Wilson, P. W., 1958, Physiology of nitrogen fixation by *Aerobacter aerogenes, J. Bacteriol.* **75:**21–25.

Pengra, R. M., and Wilson, P. W., 1959, Trace metal requirements of *Aerobacter aerogenes* for assimilation of molecular nitrogen, *Proc. Soc. Exp. Biol. Med.* **100:**436–439.

Phillips, D. H., and Johnson, M. J., 1961, Aeration in fermentations, *J. Biochem. Microbiol. Technol. Eng.* **3:**277–309.

Plotho, O. von, 1941, Die Synthese der Knollchen an den Wurzeln der Erle, *Arch. Microbiol.* **12:**1–18.

Pohlman, G. G., 1931, Nitrogen fixation by *Rhizobium meliloti* and *Rhizobium japonicum*, *J. Am. Soc. Agron.* **23:**70–77.

Pommer, E., 1959, Uber die Isolierurig dew Endophyten aus den Wurzelknollchen *Alnus glutinosa* Gaertn. und Uber erfolgreiche Re-Infektionsversuche, *Ber. Deutsch Bot. Gesell.* **72:**138–150.

Postgate, J. R., 1982, *The Fundamentals of Nitrogen Fixation,* Cambridge University Press, New York.

Prakash, R. K., Schilperoort, R. A., and Nuti, M. P., 1981, Large plasmids of fast-growing rhizobia: Homology studies and location of structural nitrogen fixation (nif) genes, *J. Bacteriol.* **145:**1129–1136.

Prantl, K., 1889, Die Assimilation freien Stickstoffs und der Parasitismus von Nostoc, *Hedwigia* **28:**135–136.

Prazmowski, A., 1890, Die Wurzelknollchen der Erbse, *Landw. Vers. Sta.* **37:**161–238.

Pringsheim, E. G., 1913, Kulturversuche mit chlorophyllfuhrenden Mikroorganismen, III. Mitteilung. Zur Physiologie der Schizophyceen, *Beit. Biol. Pflanzen* **12:**49–108.

Rayner, M. C., 1915, Obligate symbiosis in *Calluna vulgaris, Ann. Botany* **29:**97–123.

Rayner, M. C., 1922, Nitrogen fixation in the Ericaceae, *Bot. Gaz.* **73:**226–235.

Remy, E., 1909, Untersuchungen uber die Stickstoffsammlungsvorgange in ihrer Beiziehung zum Bodenklima, *Zentralbl. Bakteriol (II)* **22:**561–651.

Richards, B. N., and Voigt, G. K., 1964, Role of mycorrhiza in nitrogen fixation, *Nature* **201:**310–311.

Rinaudo, G., Balandreau, J., and Dommergues, Y., 1971, Algal and bacterial non-symbiotic nitrogen fixation in paddy soils, *Plant Soil, Special Vol.,* pp. 471–479.

Ruehle, J. L., and Marx, D. H., 1979, Fiber, fuel, and fungal symbionts, *Science* **206:**419–422.

Ruinen, J., 1975, Nitrogen fixation in the phyllosphere, in: *Nitrogen Fixation by Free-Living Microorganisms* (W. D. P. Stewart, ed.), Cambridge University Press, New York, pp. 85–100.

Ruvkun, G. B., and Ausubel, F. M., 1980, Interspecies homology of nitrogenase genes, *Proc. Natl. Acad. Sci. USA* **77:**191–195.

Rychert, R. C., and Skujins, J., 1974, Nitrogen fixation by blue-green algae-lichen crusts in the Great Basin desert, *Soil Sci. Soc. Am. Proc.* **38:**768–771.

Schloesing, J. T., 1888, Sur les relations de l'azote gazeux de l'atmosphérique avec la terre végétale. *C. R. Acad. Sci. (Paris)* **106:**898–902; 982–987; **107:**290–296.

Schloesing, A., and Laurent, E., 1890, Sur la fixation de l'azote libre par les Légumineuses, *C. R. Acad. Sci. (Paris)* **111:**750–753.

Schloesing, T., and Laurent, E., 1892, Recherches sur la fixation de l'azote libre par les plantes, *Ann. Inst. Pasteur* **6:**65–115.

Schloesing, J. T., and Muntz, A., 1877, Sur la nitrification par les ferments organiques, *C. R. Acad. Sci. (Paris)* **84:**301–303.

Schollhorn, R., and Burris, R. H., 1966, Study of intermediates in nitrogen fixation, *Fed. Proc.* **25:**710.

Schultz, J. E., and Breznak, J. A., 1978, Heterotrophic bacteria present in hindguts of wood-eating termites (*Reticulitermes flavipes* Kollar), *Appl. Environ. Microbiol.* **35:**930–936.

Schultz-Lupitz, A., 1881, Reinertrage auf leichtem Boden, ein Wort der Erfahrung, zur Abwehr der wirtschaftlichen Noth, *Landw. Jahrb.* **10:**777–848.

Shields, C. M., Mitchell, C., and Drouet, F., 1957, Alga- and lichen stabilized surface crusts as soil nitrogen sources, *Am. J. Bot.* **44:**489–498.

Silver, W. S., 1977, Foliar associations with higher plants, in: *A Treatise on Dinitrogen Fixation* (R. W. F. Hardy and W. S. Silver, eds.), John Wiley and Sons, New York, pp. 153–184.

Silvester, W. B., 1977, Dinitrogen fixation by plant associations excluding legumes, in: *A Treatise on Dinitrogen Fixation, Section IV: Agronomy and Ecology* (R. W. F. Hardy and A. H. Gibson, ed.), John Wiley and Sons, New York, pp. 141–190.

Silvester, W. B., and Bennett, K. J., 1973, Acetylene reduction by roots and associated soil of New Zealand conifers, *Soil Biol. Biochem.* **5:**171–179.

Singh, R. N., 1942, The fixation of elemental nitrogen by some of the commonest blue-green algae from paddy field soils of the United Provinces and Bihar, *Indian J. Agr. Sci.* **12:**743–756.

Singh, R. N., 1961, *Role of Blue-Green Algae in Nitrogen Economy of Indian Agriculture,* Indian Council of Agricultural Research, New Delhi, 175 pp.

Skinner, C. E., 1928, The fixation of nitrogen by *Bacterium aerogenes* and related species, *Soil Sci.* **25:**195–205.

Sloger, C., and Silver, W. S., 1967, Biological reductions catalyzed by symbiotic nitrogen-fixing tissues, *Bacteriol. Proc.* p. 12.

Smartt, J., 1969, Evolution of American *Phaseolus* beans under domestication, in: *The Domestication and Exploitation of Plants and Animals* (P. J. Ucko and G. W. Dimbleby, eds.), Gerald Duckworth and Co., London, pp. 451–462.

Smith, V. H., 1983, Low nitrogen to phosphorous ratios favor dominance by blue-green algae in lake phytoplankton, *Science* **221:**669–671.

Spratt, E. R., 1912, The formation and physiological significance of root nodules in the Podocarpineae, *Ann. Bot. (London)* **26:**801–814.

Starkey, R. L., 1938, Some influences of the development of higher plants upon the microorganisms in the soil. V. *Soil Science* **45:**207–249.

Stewart, W. D. P., 1965, Nitrogen turnover in marine and brackish habitats, *Ann. Bot.* **29:**229.

Stewart, W. D. P., 1966, *Nitrogen Fixation in Plants,* Athlone Press, London.

Stewart, W. D. P., 1967, Nitrogen-fixing plants, *Science* **158:**1426–1432.

Stewart, W. D. P., and Lex, M., 1970, Nitrogenase activity in the blue-green algae *Plectonema boryanum* strain 594, *Arch. Mikrobiol.* **73:**250–260.

Stewart, W. D. P., Fitzgerald, G. P., and Burris, R. H., 1967, *In situ* studies on N_2 fixation, using the acetylene reduction technique, *Proc. Natl. Acad. Sci. USA* **58:**2071–2078.

Stewart, W. D. P., Haystead, A., and Pearson, H. W., 1969, Nitrogenase activity in heterocysts of blue-green algae. *Nature (London)* **224:**226–228.

Stewart, W. D. P., Mague, T., Fitzgerald, G. P., and Burris, R. H., 1971, Nitrogenase activity in Wisconsin lakes of differing degrees of eutrophication, *New Phytol.* **70:**497–502.

Stewart, W. D. P., Sampaio, M. J., Isichei, A. O., and Sylvester-Bradley, R.,1978, Nitrogen fixation by soil algae of temperate and tropical soils, in: *Limitations and Potentials for Biological Nitrogen Fixation in the Tropics* (J. Dobereiner, R. H. Burris, and A. Hollaender, eds.), Plenum Press, New York, pp. 41–63.

Stutz, R. C., and Bliss, L. C., 1975, Nitrogen fixation in soils of Truelove Lowland, Devon Island, Northwest Territories, *Can. J. Bot.* **53:**1387–1399.

Sutton, W. D., Pankhurst, C. E., and Craig, A. S., 1981, The *Rhizobium* bacteroid state, in: *Biology of the Rhizobiaceae* (K. L. Giles and A. G. Atherly, eds.), Academic Press, New York, 149–171.

Tchan, Y. T., and Beadle, N. C. W., 1955, Nitrogen economy in semiarid plant communities. Part II. The non-symbiotic nitrogen-fixing organisms, *Proc. Linnean Soc. New South Wales* **80:**97–104.

Ternetz, C., 1907, Über die assimilation des atmospharischen stickstoffs durch Pilze, *Jahrb. Wiss. Botany* **44:**353–408.

Theophrastus, 1916, *Enquiry into Plants, trans. by Sir Arthur Hort,* Vol. II, W. Heinemann, London.

Tjepkema, J. D., and Evans, H. J., 1975, Nitrogen fixation by free-living rhizobia in a defined liquid medium, *Biochem. Biophys. Res. Commun.* **65:**625–628.

Tjepkema, J. D., and van Berkum, P., 1977, Acetylene reduction by soil cores of maize and *Sorghum* in Brazil, *Appl. Env. Microbiol.* **33:**626–629.

Torrey, J. G., and Tjepkema, J. D., 1979, Symbiotic nitrogen fixation in actinomycete-nodulated plants, Preface, *Bot. Gaz. (Chicago)* **140** (Suppl.):i–ii.

Trinick, M. J., 1973, Symbiosis between *Rhizobium* and the non-legume *Trema aspera, Nature (London)* **244:**459–460.

Truchet, G., Rosenberg, C., Vasse, J., Juillot, J.-S., Camut, S., and Denarie, J., 1984, Transfer of *Rhizobium meliloti* pSym genes into *Agrobacterium* tumefaciens: Host-specific nodulation by a typical infection, *J. Bacteriol.* **157:**134–142.

Vénkatamaran, G. S., 1981, Blue-green algae: a possible remedy to nitrogen scarcity, *Curr. Sci.* **50:**253–256.

Verma, D. P., and Long, S., 1983, The molecular biology of *Rhizobium*–legume symbiosis, *Int. Rev. Cytology (Suppl.)* **14:**211–245.

Ville, G., 1855, Rapport sur un travail de M. Georges Ville, dont l'objet est de prouver que le gaz azote de l'air s'assimilé aux végétaux. Commission composée de M. M. Dumas, Regnault, Payen, Decaisne, Peligot, Chevreul (rapporteur), *C. R. Acad. Sci. (Paris)* **41:**757–778.

Virtanen, A. I., 1945, Symbiotic nitrogen fixation, *Nature* **155:**747–748.

Virtanen, A. I., and Laine, T., 1938, Biological synthesis of amino acids from atmospheric nitrogen, *Nature* **141:**748.

Virtanen, A. I., and Laine, T., 1946, Red, brown and green pigments in leguminous root nodules, *Nature* **157:**25–29.

Virtanen, A. I., Moisio, T., Allison, R. M., and Burris, R. H., 1954, Fixation of nitrogen by excised nodules of the alder, *Acta Chem. Scand.* **8:**1730–1731.

Waksman, S. A., 1932, *Principles of Soil Microbiology,* Williams and Wilkins, Baltimore.

Ward, H. M., 1887, On the tubercular swellings on the roots of *Vicia faba, Phil. Trans. Roy. Soc. London Ser.* **178:**539–562.

Ward, H. M., 1895, *Sci. Prag.* **3:**251–271.

Warington, R., 1878–1891, On nitrification, I–IV, *J. Chem. Soc.,* **33:**44–51; **35:**429–456; **45:**637–672; **59:**484–529.

Warren, J. A., 1910, Additional notes on the number and distribution of native legumes in Nebraska and Kansas, Circ. U.S.D.A. Bureau Plant Ind. **70:**1–8.

Watanabe, I., and Brotonegoro, S., 1981, Paddy fields, in: *Nitrogen Fixation, Volume I, Ecology* (W. J. Broughton, ed.), Clarendon Press, Oxford, pp. 242–263.

Waterbury, J. B., Calloway, C. B., and Turner, R. D., 1983, A cellulolytic nitrogen-fixing bacterium cultured from the gland of Deshayes in shipworms (Bivalvai:Teredinidae), *Science* **221:**1401–1402.

Waughman, G. J., and Bellamy, D. J., 1972, Acetylene reduction in surface peat, *Oikos* **23:**353–358.

Waughman, G. J., French, J. R. J., and Jones, K., 1981, Nitrogen fixation in some terrestrial environments, in: *Nitrogen Fixation Volume I, Ecology* (W. J. Broughton, ed.), Clarendon Press, Oxford, pp. 135–192.

Way, J. T., 1856, On the composition of the waters of land drainage and of rain, *J. Roy. Agr. Soc. England (London)* **17:**123–162.

Wilkinson, C. R., and Fay, P., 1979, Nitrogen fixation in coral reef sponges with symbiotic cyanobacteria, *Nature (London)* **279:**527–529.

Wilson, E. H., 1920, A phytogeographical sketch of the ligneous flora of Formosa, *J. Arnold Arboretum* **2:**25–41.

Wilson, P. W., 1940, *The Biochemistry of Symbiotic Nitrogen Fixation,* University of Wisconsin Press, Madison.

Wilson, P. W., 1957, On the sources of nitrogen of vegetation etc., *Bacteriol. Rev.* **21:**215–226.

Wilson, P. W., 1958, Asymbiotic nitrogen fixation, in: *Handbuch der Pflanzenphysiologie, Vol. 8, Nitrogen Metabolism* (W. Ruhland, ed.), Springer-Verlag, Berlin, pp. 9–47.

Wilson, P. W., 1969, First steps in biological nitrogen fixation, *Proc. R. Soc. B* **172:**319–325.

Wilson, P. W., and Umbreit, W. W., 1937, Mechanism of symbiotic nitrogen fixation. III. Hydrogen as a specific inhibitor, *Arch. Mikrobiol.* **8:**440–457.

Winogradsky, S. N., 1893, Sur l'assimilation de l'azote gazeux de l'atmosphère par les microbes, *C. R. Acad. Sci. (Paris)* **116:**1385–1388.

Winogradsky, S., 1930, Sur la synthèse de l'ammoniac par les *Azotobacter* du sol, *C. R. Acad. Sci. (Paris)* **190:**661–665.

Winogradsky, S., 1932, Sur la synthèse de l'ammoniac par les *Azotobacter* du sol, *Ann. Inst. Pasteur* **48:**269.

Winter, G., 1935, Uber die Assimilation des Luftstickstoff durch endophytische Blaualgen, *Beitr. Biol. Pflanzaen* **83:**295–335.

Witty, J. F., Day, J. M., and Dart, P. J., 1977, The nitrogen economy of the Broadbalk experiments: II. Biological nitrogen fixation, in: *Rothamstead Experimental Station, Report for 1976, Part 2,* Harpenden: Lawes Agricultural Trust, pp. 111–118.

Wolff, H., 1927, *Jahrb. Wiss. Botan.* **66:**1–34.
Woronin, M., 1866, Ueber die bei der Schwarzerle (*Alnus glutinosa*) und der gewohnlichen Gartenlupine (*Lupinus mutabilis*) auftretenden Wurzelanschwellungen, *Mem. Acad. Imp. Sci., St. Petersbourgh* **10** (Ser. 7)**:**1–13.
Woronine, M., 1867, Observations sur certaines excroissances que présentent les racines de l'aune et du lupin des jardins, *Ann. Sci. Nat. Bot.* **7** (Ser. 5)**:**73–86.
Wyatt, J. T., and Silvey, J. K. G., 1969, Nitrogen fixation by *Gloeocapsa, Science* **165:**908–909.
Yates, M. G., and Jones, C. W., 1974, Respiration and nitrogen fixation in *Azotobacter, Adv. Microb. Physiol.* **11:**97–135.
Youngken, H. W., 1919, The comparative morphology, taxonomy and distribution of the *Myricaceae* of the Eastern United States, *Contrib. Bot. Lab. Univ. Pa.* **4:**339–400.
Zavitkovski, J., and Newton, M., 1968, Effect of organic matter and combined nitrogen on nodulation and nitrogen fixation in red alder, in: *Biology of Alder*, (J. M. Trappe, J. F. Franklin, R. F. Tarrant, and G. M. Hansen, eds.), Pacific Northwest Forest and Range Experiment Station, Forest Service USDA, Portland, pp. 209–223.
Zelitch, I., Rosenblum, E. D., Burris, R. H., and Wilson, P. W., 1951, Isolation of the key intermediate in biological nitrogen fixation by *Clostridium pasteurianum, J. Biol. Chem.* **191:**295–298.
Ziegenspeck, H., 1929, Die cytologischen Vorgange in den Knollchen von *Hippophae rhamnoides* (sanddorn) und *Alnus glutinosa* (Erle), *Ber. Dtsch. Bot. Gea.* **47:**50–58.
Zohary, D., and Hopf, M., 1973, Domestication of pulses in the Old World, *Science* **182:**887–894.

7

THE POSITION OF BACTERIA AND THEIR PRODUCTS IN FOOD WEBS

Henry L. Ehrlich

INTRODUCTION

When examining a drop of pond water under the microscope, an observer may see in it different kinds of bacteria, algae, protozoa, and even occasional invertebrates, such as rotifers, cladocerans, copepods, nematodes, turbellarians, and others. The presence of these very different organisms together in the pond water sample is not mere coincidence. They are the members of a food web, in which each organism, its secretions and excretions, or its dead remains represent food for one or more of the other organisms in the assemblage.

CONCEPTS

The concept of the *food web* can be traced back at least as far as 1927 when Elton used it in his book on animal ecology (pp. 55–56). He called it *food cycle* instead of food web and related it to feeding interdependencies among members of a biological community. In any such community that includes many different types of food interdependencies, one or more *food chains* (Elton, 1927, pp. 55–56; see also Lapedes, 1974) exist, which consist of several *trophic levels* (Lindeman, 1942, according to Dice, 1952). Food chains may be divided into two categories, *assimilatory* and *dissimilatory* (Figure 1). In assimilatory food chains, the emphasis is on growth and reproduction of organisms at each trophic level with direct dependence on the organisms of the preceding trophic level. In dissimilatory food chains, the emphasis is on stepwise decomposition of complex organic molecules to simpler organic molecules by different organisms at successive trophic levels. The ultimate chem-

Henry L. Ehrlich • Department of Biology, Rensselaer Polytechnic Institute, Troy, New York 12181.

ASSIMILATORY FOOD CHAIN	DISSIMILATORY FOOD CHAIN
secondary feeder	secondary decomposer
↑	↑
primary feeder	primary decomposer
↑	↑
primary producer	depolymerizer

FIGURE 1. Food chains. Primary feeders may consume live primary producers or their photosynthate. Secondary feeders may consume live primary producers or their secretions or excretions. Depolymerizers and primary decomposers may be the same or different organisms. Secondary decomposers mineralize the metabolic excretions of primary decomposers. Primary decomposers may be sufficient to mineralize completely some forms of organic matter.

ical products in dissimilatory food chains may be inorganic, in which case the total dissimilatory process is called *mineralization.* The trophic levels in assimilatory food chains were first named by Thienemann (1926) according to Dice (1952). They include *primary producers,* which transform fully oxidized carbon (CO_2) into reduced carbon (organic matter) by *photolithotrophy* (photosynthetic autotrophy) or *chemolithotrophy* (chemosynthetic autotrophy). The primary producers, in turn, may be preyed upon by *primary feeders,* and they, in turn, may be preyed upon by *secondary feeders* (Alexander, 1971). Some of the primary producers may nourish primary feeders by secretions or with their dead remains, and the primary feeders may similarly nourish secondary feeders by secretions, excretions, or their dead remains.

The trophic levels in dissimilatory food chains have not been specifically named before. In the present discussion the organisms at the different levels will be identified as depolymerizers and primary and secondary decomposers, respectively (Figure 1). The depolymerizers and primary decomposers may be the same organisms, but their activities at the two trophic levels are different. Depolymerizers produce exoenzymes which catalyze hydrolysis of polymers outside the cells, whereas decomposers metabolize oligomers and monomers, which may have resulted from depolymerization or are products of metabolism of other decomposers, primary or secondary feeders, or even primary producers (Figure 2). Both depolymerizers and decomposers may function aerobically or anaerobically. Under anaerobic conditions, at least two trophic levels of decomposition may be required to achieve complete mineralization, whereas under aerobic conditions, only one level of decomposition may be required.

Food chains define a hierarchy of trophic levels, i.e., who feeds on whom, or who is fed by whom in a biological community. The term food web rec-

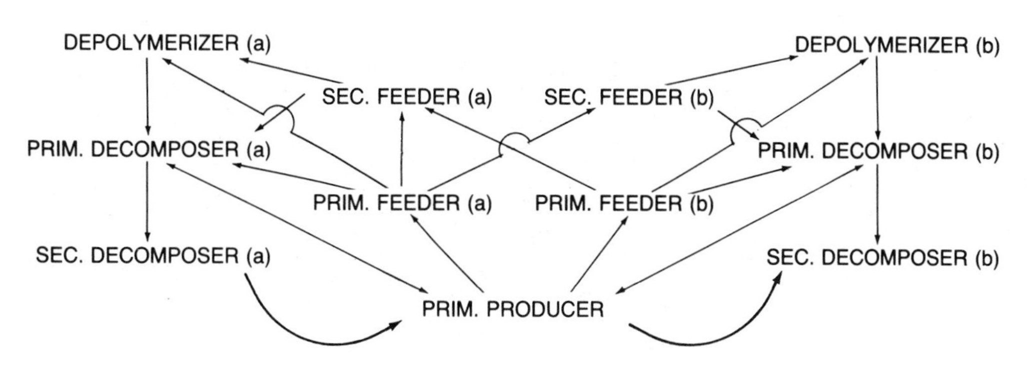

FIGURE 2. Food web or food cycle. Arrows indicate flow of nutrients.

ognizes that food chains may branch, i.e., an organism at one trophic level may be the source of food for more than one type of organism at the next trophic level, and indeed, the concept recognizes that a primary or secondary feeder may nourish primary producers by the CO_2 it produces in its respiratory metabolism, or by the organic matter it secretes or excretes and which is then mineralized by decomposers (Figure 2). A food chain, whether assimilatory or dissimilatory, traces the main line of feeding hierarchy of a group of organisms in a biological community. In a complex community many such food chains exist. A food web traces the interrelationships of these different food chains.

Elton (1927) also recognized a *pyramid of numbers* in biological communities by which he meant that the number of individuals at successive trophic levels in an assimilatory food chain decreases markedly. This concept has been confirmed many times and is usually attributed to energy limitations (Figure 3). Biochemistry has established that to assimilate food constituents, an organism has to do work, i.e., it has to expend chemical energy. Primary and secondary feeders derive this energy from a portion of the total food they consume. Furthermore, they are only able to use part of all the potentially available energy in the food they consume for assimilation (Figure 3). The rest of the energy is lost as heat, and in a few organisms also as light. Because only a portion of the total food consumed is assimilated, the amount of food available at each successive trophic level becomes less than at the preceding level. Consequently, the number of individuals it can support at each trophic level becomes progressively smaller.

Although thermodynamic inefficiency has been the most widely accepted explanation for the pyramid of numbers, Pimm and Lawton (1977, 1978) have suggested that it is population dynamics and not ecological energetics that accounts for it. They have argued that if food chains were primarily limited by available energy, they should be longer than they are. Pimm and Lawton developed a mathematical model to support their contention that population dynamics is the controlling factor which determines the length of

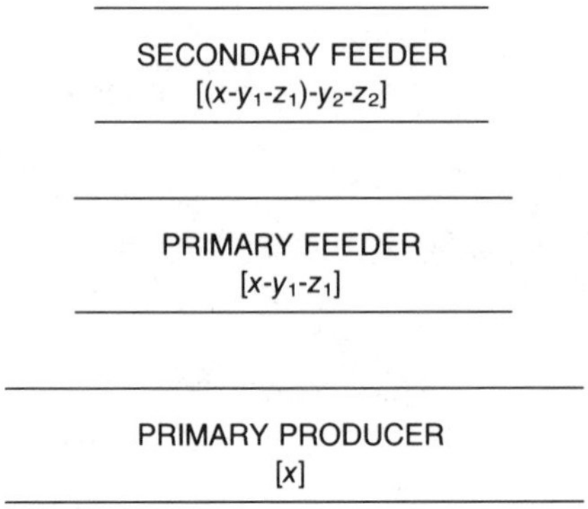

SECONDARY FEEDER
$[(x-y_1-z_1)-y_2-z_2]$

PRIMARY FEEDER
$[x-y_1-z_1]$

PRIMARY PRODUCER
$[x]$

FIGURE 3. Pyramid of numbers. Symbols: x denotes total biologically available energy at the level of primary producers; y_1 denotes the portion of biologically available energy in x consumed by primary feeders for assimilation (endogonic reactions); z_1 denotes the portion of biologically available energy in x lost as heat by primary feeders in exogonic reactions; y_2 denotes biologically available energy in $(x - y_1 - z_1)$ consumed by secondary feeders for assimilation; z_2 denotes the portion of biologically available energy in $(x - y_1 - z_1)$ lost as heat by secondary feeders in exogonic reactions.

food chains. Probably energy limitation and population dynamics must both be taken into account. Which of these has the dominant influence must depend on the complexity of the biological community which is analyzed.

The members of a food chain and food web can be classified according to the nature of the food they consume or how they obtain it. Thus, primary producers consume inorganic nutrients, obtaining their carbon from CO_2 (including HCO_3^- and CO_3^{2-}) and their nitrogen from inorganic forms (N_2, NH_3, NO_3^{2-}). Most primary producers obtain the energy for assimilating these nutrients from sunlight, i.e., they photosynthesize. In aquatic environments, primary producers are mainly photosynthetic microorganisms (cyanobacteria, algae, photosynthetic dinoflagellates), but in shallow freshwater environments, they may be plants. As will be discussed in more detail later, in some special deep-sea locations at sea-floor spreading centers, the primary producers are mainly chemolithotrophic bacteria, which obtain their energy from the oxidation of inorganic compounds such as H_2, Fe(II), NH_3, NO_2^- reduced forms of sulfur, and some other oxidizable inorganic compounds (Baross et al., 1982; Ehrlich, 1981). On land, the primary producers are mainly plants, although algae and photosynthetic prokaryotes, as well as chemolithotrophic bacteria, may also make small contributions. A major exception are emergent soils such as those developing from cooled lava beds some time after volcanic eruption. They may be pioneered by cyanobacteria, beggiatoas, and mosses or lichens (e.g., Henriksson and Henriksson, 1981; Rodgers and Henriksson, 1976). The primary producers form the base of assimilatory food chains.

Decomposers are organisms, usually bacteria, fungi, and some protozoa which break down nonliving organic matter, ultimately into its inorganic constituents (e.g., CO_2, H_2O, NH_3, H_2S, SO_4^{2-}, PO_4^{3-} etc.) (Figure 1). Depending on the nature of the organic matter, complete mineralization may be accomplished by a single type of organism or by a succession of different organisms. Organisms which feed on insoluble dead organic matter are known as *saprobes,* a term derived from *saprophyte,* which denotes any vegetable organism that lives on decaying organic matter. The term and hence the concept was used as far back as 1875 (Murray et al., 1933) and was applied to molds and other fungi.

Some primary and secondary feeders, microbial as well as metazoan, consume living food. A feeder that captures live food is known as a *predator,* and its live food is known as *prey.* A few predators are found among bacteria and fungi, but most are found among protozoa and metazoa. Protozoan and metazoan feeders which consume microbes living on decaying organic matter itself are known as *detritivores* (Findley and Tenore, 1982). Metazoa that feed on microbial mats formed by filamentous microbes, especially cyanobacteria and algae, are known as *grazers.* A feeder may establish a more permanent and intimate relationship with its live food source. Such a feeder is known as a *mutualistic symbiont* if mutual benefit derives from this relationship, as a *commensal symbiont* if only the feeder derives benefit and the host is not benefited or harmed, or as a *parasitic symbiont* if the feeder benefits and the host is harmed. Parasites that cause disease are known as *pathogens.* For further discussion of this classification see Alexander (1971).

Microorganisms growing in mixed culture may depend on a cross-feeding relationship in which each of the partners releases a metabolite that is essential to the growth of one or more of the others, or in which the consumption by one of the partners of a metabolic product formed by another in the mixture is essential to the growth of both. Under otherwise similar growth conditions, none of the organisms in the mixed culture may be able to grow alone, or if they can, their energy metabolism may be altered. Such cross-feeding relationships in a mixed culture is known as *syntrophy.* In some cases, a consortium of bacteria may form a tight aggregate with a nonrandom organization of the members, as was found, for instance, with an assemblage which formed methane from acetate (Bochem et al., 1982). Other consortia of tight bacterial aggregates, sometimes including other types of organisms, are known. Kefir grains, which are cauliflowerlike clumps which may measure one or two centimeters in diameter and which are used in the fermentation of milk to produce the beverage kefir, consist of *Streptococcus lactis, S. cremoris, Leuconostoc dextranicus, Bacillus kefir,* and several yeasts (Frazier, 1967). *Mixotrichia paradoxa,* a protozoan, whose habitat is the gut of the termite *Mastotermis darwiniensis,* has spirochetes and bacterial rods attached to special brackets on its cell surface. The spirochetes provide the locomotion of the protozoan by rhythmic, coordinated undulations under the control of the protozoan (Canale-Parola, 1978). The gutless vestimentiferan tube worm *Riftia pachyptila* (Jones) contains trophosomes in its coelomic cavity which are dense spherical aggregates of sulfur- and methane-oxidizing bacteria. The trophosomes are 3–5 μm in diameter. They play a central role in the nutrition of the worm (Cavannaugh et al., 1981; Jannasch, 1984) (see below). Lichens consisting of cyanobacteria or algae and fungi are still another example of non-randomly organized microbial consortia (Ahmadjian and Hale, 1973) (see below).

Some microorganisms produce *secondary metabolites,* which are special substances produced late in the growth phase. They have a growth regulating effect (Rose, 1979). The concept of secondary metabolites as it relates to microbes was developed in the 1960s and 1970s by Bu'lock in England and Demain in the U.S.A. (Rose, 1979). These substances fall chiefly into four groups (Rose, 1979): (1) penicillins, cephalosporins, peptide antibiotics, alkaloids, and toxins; (2) aminoglycoside antibiotics; (3) tetracyclines; and (4) gibberellins.

BACTERIAL INTERACTIONS IN FOOD WEBS

Primary Producers

Among prokaryotes, cyanobacteria are a common point of entry for the external energy that is needed to drive the food cycle in freshwater environments. Photolithotrophic purple and green bacteria and chemolithotrophic bacteria, on the other hand, are only rarely a dominant point of entry for external entry. Two locales where the latter two types of bacteria are dominant primary producers are the Cyrenaican lakes in Libya and Lake Sernoye in the U.S.S.R. (Ivanov, 1968). All primary producers, as pointed out earlier, transform solar energy, or chemical energy from the oxidation of inorganic compounds into biochemically useful energy, part of which is used for synthesizing organic matter from CO_2. The energy required by successive trophic levels derives ultimately from the energy stored by the primary producers. While cyanobacteria and algae are dominant as primary producers in the photic zone of freshwater systems, diatoms and dinoflagellates are dominant in the photic system of oceans (Alexander, 1971; Sieburth, 1979, p. 196). When these photosynthetic organisms form the base of microbial food chains, it may be through their excretions of *photosynthate* by which they nourish primary microbial feeders. The photosynthate may include glycollate, sugars, fatty acids, polypeptides, amino acids, vitamins, and nucleotides (Alexander, 1971, p. 228; Litchfield and Prescott, 1970; Sieburth, 1979), which may serve as carbon, nitrogen, energy, and growth factor sources for the heterotrophs. The remains of the photosynthesizers after their death are fed upon by decomposers, which mineralize them.

Primary producers may form intimate relationships with primary feeders. Cyanobacteria, algae, and dinophyceae may reside as mutualistic symbionts in protozoa and metazoa (Sieburth, 1979). They release photosynthate within their hosts and thus feed them. Parasitization of cyanobacteria and algae by fungi such as hyphochytridiales and chytridiales is also known (Sieburth, 1979;

Sparrow, 1973). In these instances, the primary feeder, a fungus, abuses its host by absorbing cytoplasmic constituents via haustoria, which are cellular extensions of the fungus with which it penetrates the host cell. In addition, cyanobacteria, algae, and dinoflagellates may be grazed upon by metazoa in a form of predation. In anaerobic environments, such as those of salt marshes, dystrophic lakes, and regions just above the chemocline in meromictic lakes, purple and green photosynthetic bacteria and some cyanobacteria may form syntrophic relationships with sulfate-reducing and sulfur-reducing bacteria by furnishing the latter two types with electron acceptors of sulfate and sulfur, respectively (Cohen et al., 1975; Gray et al., 1973; Parkin and Brock, 1981; Pfennig and Biebl, 1976, 1978). The sulfate- and sulfur-reducing bacteria, in turn, regenerate the H_2S needed by the photosynthetic bacteria. Moreover, the photosynthate of the purple and green bacteria may feed anaerobic heterotrophs. The dead remains of all these organisms will be mineralized by decomposers.

Lichens, which are an intimate association of fungi and an oxygen-producing photosynthesizer, may contain cyanobacteria as the latter component (Ahmadjian and Hale, 1973). As primary producers, lichens flourish in arid environments on the surfaces of rocks or in porous spaces within them; they also may grow on soil, or epiphytically on plants (Ahmadjian and Hale, 1973; Friedmann, 1982). Although representing a self-contained food chain, lichens may also form the base of more complex food chains (Ahmadjian and Hale, 1973).

In environments where light does not penetrate, chemolithotrophs rather than photolithotrophs may form the base of food chains and be the point of entry of external energy to drive the food cycle. Thus, in caves, deep rock fissures, and other such sites on land, bacteria which can oxidize one or more of the following: H_2S, S^0, metal sulfides, Fe(II), As(III), U(IV), H_2, Mn(II), etc., and use some of the available energy to assimilate CO_2, may be the important primary producers (DiSpirito and Tuovinen, 1982; Ehrlich, 1963, 1981). In the deep sea, at about 2500 m, at sea-floor spreading centers where hydrothermal solutions rich in H_2S, H_2, Mn(II), Fe(II), etc. are discharged, chemolithotrophs abound which can use these oxidizable substances as energy sources to assimilate CO_2, and thus form the base of a food chain on which a variety of metazoa depend (Jannasch and Wirsen, 1979). Although the immediate source of energy to drive the food cycle surrounding hydrothermal vents at sea-floor spreading centers is chemical, indirectly it is geothermal. This is because H_2S, H_2, Mn(II), Fe(II), etc., in the hydrothermal solution originate when seawater penetrates porous basalt to depths of more than 1 km; as the water is heated to as high as 350–400°C by thermal energy from underlying magma chambers, these solutes react with the basalt. In this reaction, sulfide is generated and heavy metals are mobilized by acid resulting from the formation of $Mg(OH)_2$ in basalt modified through reaction with Mg^{2+} from seawater. The chemically altered seawater is forced upward through porous channels and fissures in the basalt by hydrostatic pressure and is discharged as hydrothermal solution from vents on the sea-floor (Seyfried and Bischoff, 1981; Seyfried and Mottl, 1982; Mottl et al., 1979).

Decomposers

A wide range of aerobic, facultative, and anaerobic heterotrophic bacteria play a central role in breaking down and mineralizing organic matter. This activity was recognized as decay and putrefaction long before bacteria were recognized as living entities (17th century by Antonie van Leeuwenhoek) and as the cause of these activities (19th century by Pasteur and others) (see Chapters 1 and 2, this volume). Despite the central role of bacteria as decomposers, some natural polymers of plant origin such as lignin and to some extent cellulose are more efficiently attacked by fungi (Alexander, 1977; Zeikus, 1982). Fungi depolymerize these substances and chemically modify a portion or all of the resultant oligomers and monomers to make them more readily usable by bacteria, which for the most part lack the capacity to form the exoenzymes necessary for lignin and cellulose depolymerization. The fungi are examples of saprobes (see Chapter 8, this volume).

As previously explained, decomposers may form a trophic hierarchy of their own with two or more trophic levels, depending on the complexity and concentration of the organic matter and whether or not oxygen is available. The more anaerobic, the more complex the trophic hierarchy of decomposers is likely to be. The need for a hierarchy of decomposers to achieve complete mineralization is that some aerobic and many anaerobic bacteria break organic matter down incompletely, and that anaerobes often are limited in the range of substrates they can attack, different groups of anaerobes attacking different types of compounds. Until it was realized that anaerobically respiring bacteria such as sulfate-reducers, denitrifiers, nitrate reducers (to ammonia), and acetate-utilizing methanogens could mineralize a narrow range of organic compounds (Fenchel and Blackburn, 1979; Jørgensen, 1982; Pfennig and Biebl, 1976; Sørensen, 1978; Sørensen et al., 1979; Widdel and Pfennig, 1977, 1981; Zeikus, 1977), it was thought that complete mineralization of organic matter under anaerobic conditions was not possible. Fermentation was thought to be the dominant process in anaerobic decomposition. Fermentation, in a physiological sense, always generates new organic compounds as end-products, although generally simpler and often more diverse than the original substrates, because fermentation is a process in which organic compounds are the terminal electron acceptors rather than oxygen or some other reducible inorganic compound. Products of fermentation such as caproate, butyrate, propionate, acetate, ethanol, etc., were thought not to be mineralizable by anaerobic respiration.

The point of departure for a food chain featuring decomposers may be at any trophic level of a food chain involving assimilators such as microbes and metazoa, or plants, herbivores, and carnivores or omnivores (Figure 2). The decomposers insure a continuing supply of CO_2, inorganic nitrogen, sulfur, phosphorus and other essential inorganics, which might otherwise be permanently fixed in organic matter. The decomposers are an essential link in the cycling of matter in nature.

Bacteria as Predators and Prey

The preying upon live bacteria as food by phagotrophic protozoa such as most amoebae and ciliophores, and some flagellates (Cloudsley-Thompson, 1972; Guiterrez and Davis, 1959; Habte and Alexander, 1975; Ramirez and Alexander, 1980; Sieburth, 1979), and by some metazoa, such as some worms, molluscs, crustaceans, etc. (Wood, 1965), is not uncommon among the feeding habits of biological communities. More unusual is the preying upon live bacteria by other bacteria. The currently best known example of a predatory bacterium is *Bdellovibrio,* but other predatory bacteria have been described (Casida, 1980a,b; Varon and Shilo, 1980). *Bergey's Manual of Determinative Bacteriology* (Buchanan and Gibbons, 1974) lists three species of *Bdellovibrio: B. bacteriovorus, B. stolpii,* and *B. starrii. Bdellovibrio* was first described in 1963 by Stolp and Starr. Free-living, it is a small vibrioid to rod-like organism (0.3 × 1.0 μm) which, when preying on a susceptible bacterium attaches to the cell envelope of its prey and penetrates into its periplasmic space. There the predatory bacterium develops into a spiral-shaped cell, several times as long as the free-living cell. It grows at the expense of the cytoplasmic content of the host bacterium by making the plasma membrane of the host very leaky and causing it to draw away from the wall of the host cell. The host cell loses viability within a few minutes after *Bdellovibrio* attachment. Enzymatic activities of the host cell seem not to be required by *Bdellovibrio* for its growth and reproduction in the periplasmic space. Instead it seems to feed on the protein, nucleic acids, lipids, and cell wall components of the host cell, degrading them into monomeric forms for assimilation. The advantage to this mode of nutrition by *Bdellovibrio* seems to be in energy conservation. By assimilating preformed precursors for the major cell polymers (proteins, nucleic acids, polysaccharides, and lipids), much expenditure of energy for the synthesis of the monomers is eliminated (Rittenberg and Hespell, 1975). The host range of *Bdellovibrio* includes various gram-negative bacteria, such as enterobacteria, pseudomonads, *Azotobacter chroococcum, Rhizobium* spp., *Agrobacterium tumefaciens,* and *Sphaerotilus natans.* For a more detailed discussion of *Bdellovibrio* refer to Stolp (1981) and Varon and Shilo (1980).

Bacteria as Mutualistic, Commensal, or Parasitic Symbionts

Some bacteria may form relatively permanent and mutually beneficial relationships with each other. Other bacteria may form relationships in which only one of the partners benefits but the other is not harmed, or in which one of the partners benefits by harming the other. Many if not all symbiotic relationships involving bacteria are clearly dependent on their biochemical activity. Some release exoenzymes in the gut of herbivorous animals to depolymerize macromolecules such as cellulose into soluble oligomers, dimers, and monomers, which they then share directly or after further biochemical modification with their animal host (Stanier et al., 1976; Taylor, 1982). Others fix nitrogen in the gut of their animal host (for instance, workers of dry-rot termites according to Benemann, 1973) and share the products of fixation with their host. *Escherichia coli* shares the vitamin K which it produces in the human gut with its host, which cannot form it (Brock, 1979). *E. coli* benefits in this relationship with its human host from a rich general nutrient supply and from an optimal growth temperature in the gut.

Lichens, which are stable associations of a cyanobacterium or a green or yellow-green alga and a fungus, together forming a thallus in which the cyanobacterial or algal and fungal partners are differentially positioned, are an example of a mutualistic symbiosis (Ahmadjian and Hale, 1973). The association of cyanobacterium or alga and fungus is stable only under conditions of severe nutritional limitation. In the presence of a plentiful nutrient supply, a lichen dissociates into its component organisms. The algal component of a lichen shares its photosynthate with the fungal partner. If the lichen includes a cyanobacterial partner, the cyanobacterium may fix dinitrogen and share the products of the fixation with its fungal partner. The fungus, which usually covers the cells of the photosynthetic partner in the thallus, shares water and minerals which it absorbs from the surround with its photosynthetic partner and may also shade it from excessive sunlight to optimize its photosynthetic activity. Mineral uptake by the fungus is greatly facilitated by lichen acids which act as chelators (see Stanier et al., 1976). For a complete review of lichens refer to Ahmadjian and Hale (1973).

Bacteria may even form endosymbiotic relationships with protozoan and metazoan cells (Breed et al., 1957; Buchanan and Gibbons, 1974). Some paramecia, for instance, harbor gram-negative rods related to *Proteus, Haemophilus, Acinetobacter,* or *Cytophaga* (Preer et al., 1974). *Paramecium* requires a special genetic endowment to accommodate any of these endosymbionts. The basis for a nutritional dependency has so far not been elucidated in the case of *Paramecium,* except in one instance in which the endosymbiont was shown to synthesize the folic acid needed by the host (Stanier et al., 1976). In *Crythidia oncopelti,* a trypanosome, the bacterial endosymbiont apparently furnishes its host with amino acids as well as other growth factors (Stanier et al., 1976). Since many of these endosymbionts have been cultured in the laboratory outside their host, nutritional interdependency cannot be absolute. The presence of endosymbionts in a paramecium host seems to render the host more competitive in the presence of uninfected paramecia. Toxins produced by the endosymbiont, to which the host is resistant, will kill uninfected, susceptible cells if the toxin is liberated by the host.

Cyanobacterial symbionts (cyanellae) have been found in a few protozoa (*Cyanophora, Peliama,* and *Panlinella*) (Stanier et al., 1976). Presumably, they share their photosynthate with the protozoan host. The contribution of the protozoan host to the cyanellae other than protection is not clear.

Members of a group of small, gram-negative bacteria known as rickettsias are examples of intracellular parasitic symbionts, which may or may not cause disease in their host. For instance, *Rickettsia* spp. and *Coxiella burnetii* parasitize arthropods and man (Weiss, 1981), but are only pathogenic for man (Stanier et al., 1976). *Rickettsia* multiplies in the cytosol of host cells, whereas *Coxiella* multiplies in phagolysosomes of phagosomes of host cells (Weiss, 1981). Many of this group are obligate parasites, but a few like *Bartonella* can be cultivated on rich semisolid agar media (Kreier et al., 1981). Whether these parasites contribute anything to their host is unknown. Some, like Chlamydiales, derive major benefit from their host in the form of cofactors and high-energy phosphate such as adenosine 5′-triphosphate, which they cannot synthesize themselves (Page, 1981). When these intracellular parasites cause the death of their host, they clearly initiate a decomposer food chain.

Mutualism, commensalism, and parasitism by bacteria have also been noted in plant cells (Buchanan and Gibbons, 1974). This involves walled bacteria (Davis et al., 1981; Gardner et al., 1982; Mundt and Hinkle, 1976) as well as wall-less bacteria known as mycoplasmas. *Spiroplasma* is a recently intensively studied example of mycoplasmas (Maramarosch, 1981; Tully and Whitcomb, 1981). These organisms are pathogenic for their plant host. Spiroplasmas interfere with plant hormonal balances and thereby affect plant development (Maramarosch, 1981). They may also produce toxins which shorten the life-span of the affected host plant and lead to its death, thereby initiating another decomposer food chain.

In order for bacteria to be parasitic in a multicellular host, they need not invade cells but may reside in tissue spaces, i.e., among the cells of the tissue, in extracellular fluids or, in the case of metazoa, in the lumen of the digestive, excretory, or respiratory systems. In such a relationship, the bacteria probably do not contribute any metabolic products of value to the host, but, when pathogenic, contribute products that are injurious. Such products may be proteins, as for example in the case of botulinum toxin generated by *Clostridium botulinum,* tetanus toxin generated by *Cl. tetani,* diphtheria toxin generated by *Corynebacterium diphtheriae,* or enterotoxin generated by *Shigella dysenteriae, Staphylococcus aureus,* or *Vibrio cholerae* (Davis et al., 1967; Boyd and Hoerl, 1977). Such products may also be lipopolysaccharides produced by such organisms as *Salmonella, Shigella,* and some strains of *Escherichia* (Boyd and Hoerl, 1977; Bradley, 1979; Davis et al., 1967). The lipopolysaccharides are derived from the cell walls of the parasitic bacteria and released only after their death. These toxins may weaken the host and hasten its death. The parasites may thereby initiate a decomposer food chain.

Syntrophy

The basis for some mutualistic relationships may be syntrophy. In such relationships, one or more of the interacting organisms are a source of one or more needed major or minor metabolites which they share with their partners in a metabolite exchange. The partners may be exclusively bacteria, or one or more may be bacterial and the other nonbacterial. A number of examples of syntrophy will be cited.

In anaerobic environments as diverse as freshwater and saline reducing sediments and the rumen of cows, sheep, goats, deer, and other cloven-hoofed, horned quadrupeds, a phenomenon of interspecies hydrogen transfer occurs. Here one group or a hierarchy of two or more groups of bacteria ferment carbohydrates or other organic energy sources, producing H_2, CO_2, volatile acids, and some other products, while anaerobic hydrogen-utilizing bacteria such as methanogens, acetogens, or sulfate-reducers utilize the H_2 to reduce CO_2, respectively, to methane or acetate, or sulfate or sulfur to hydrogen sulfide (Balch et al., 1982; Braun et al., 1981; Fenchel and Blackburn, 1979; Ianotti et al., 1973; Laube and Martin, 1981; Leigh et al., 1981; McInerney et al., 1979; Pfennig and Biebl, 1981; Pfennig et al., 1981; Wiegel et al., 1981; Winter and Wolfe, 1980; Wolin, 1979; Wolin and Miller, 1982). Some H_2-producing organisms cannot continue their fermentation unless the H_2 they produce is removed from their environment, as by bacterial H_2-consumption. One example is the S organism which ferments ethanol to H_2 and acetate (Bryant et al., 1967; Reddy et al., 1972). Another example is represented by certain desulfovibrios growing in the absence of sulfate as terminal hydrogen acceptor (Bryant et al., 1977). The composition of end-products from H_2-producing organisms like *Ruminococcus albus* may be changed if the H_2 produced is consumed by another organism. In the absence of a H_2-consumer, *R. albus* forms acetate, ethanol, CO_2 and H_2 from cellulose, whereas in the presence of a H_2-consumer, *R. albus* forms only acetate and CO_2, the H_2 being removed as soon as it is formed and therefore not available to *R. albus* for ethanol formation. The external H_2 removal prevents reduction of any acetyl-SCoA by *R. albus* to ethanol (Wolin, 1979). Other such examples involve *Ruminococcus flavefaciens* (Latham and Wolin, 1977) and *Selenomonas ruminantium* (Chen and Wolin, 1977). For a more complete discussion of hydrogen metabolism in the rumen see, for instance, Hungate (1967; also, this volume) and Wolin and Miller (1982).

In saline reducing sediments, a competition for fermentatively produced H_2 may exist between sulfate-reducers and methanogens of the type that form methane from CO_2 and H_2. In such competition, the sulfate-reducers outcompete the methanogens (see, for instance, Abram and Nedwell, 1978; Fenchel and Blackburn, 1979; Mountford et al., 1980; Oremland and Taylor, 1978). The explanation for the more successful competition for H_2 by the sulfate-reducers resides in their five-fold higher affinity for hydrogen compared to methanogens (Kristjansson et al., 1982). Bacterial sulfate-reduction and methanogenesis are compatible when methanogens are not dependent on H_2 but form methane from methanol or trimethyl amine (Oremland et al., 1982).

In reducing environments such as anaerobic sediments, syntrophic relationships may exist not only around H_2 production but also around acetate production. In anaerobic marine sediments, acetate may be produced in the bacterial fermentation of carbohydrate and in the anaerobic respiration of sulfate-reducers such as *Desulfovibrio, Desulfomonas,* and some *Desulfotomaculum* species. This acetate may then be completely oxidized to CO_2 by other sulfate-reducers such as *Desulfotomaculum acetoxidans, Desulfobacter postgatei, Desulfococcus multivorus, Desulfomonas limicola, Desulfomonas magnum,* and *Desulfonema variabilis.* In these sediments, the sulfate-reducers out-compete methanogens, which can form methane from acetate because the former have a higher affinity for acetate than the methanogens (Fenchel and Blackburn, 1979; Pfennig et al., 1981; Schönheit et al., 1982). However, in freshwater anaerobic sediments in which only low amounts of sulfate occur, acetate formed by fermentation is converted mostly to methane and CO_2 by appropriate methanogens and without the use of H_2.

An example of syntrophy involving a bacterium and a plant is symbiotic nitrogen fixation (Quispel, 1974; see Chapter 6, this volume). In the classical example, *Rhizobium* in nodules of the root system of their leguminous host convert dinitrogen to ammonia using products of photosynthesis of their plant host as sources of reducing power in the dinitrogen fixation process. The rhizobia, in turn, share the fixed dinitrogen in the form of amino acids with their leguminous host (Brock, 1979; Quispel, 1974). Neither the legume nor its *Rhizobium* symbiont can fix dinitrogen as separate individuals under natural conditions (Brill, 1979). Under certain artificial conditions in the laboratory, free-living *Rhizobium* may fix dinitrogen if, for instance, it has been induced by plant cells not necessarily of legume origin and not invaded by the *Rhizobium* (Child, 1976).

Symbiotic dinitrogen fixation may also occur in the nodules of the root system of nonlegumes (Torrey, 1978) such as the alder (*Alnus* sp.) if infected with the actinomycete *Frankia* sp. (Benson, 1982), or in a pteridophyte, *Azolla,* infected with a cyanobacterium, *Anabaena azollae* (Peters, 1978). In the *Azolla-Anabaena* association, ammonia rather than amino acids appears to be shared by the symbionts (Orr and Haselkorn, 1982). They are an important source of fixed nitrogen in the cultivation of rice (Peters, 1978).

A looser association in symbiotic dinitrogen fixation occurs between some grasses and the bacterium *Azospirillum lipoferum* (Day et al., 1975; von Bülow and Döbereiner, 1975). In this instance, the dinitrogen-fixing bacterium does not invade the root tissue of the plant but grows in close proximity to it. The dinitrogen-fixer benefits from root secretions which appear to contain compounds that can serve as energy sources, while the dinitrogen-fixer secretes some of the fixed nitrogen which may be absorbed by the roots of the grasses (Brill, 1979).

The grass–*Azospirillum* relationship is a special example of syntrophic rhizosphere interaction. The rhizosphere is a zone in the soil surrounding the roots of plants (Alexander, 1977). The root surface and the adherent soil constitute the rhizoplane (Alexander, 1977). It is in the rhizoplane that *Azospirillum lipoferum* is active. Since the roots of plants not only absorb solutes from soil solution but also release exudates of a variety of organic compounds (see Alexander, 1977, p. 428) and slough off dead tissue, the rhizosphere forms a habitat for microbes, including bacteria, which benefit from these plant products nutritionally. The microbial activity, in turn, may contribute to ammonification, nitrification, phosphate, sulfate, and cation (e.g., Na^+, K^+, Mg^{2+}, Ca^{2+}, Fe^{2+}) mobilization, and formation of plant-growth regulators like indolacetic acid, gibberellins, and cytokinins, all of which are of importance to plant development (Alexander, 1977; Rovira and McDougall, 1967).

Food chains whose base rests on CO_2-fixing organisms generally involve photosynthetic organisms as primary producers, as pointed out earlier. Recently it has been found that in very special environments where sunlight cannot penetrate, chemolithotrophic bacteria are the primary producers. In at least one such case, syntrophy is involved. Large, up to 10-m-long, gutless vestimentiferan worms have been discovered living on the ocean floor at approximately 2500 m below sea-level around hydrothermal vents at sea-floor spreading-centers (see above). The vestimentiferan worms absorb H_2S discharged from the vents along with O_2 and CO_2 dissolved in the seawater by the plumes at their anterior ends. The blood in their circulatory system carries these gases to trophosomes in their coelomic cavity. A special protein in their blood appears to be involved in transporting the H_2S from the plumes to the trophosomes. This protein binds the H_2S and prevents it from poisoning the worm. At the site of the trophosomes, the H_2S is released from the carrier protein and oxidized by sulfide-oxidizing chemolithotrophic bacteria that fill the trophosomes. The H_2S furnishes energy and reducing power for the assimilation of CO_2 by the bacteria, which share the reduced carbon with their host. The bacteria have so far not been cultivated outside their host (Arp and Childress, 1981, 1983; Cavannaugh et al., 1981; Felbeck, 1981; Felbeck et al., 1982; Jones, 1981; Powell and Somero, 1983; Rau, 1981; Wittenberg et al., 1981).

Less intimate syntrophic associations between chemolithotrophic H_2S-oxidizers and molluscs have been found at these hydrothermal discharge areas as well as in more conventional shelf areas at the edge of the oceans (e.g., Buzzard's Bay, Cape Cod, Massachusetts, U.S.A.) (Cavannaugh, 1982; Felbeck et al., 1982; Rau and Hedges, 1979).

An example of bacterial symbiosis that may involve syntrophy or may be merely a commensal relationship, and which may be the basis of a public health problem involves cyanobacteria and green algae and *Legionella pneumophila.* The cyanobacteria and algae have been shown to produce a photosynthate which supports the growth of *Legionella* (Pope et al., 1982; Tison et al., 1980). Tison et al. (1980) showed that spent culture medium from *Fischerella* sp., a cyanobacterium, could support the growth of *Legionella.* Similar findings were made by Pope et al. (1982) with respect to the green algae *Chlorella, Scenedesmus,* and *Chlamydomonas,* and the cyanobacteria *Synechocystis, Phormidium, Fischerella,* and *Nostoc.* The active components in the photosynthates have not yet been identified. In other instances, heterotrophic bacteria growing in water may generate nutrients that can support the growth of *L. pneumophila* (Yee and Wadowsky, 1982). The identity of these nutrients also remains to be determined. These cross-feeding phenomena may be an explanation for the persistence of *L. pneumophila* in bodies of freshwater and in cooling tower water.

Although discussed here for special situations, it is quite likely that syntrophy involving bacteria is common in densely populated environments such as soil, sediment, and the animal gut.

Antibiosis

Many different kinds of bacteria as well as other microorganisms have been shown under laboratory conditions to produce secondary metabolites that have an inhibitory or even lethal effect on the growth of some other bacteria and some eukaryotic microorganisms. These metabolites were named antibiotics by S.A. Waksman (Rose, 1979). They are effective at low concentration. The microorganisms that produce these substances may regulate with them the passage of nutrients through the food web in both quantity and kind by forcing a selection of microorganisms that are resistant to the prevalent antibiotics and that carry out special catabolic processes. Large-scale effects of antibiotics in nature have, however, not been demonstrated so far (see, for instance Doetsch and Cook, 1973). Indeed, the effects may not occur on a large scale, if at all, either because they are not produced in nature or because of the ease with which microbes can develop resistance to antibiotics by the acquisition of R-factors (plasmids). Possibly, on a small scale, in microenvironments, antibiotics are produced and have an effect (Alexander, 1977). Rasool and Wimpenny (1982) have calculated from experimental data that *Streptomyces aureofaciens* should be able to produce enough tetracycline in a 10 μm radius around the hyphae to prevent the growth of susceptible bacteria, such as *Escherichia coli* and *Bacillus pumilus,* under conditions similar to those used in their experiments.

CONCLUSION

The foregoing makes it clear that bacteria play a central role in the food web of biological communities as primary producers, as primary and secondary feeders, and as decomposers. As suppliers of nutriment, they make contributions not only in the form of living and dead cells, but also through their metabolic products, in some cases by intimate association with other organisms. Without their participation, the food cycle of biological communities would be incomplete.

REFERENCES

Abram, J. W., and Nedwell, D. B., 1978, Inhibition of methanogens by sulfate-reducing bacteria competing for transferred hydrogen, *Arch. Microbiol.* **117**:89–92.

Ahmadjian, V., and Hale, M. E. (eds.), 1973, *The Lichens,* Academic Press, New York, London.

Alexander, M., 1971, *Microbial Ecology,* John Wiley and Sons, New York.

Alexander, M., 1977, *Introduction to Soil Microbiology,* 2nd ed., John Wiley and Sons, New York.

Arp, A. J., and Childress, J. J., 1981, Blood function in the hydrothermal vent vestimentiferan tube worm, *Science* **213**:342–344.

Arp, A. J., and Childress, J. J., 1983, Sulfide binding by the blood of the hydrothermal vent tube worm *Riftia pachyptila, Science* **219**:295–297.

Balch, W. E., Schoberth, S., Tanner, R. S., and Wolfe, R. S., 1977, *Acetobacterium,* a new genus of hydrogen-oxidizing, carbon dioxide reducing bacteria, *Int. J. Syst. Bacteriol.* **27**:355–361.

Baross, J. A., Lilley, M. D., and Gordon, J. L., 1982, Is the CH_4, H_2 and CO venting from submarine hydrothermal systems produced by thermophilic bacteria? *Nature (London)* **298**:366–368.

Benemann, J. R., 1973, Nitrogen fixation in termites, *Science* **181**:164–165.

Benson, D. R., 1982, Isolation of *Frankia* strains from alder actinorhizal root nodules, *Appl. Environ. Microbiol.* **44**:461–465.

Bochem, H. P., Schoberth, S. M., Sprey, B., and Wengler, P., 1982, Thermophilic biomethanation of acetic acid: morphology and ultrastructure of a granular consortium, *Can. J. Microbiol.* **28**:500–510.

Boyd, R. F., and Hoerl, B. G., 1977, *Basic Medical Microbiology,* Little, Brown and Company, Boston.

Bradley, S. G., 1979, Cellular and molecular mechanisms of action of bacterial endotoxins, *Ann. Rev. Microbiol.* **33**:67–94.

Braun, M., Mayer, F., and Gottschalk, G., 1981, *Clostridium aceticum* (Wieringa), a microorganism producing acetic acid from molecular hydrogen and carbon dioxide, *Arch. Microbiol.* **128**:288–293.

Breed, R. S., Murray, E. G. D., and Smith, N. R. (eds.), 1957, *Bergey's Manual of Determinative Bacteriology,* 7th ed., The Williams and Wilkins Company, Baltimore.

Brill, W. J., 1979, Nitrogen fixation: basic to applied, *Am. Scientist* **67**:458–466.

Brock, T. D., 1979, *Biology of Microorganism,* 3rd ed., Prentice-Hall, Englewood, New Jersey.

Bryant, M. P., Wolin, E. A., Wolin, M. J., and Wolfe, R. S., 1967, *Methanobacillus omelianskii,* a symbiotic association of two species of bacteria, *Arch. Mikrobiol.* **59**:20–31.

Bryant, M. P., Cambell, L. L., Reddy, C. A., and Crabill, M. A., 1977, Growth of *Desulfovibrio* in lactate or ethanol media low in sulfate in association with hydrogen utilizing methanogenic bacteria, *Appl. Environ. Microbiol.* **33**:1162–1169.

Buchanan, R. E., and Gibbons, N. E., (eds.), 1974, *Bergey's Manual of Determinative Bacteriology,* 8th ed., The Williams and Wilkins Company, Baltimore.

Canale-Parola, E., 1978, Motility and chemotaxis of spirochetes. *Ann. Rev. Microbiol.* **32**:69–99.

Casida, L. E., Jr., 1980a, Death of *Micrococcus luteus* in soil, *Appl. Environ.* **39**:1031–1034.

Casida, L. E., 1980b, Bacterial predators of *Micrococcus luteus* in soil, *Appl. Environ. Microbiol.* **39**:1035–1041.

Cavannaugh, C. M., 1982, Prokaryotic cells in marine invertebrates: possible chemoautotrophic symbionts, Abstracts, *Colloquium on Microbial Chemoautotrophy,* Ohio State University, College of Biological Sciences, October 4–5, 1982.

Cavannaugh, C. M., Gardiner, S. L., Jones, M. L., Jannasch, H. W., and Waterbury, J. B., 1981, Prokaryotic cells in the hydrothermal vent tube worm *Riftia pachyptila* Jones: Possible chemoautotrophic symbionts, *Science* **213**:342–344.

Chen, M., and Wolin, M. J., 1977, Influence of CH_4 production by *Methanobacterium ruminantium* on the fermentation of glucose and lactose by *Selenomonas ruminantium, Appl. Environ. Microbiol.* **34**:756–759.

Child, J. J., 1976, New developments in nitrogen fixation research, *BioScience* **26**:614–617.

Cloudsley-Thompson, J., 1972, The habitat and its influence on the evolutionary development of life forms, in: *Biology of Nutrition, The Evolution and Nature of Living Systems, the Organization and Nutritional Methods of Life Forms,* R.N. T-W-Fiennes (ed.), Pergamon Press, Oxford, pp. 351–372.

Cohen, Y., Jørgensen, B. B., Paden, E., and Shilo, M., 1975, Sulfide-dependent anoxygenic photosynthesis in the cyanobacterium *Oscillatoria limnetica, Nature (London)* **257:**489–492.

Davis, B. D., Dulbecco, R., Eisen, H. N., Ginsberg, H. S., and Wood, W. B., Jr., 1967, *Principles of Microbiology and Immunology,* Harper and Row, New York.

Davis, M. J., Whitcomb, R. T., and Graves Gillaspie, Jr., A., 1981, Fastidious bacteria of plant vascular tissue and invertebrates (including so-called rickettsia-like bacteria), in: *The Prokaryotes, a Handbook of Habitats, Isolation and Identification of Bacteria,* Vol. 2, (M. P. Starr, H. Stolp, H. G. Truper, A. Balows, and H. C. Schlegel, eds.), Springer Verlag, Berlin, pp. 2172–2188.

Day, J. M., Neves, M. C. P., and Dobereiner, J., 1975, Nitrogenase activity on the roots of tropical forage grasses, *Soil Biol. Biochem.* **7:**107–112.

Dice, L. R., 1952, *Natural Communities,* The University of Michigan Press, Ann Arbor.

Dispirito, A. A., and Tuovinen, O. H., 1982, Uranous ion oxidation and carbon dioxide fixation by *Thiobacillus ferrooxidans, Arch. Microbiol.* **133:**28–32.

Doetsch, R. N., and Cook, T. M., 1973, *Introduction to Bacteria and Their Ecobiology,* University Park Press, Baltimore.

Ehrlich, H. L., 1963, Microroganisms in acid mine drainage from a copper mine, *J. Bacteriol.* **86:**350–352.

Ehrlich, H. L., 1981, *Geomicrobiology,* Marcel Dekker, New York.

Elton, C., 1927, *Animal Ecology,* Sidgwick and Jackson, Ltd., London.

Felbeck, H., 1981, Chemoautotrophic potential of the hydrothermal vent tube worm *Riftia pachyptila* Jones (Vestimentifera), *Science* **213:**336–338.

Felbeck, H., Powell, M., Wienhausen, G., Hand, S., and Somero, G., 1982, Chemoautotrophic bacteria as symbionts in animals from sulfide-rich environments, Abstract, *Colloquium on Microbial Chemoautotrophy,* Ohio State University, College of Biological Sciences, October 4–5, 1982.

Fenchel, T., and Blackburn, T. H., 1979, *Bacteria and Mineral Cycling,* Academic Press, New York.

Findley, S., and Tenore, K., 1982, Nitrogen source for a detritivore: detritus substrate versus associated microbes, *Science* **218:**371–373.

Frazier, W. C., 1967, *Food Microbiology,* 2nd ed., McGraw-Hill Book Company, New York.

Friedmann, E. I., 1982, Endolithic microorganisms in the Antarctic cold desert, *Science* **215:**1045–1053.

Gardner, J. M., Feldman, A. W., and Zablotowicz, R. M., 1982, Identity and behavior of xylem-residing bacteria in rough lemon roots of Florida Citrus trees, *Appl. Environ. Microbiol.* **43:**1335–1342.

Gray, B. H., Powler, C. F., Nugent, N. A., Rigopoulos, N., and Fuller, R. C., 1973, Reevaluation of *Chloropseudomonas ethylica* strain 2-K, *Int. J. Syst. Bacteriol.* **23:**256–264.

Guiterrez, J., and Davis, R. E., 1959, Bacterial ingestion by rumen ciliates *Entodinium* and *Diplodinium. J. Protozool.* **6:**222–226.

Habte, M., and Alexander, M., 1975, Protozoa as agents responsible for the decline of *Xanthomonas campestris* in soil, *Appl. Microbiol.* **29:**159–164.

Henriksson, E., and Henriksson, L. E., 1981, Beggiatoas as pioneers in volcanic soils of Surtsey, Iceland, *Acta Bot. Isl.* **6:**11–14.

Hungate, R. E., 1967, Hydrogen as an intermediate in the rumen fermentation, *Arch. Mikrobiol.* **59:**158–164.

Ianotti, E. L., Kaflewitz, D., Wolin, M. J., and Bryant, M. P., 1973, Glucose fermentation products of *Ruminococcus albus* grown in continuous culture with *Vibrio succinogenes:* changes caused by interspecies transfer of hydrogen, *J. Bacteriol.* **114:**1231–1240.

Ivanov, M. V., 1968, *Microbiological Processes in the Formation of Sulfur Deposits,* Israel Program for Scientific Translation, Jerusalem, U.S. Dept. Commerce, Clearinghouse, Fed. Sci. Tech. Inf., Springfield, Va.

Jannasch, H. W., 1984, Microbial processes at deep sea hydrothermal vents, in: *Hydrothermal Processes at Seafloor Spreading Centers,* (P. A. Rona, K. Bostrom, L. Laubier, and K. L. Smith, Jr., eds.), Plenum Press, New York, pp. 677–709.

Jannasch, H. W., and Wirsen, C. O., 1979, Chemosynthetic primary production at East Pacific sea floor spreading centers. *BioScience* **29:**592–598.

Jones, M. L., 1981, Riftia pachyptila Jones: observations on the vestimentiferan worm from the Galapagos Rift. *Science* **213:**333–336.

Jørgensen, B. B., 1982, Mineralization of organic matter in the seabed-role of sulfate reduction. *Nature (London)* **296:**643–645.

Kreier, J. P., Dominguez, N., Krampitz, H. E., Gothe, R., and Ristic, M., 1981, in: *The Prokaryotes, a Handbook of Habitats, Isolation and Identification of Bacteria,* Volume 2 (M. P. Starr, H. Stolp, H. G. Trüper, A. Balows, and H. C. Schlegel, eds.), Springer Verlag, Berlin, pp. 2189–2209.

Kristjansson, J. K., Schönheit, P., and Thauer, R. K., 1982, Different K_s values for hydrogen of methanogenic bacteria and sulfate reducing bacteria: an explanation for the apparent inhibition of methanogenesis by sulfate, *Arch. Microbiol.* **131:**278–282.

Lapedes, D. N. (ed.), 1974, *McGraw-Hill Encyclopedia of Environmental Science,* McGraw-Hill Book Company, New York.

Latham, M. J., and Wolin, M. J., 1977, Fermentation of cellulose by *Ruminococcus flavefaciens* and in the presence and absence of *Methanobacterium ruminantium, Appl. Environ. Microbiol.* **34:**297–301.

Laube, O. M., and Martin, S. M., 1981, Conversion of cellulose to methane and carbon dioxide by triculture of *Acetivibrio cellolyticus, Desulfovibrio* sp., *Methanosarcina barkeri, Appl. Environ. Microbiol.* **42:**413–420.

Leigh, J. A., Mayer, F., and Wolfe, R. S., 1981, *Acetogenum kivui,* a new hydrogen-oxidizing, acetogenic bacterium, *Arch. Microbiol.* **129:**275–280.

Lindeman, R. L., 1942, The trophic-dynamic aspect of ecology, *Ecology* **23:**399–418.

Litchfield, D. C., and Prescott, J. M., 1970, Analysis of dansylation of amino acids dissolved in marine and freshwaters, *Limnol. Oceanogr.* **15:**250–256.

Mandels, R. I., 1975, Microbial sources of cellulose, *Biotechnol. Bioeng. Symp.* **5:**81–105.

Maramarosch, K., 1981, Spiroplasma: Agents of animal and plant diseases, *BioScience* **31:**374–380.

McInerney, M. J., Bryant, M. P., and Pfennig, N., 1979, Anaerobic bacterium that degrades fatty acids in syntrophic association with methanogens, *Arch. Microbiol.* **122:**129–135.

Mottl, M. J., Holland, H. D., and Corr. R. F., 1979, Chemical exchange during hydrothermal alteration of basalt by seawater. II. Experimental results for Fe, Mn, and sulfur species, *Geochim. Cosmochim. Acta* **43:**869–884.

Mountford, D. O., Asher, R. A., Mays, E. L., Tiedje, J.M., 1980, Carbon and electron flow in mud and sandflat intertidal sediments of Delaware inlet, Nelson, New Zealand, *Appl. Microbiol.* **39:**686–694.

Mundt, J. O., and Hinkle, N. F., 1976, Bacteria within ovules and seeds, *Appl. Environ. Microbiol.* **32:**694–698.

Murray, J. A. H., Bradley, H., Craigie, W. A., and Onions, C. T. (eds.), 1933, *The Oxford English Dictionary,* Volume IX, Oxford at the Clarendon Press.

Oremland, R. S., and Taylor, B. F., 1978, Sulfate reduction and methanogenesis in marine sediments, *Geochim. Cosmochim. Acta* **42:**209–214.

Oremland, R. S., Marsh, L. M., and Polcin, S., 1982, Methane production and simultaneous sulfate reduction in anoxic, salt marsh sediments, *Nature (London)* **296:**143–145.

Orr, J., and Haselkorn, R., 1982, Regulation of glutamic synthetase activity and synthesis in free-living and symbiotic *Anaboena* spp., *J. Bacteriol.* **152:**626–635.

Page, L. A., 1981, Obligatory intracellular bacteria: the genus Chlamidia, In: *The Prokaryotes, a Handbook on Habitats, Isolation and Identification of Bacteria,* Volume 2 (M. P. Starr, H. Stolp, H. G. Trüper, A. Balows, and H. G. Schlegel, eds.), Springer Verlag, Berlin, pp. 2210–2222.

Parkin, T. B., and Brock, T. D., 1981, The role of phototrophic bacteria in the sulfur cycle of a meromictic lake. *Limnol. Oceanogr.* **26:**880–890.

Peters, G. A., 1978, Blue-green algae and algal associations, *BioScience* **28:**580–585.

Pfennig, N., and Biebl, H., 1976, *Desulfomonas acetoxidans* gen. nov. and sp. nov., a new anaerobic, and sulfur-reducing, acetate-oxidizing bacterium, *Arch.Microbiol.* **110:**3–12.

Pfennig, N., and Biebl, H., 1978, Growth yields of green sulfur bacteria in mixed cultures with sulfur and sulfate reducing bacteria, *Arch. Microbiol.* **117:**9–16.

Pfennig, N., and Biebl, H., 1981, The dissimilatory sulfur-reducing bacteria, in: *The Prokaryotes, a Handbook of Habitats, Isolation and Identification of Bacteria,* Volume 1 (M. P. Starr, H. Stolp, H. G. Trüper, A. Balows, and H. G. Schlegel, eds.), Springer Verlag, Berlin, pp. 941–947.

Pfennig, N., Widdel, F., and Trüper, H. G., 1981, The dissimilatory sulfate-reducing bacteria, in: *The Prokaryotes: a Handbook of Habitats, Isolation and Identification of Bacteria,* Volume 1 (M. P. Starr, H. Stolp, H. G. Trüper, A. Balows, and H. G. Schlegel, eds.), Springer Verlag, Berlin, pp. 926–940.

Pimm, S. L., and Lawton, J. H., 1977, Number of trophic levels in ecological communities, *Nature (London)* **268:**329–331.

Pimm, S. L., and Lawton, J. H., 1978, On feeding on more than one trophic level, *Nature (London)* **275:**542–544.

Pope, D. H., Soracco, R. J., Gill, H. G., and Fliermans, C. B., 1982, Growth of *Legionella pneumophila* in two-membered cultures with green algae and cyanobactreria, *Curr. Microbiol.* **7:**319–322.

Powell, M. A., and Somero, G. N., 1983, Sulfide binding by the bolood of the hydrothermal vent tube worm, *Riftia pachyptila, Science* **219:**295–297.

Preer, J. R., Jr., Preer, L. B., and Jurand, A., 1974, Kappa and other endosymbionts in *Paramecium aurelia, Bacteriol. Rev.* **38:**113–163.

Quispel, A. (ed.), 1974, *The Biology of Nitrogen Fixation,* North Holland Publishing Company, Amsterdam.

Ramirez, C., and Alexander, M., 1980, Evidence suggesting protozoan predation of *Rhizobium* associated with germinating seeds in the rhizosphere of beans (*Phaseolus vulgaris* L.), *Appl. Environ. Microbiol.* **40:**492–499.

Rasool, K. A., and Wimpenny, J. W. T., 1982, Mixed continuous culture experiments with an antibiotic-producing streptomycete and *Escherichia coli, Microb. Ecol.* **8:**267–277.

Rau, G. H., 1981, Hydrothermal vent clam and tube worm $^{13}C/^{12}C$: further evidence on non-photosynthetic food sources, *Science* **213:**338–340.

Rau, G. H., and Hedges, J. I., 1979, Carbon-13 depletion in a hydrothermal vent mussel: suggestion of a chemosynthetic food source, *Science* **203:**648–649.

Reddy, C. A., Bryant, M. P., and Wolin, M. J., 1972, Characteristics of S organism isolated from *Methanobacillus omelianskii, J. Bacteriol.* **109:**539–545.

Rittenberg, S. C., and Hespell, R. B., 1975, Energy efficiency of interperiplasmic growth of *Bdellovibrio bacteriovorus, J. Bacteriol.* **121:**1158–1165.

Rodgers, G. A., and Henriksson, E., 1976, Associations between the blue-green algae *Anaboena variabilis* and *Nostoc muscorum* and the moss *Funaria hygrometrica* with reference to the colonization of Surtsey, *Acta Bot. Isl.* **4:**10–15.

Rose, A. H., 1979, Production and industrial importance of secondary products of metabolism, in: *Economic Microbiology,* Volume 3, *Secondary Products of Metabolism* (A. H. Rose, ed.), Academic Press, London, pp. 1–33.

Rovira, A. D., and McDougall, B. M., 1967, Microbial and biochemical aspects of the rhiosphere, in: *Soil Biochemistry,* Volume 1 (A. D. McLaren and J. M. Petersen, eds.), Marcel Dekker, New York, pp. 417–463.

Schönheit, P., Kristjansson, J. K., and Thauer, R. K., 1982, Kinetic mechanism for the ability of sulfate reducers to out-compete methanogens for acetate, *Arch. Microbiol.* **132:**285–288.

Seyfried, W. E., Jr., and Bischoff, J. L., 1981, Experimental seawater-basalt interactions at 300°C, 500 bars, chemical exchange, secondary mineral formation and implication for the transport of heavy metals, *Geochim. Cosmochim. Acta* **45:**135–147.

Seyfried, W. E., Jr., and Mottl, M. J., 1982, Hydrothermal alteration of basalt by seawater under seawater-dominated conditions, *Geochim. Cosmochim. Acta* **46:**985–1002.

Sieburth, J. McN., 1979, *Sea Microbes,* Oxford University Press, New York.

Sørensen, J., 1978, Capacity for denitrification and reduction of nitrate to ammonia in a coastal marine sediment, *Appl. Environ. Microbiol.* **35:**301–305.

Sørensen, J., Jørensen, B. B., and Revsbeck, N. P., 1979, A comparison of oxygen, nitrate, and sulfate respiration in coastal marine sediments, *Microb. Ecol.* **5:**105–115.

Sparrow, F. K., 1973, Chytridiomycetes, Hyphochytridiomycetes, chapter 6, in: *The Fungi, an Advanced Treatise* (G. C. Ainsworth, F. K. Sparrow, and A. S. Sussman, eds.), Academic Press, New York, London, pp. 85–110.

Stanier, R. Y., Adelberg, E. A., and Ingraham, J. L., 1976, *The Microbial World,* Prentice-Hall, Inc., Englewood Cliffs, New Jersey.

Stolp, H. J., 1981, The genus *Bdellovibrio,* in: *The Prokaryotes, a Handbook on Habitats, Isolation and Identification of Bacteria,* Volume 2 (M. P. Starr, H. Stolp, H. G. Trüper, A. Balows, and H. G. Schlegel, eds.), Springer Verlag, Berlin, pp. 618–629.

Stolp, H. J., and Starr, M. P., 1963, *Bdellovibrio bacteriovorus* gen. et sp. n., a predatory ectoparasitic and bacteriolytic microorganism, *Antonie van Leeuwenhoek J. Microbiol. Serol.* **29:**217–248.

Taylor, E. C., 1982, Role of aerobic microbial popualtions in cullulose digestion by desert millipedes, *Appl. Environ. Microbiol.* **44:**281–291.

Thienemann, A., 1926, Der Nahrungskreislauf im Wasser, *Verh. deutsch. Zool. Ges.* **31:**29–79.

Tison, D. L., Pope, D. H., Cherry, W. B., and Fliermans, C. B., 1980, Growth of *Legionella pneumophila* in association with blue-green algae (cyanobacteria), *Appl. Environ. Microbiol.* **39:**456–459.

Torrey, J. G., 1978, Nitrogen fixation by actinomycete-nodulated angiosperms, *BioScience* **28:**586–592.

Tully, J. G., and Whitcomb, R. F., 1981, The genus *Spiroplasma,* in: *The Prokaryotes, a Handbook on Habitats, Isolation and Identification of Bacteria,* Volume 2 (M. P. Starr, H. Stolp, H. G. Trüper, A. Balows, and H. G. Schlegel, eds.), Springer Verlag, Berlin, pp. 2271–2284.

Tuovinen, O. H., 1981, Uranous ion oxidation and carbon dioxide fixation by *Thiobacillus ferrooxidans, Arch. Microbiol.* **133:**28–32.

Varon, M., and Shilo, M., 1980, Ecology of aquatic bdellovibrios, *Adv. Aquat. Microbiol.* **2:**1–48.

von Bülow, J. F. W., and Döbereiner, J., 1975, Potential for nitrogen fixation in maize genotypes in Brazil, *Proc. Natl. Acad. Sci. USA* **72:**2389–2393.

Weiss, E., 1981, The family Rickettsiaceae: human pathogens, in: *The Prokaryotes, a Handbook of Habitats, Isolation and Identification of Bacteria,* Volume 2 (M. P. Starr, H. Stolp, H. G. Trüper, A. Balows, and H. G. Schlegel, eds.), Springer Verlag, Berlin, pp. 2137–2160.

Widdel, F., and Pfennig, N., 1977, A new anaerobic, sporing, acetate-oxidizing sulfate-reducing bacterium, *Desulfotomaculum* (emend.) *Acetoxidans, Arch. Microbiol.* **112:**119–122.

Widdel, F., and Pfennig, N., 1981, Studies on dissimilatory sulfate-reducing bacteria that decompose fatty acids. I. Isolation of new sulfate-reducing bacteria enriched with acetate from saline environments. Description of *Desulfobacter postgatei* gen. nov. sp. nov., *Arch. Microbiol.* **129:**395–400.

Wiegel, J., Braun, M., and Gottschalk, G., 1981, *Clostridium thermoautotrophicum* species novum, a thermophile producing acetate from molecular hydrogen and carbon dioxide, *Curr. Microbiol.* **5:**255–260.

Winter, J. U., and Wolfe, R. S., 1980, Methane formation from fructose by syntrophic association of *Acetobacter woodii* and different strains of methanogens, *Arch. Microbiol.* **124:**73–79.

Wittenberg, J. B., Morris, R. J., Gibson, Q. H., and Jones, M. L., 1981, Hemoglobin kinetics of the Galapagos Rift vent tube worm *Riftia pachyptila* Jones (Pogonophora, Vestimentifera), *Science* **213:**344–346.

Wolin, M. J., 1979, The rumen fermentation: a model for microbial interactions in anaerobic exosystems, *Adv. Microb. Ecol.* **3:**49–77.

Wolin, M. J., and Miller, T. L., 1982, Interspecies hydrogen transfer: 15 years later, *ASM News* **48:**561–565.

Wood, E. J. F., 1965, *Marine Microbial Ecology,* Chapman and Hall, Ltd., London and Reinholt Publishing Corporation, New York.

Yee, R. B., and Wadowsky, R. M., 1982, Multiplication of *Legionella pneumophila* in unsterilized tapwater, *Appl. Environ. Microbiol.* **43:**1330–1334.

Zeikus, J. G., 1977, The biology of methanogenic bacteria, *Bacteriol. Rev.* **41:**514–541.

Zeikus, J. G., 1982, Lignin metabolism and the carbon cycle. Polymer biosynthesis, biodegradation, and environmental recalcitrance, *Adv. Microb. Ecol.* **5:**211–243.

8

A COMPARISON OF THE ROLES OF BACTERIA AND FUNGI

D.M. GRIFFIN

INTRODUCTION

Most bacteriologists have little knowledge of fungi and probably harbor doubts that they are proper subjects for microbiological research. Most mycologists know little about bacteria and tend to regard them as organisms with bizarre metabolisms and lacking in morphological charm. Even those whose trade should encompass both bacteria and fungi, such as pathologists and ecologists, tend to lean strongly toward one or other of these two groups. My own leaning towards fungi will probably be all too obvious to readers of this chapter.

Making a comparison of the roles of bacteria and fungi in nature is a worthwhile but daunting task. To be done comprehensively, with a proper regard for underlying metabolic and physiological factors, a review of much of the literature of microbiology *sensu lato* would be necessary. I have not had time nor have I the breadth of expertise to conduct such a review authoritatively, so my treatment will inevitably be selective. Further, I have assumed that readers will be more familiar with bacteria than with fungi, and I have therefore given greater emphasis to fungi. Those requiring more detailed information on bacterial ecology will find it in other chapters and volumes of this treatise. The recent articles by Starr and Schmidt (1981) and Schlegel and Jannasch (1981) also provide excellent introductions to prokaryotic diversity and ecology.

Carlile (1980) has provided a valuable overview in an article entitled "From prokaryote to eukaryote: gains and losses." He pointed out that prokaryotes can maintain a biosphere without any assistance from eukaryotes, but that the reverse is not possible. Indeed, at the biochemical level, the capabilities of prokaryotes are those of life itself: limitations are imposed only by the restricted morphological diversity of prokaryotes associated with con-

D. M. Griffin • Department of Forestry, Australian National University, Canberra 2601, Australia.

straints on their size and structural complexity. Small size has its compensations, however, as evidenced by the extremely high specific growth rates of many bacteria, no doubt associated with high surface area to volume ratios and efficient diffusion within the very small cell.

Within the prokaryotes, all four major forms of metabolism (chemoautotrophism, photoautotrophism, chemoheterotrophism and photoheterotrophism) can be found, although chemoheterotrophism is the most common. Fungi, however, are all chemoheterotrophs. The literature abounds with statements that fungi also require oxygen for continued growth. Although many species are capable of anaerobic fermentation, there has been a general assumption that the energy yield from this process is insufficient to permit continued growth in anaerobic environments in competition with other specialized microbes. All this has been vigorously challenged by Emerson and Natvig (1981). They deny that all fungi are strictly aerobic and on the contrary have gathered evidence that (1) anaerobic fermentation with the production of lactic acid or ethanol is widespread in the lower fungi, (2) strict facultative anaerobiosis is widespread, (3) facultative anaerobiosis coupled with obligate fermentation occurs in a few species, and (4) obligate anaerobiosis occurs among some chytridlike fungi characteristic of the sheep rumen. The evidence now seems clear that some fungi are able to grow in nature in anaerobic environments such as stagnant waters and the rumen, although all may be auxotrophic for essential metabolites that can be formed only oxidatively. Because of the extremely limited knowledge of the ecology of fungi in anaerobic sites, however, I shall limit myself to a comparison of the roles of bacteria and fungi as aerobic decomposers (saprophytes) or parasites.

SIGNIFICANCE OF GROWTH HABIT

Unicellular and Filamentous Forms

In the previously cited article, Carlile (1980) suggested that prokaryotes have a limited capacity for specific cellular interactions and linked this with their slight development of colonial differentiation:

> In a population the arrangement of cells is random, cellular interaction fortuitous, and competition between cells total. In a colony there is a fairly well-defined distribution of cells, significant cell-cell integration and cooperation, but also some competition. In a multicellular organism, form is well-defined and the suppression of cellular autonomy and competition between cells is virtually complete. A spectrum of forms occurs between what is indisputably a population of unicellular individuals through the colonial condition to undoubtedly multicellular organism.

On these criteria, most bacterial "colonies" are not colonies at all: they are mere accumulations of cells. Colony formation however, occurs in the actinomycetes, cyanobacteria and myxobacteria, and colonies of *Streptomyces* have many of the organizational and morphological features of the simpler eukaryotes (Kalakoutskii and Agre, 1976). Such genera therefore provide a morphological bridge between the unicellular prokaryotes and the well-organized colonies of filamentous fungi. The hypha appears to be a very simple morphological advance on the unicellular condition (and the same organism may be able to exhibit both forms), yet it is in fact an innovation of the greatest ecological significance. The comparison between the unicellular and filamentous conditions, in terms of ecology, will be a major theme of this article.

Most bacteria are unicellular; only in the Actinomycetales are hyphae produced, perhaps being best developed in *Streptomyces*. In contrast, the hypha is the vegetative form of most fungi (Division Eumycota); only two groups of quite different morphologies are essentially unicellular in nature. The first is the order Chytridiales within the subdivision Mastigomycotina. Some members of this primitive group consist of a single spherical cell whilst others also produce fine rhizoids from the spherical body. These rhizoids (walled, aseptate, anucleate branching extensions of the thallus) may perhaps be seen as forerunners of hyphae. While the unicellular form is undoubtedly original in the chytrids, it is probably a secondary development from mycelial ancestors in the second unicellular group, the yeasts. The simple unicellular yeast form occurs during saprophytic growth mainly in members of the Endomycetaceae (Class Hemiascomycetes, subdivision Ascomycotina), Sporobolomycetaceae (subdivision Basidiomycotina) and Cryptococcaceae (subdivision Deuteromycotina). Further, some fungi that cause systemic mycoses grow saprophytically in the mycelial form but become unicellular when phagocytized within human or animal tissues. Vascular wilt fungi, e.g. *Ceratocystis* spp. and *Verticillium* spp., also form a yeastlike phase within host xylem although their more general growth is by hyphae. The yeast phase can be induced in culture in other groups that exist in nature exclusively in the mycelial form.

Fungi able to adopt both the mycelial and yeast forms are said to be dimorphic, and the factors determining form have been studied extensively (Romano, 1966; Stewart and Rodgers, 1978). The development of the yeast phase is associated with various ultrastructural changes that affect cell wall morphogenesis. These changes can be induced by a wide variety of environmental factors, of which temperature elevation is one of the most common (Anderson, 1978). Earlier interpretations of the experimental data pointed to a causal relationship between the adoption of the yeast form and impairment of oxidative metabolism. More recently it has been argued that cyclic adenosine-3′,5′-monophosphate (cyclic AMP) is a major unifying factor in dimorphism, high endogenous concentrations being associated with the yeast phase.

Translocation

That filamentous fungi are able to move substances from sites of absorption to sites of utilization is shown by the growth of hyphae into air away from the nutrient substrate. In hyphae, growth is restricted to the apex, where new cell wall and membranes are generated, and the processes whereby nutrients are moved from the sites of absorption to the apex is one aspect of translocation.

Translocation has been studied and reviewed extensively (Schutte, 1956; Burnett, 1968; Thrower and Thrower, 1968a,b; Wilcoxson and Sudia, 1968; Jennings et al., 1974; Eze, 1975; Read and Stribley, 1975; Jennings, 1976; Lucas, 1977; Howard, 1978; Nesbitt et al., 1981). Confusion occurs because of conflicting definitions of translocation. Some writers adopt a restricted meaning, limiting translocation to the movement of chemicals within hyphae independent of growth, e.g., through existing hyphae remote from apices or in directions contrary to that of hyphal extension. Other writers use a more liberal definition that includes nutrient movements associated with growth. Thus what is a translocating fungus to one group is nontranslocating to another. Translocation also depends on experimental conditions.

I shall use the concept of translocation in its wider connotation, and it can then be said that all filmentous fungi are likely to translocate. There is ample evidence for the translocation in many fungi not only of carbohydrates, and nitrogen and phosphorous compounds, but also of a wide range of other chemicals.

In many members of the Ascomycotina and Basidiomycotina, the movement of nuclei through pre-existing mycelia can also be demonstrated, and this is of importance in sexual reproduction. It is of interest that translocation of chemicals and nuclei occurs as much in septate as in aseptate hyphae. In most species, septa within active hyphae contain pores of various morphologies (Gull, 1978) which permit the passage of materials. In moribund hyphae, these pores become plugged, with obvious effects on ability to translocate. During nuclear migration, part of the complex pore of *Corprinus lagopus* is dissolved, resulting in pore enlargement (Giesy and Day, 1965).

There is still much uncertainty as to the mechanisms of translocation of chemicals, and their relative significance. Diffusion, cytoplasmic streaming and pressure-generated bulk flow have all been suggested but the first two appear inadequate to account for all the observational data and the third needs more investigation.

In unicellular bacteria and fungi, translocation, virtually by definition, does not occur. Indeed, in most prokaryotes the movement of metabolites within the cell must be entirely by diffusion since the cytoplasm is more viscous than in eukaryotes and streaming has not been reported (Starr and Schmidt, 1981). The significance of translocation in filamentous prokaryotes is unknown. Species of *Streptomyces* produce limited aerial hyphae and aerial sporulating structures, so presumably some translocation *sensu lato* must occur. Actinomycete mycelium is septate (Williams et al., 1973; Kutzner, 1981), but I am not aware whether pores exist. In the fungus *Geotrichum candidum,* the septa are without pores but adjacent "cells" are connected by plasmodesmata-like structures (Wilsenach and Kessel, 1965); it is perhaps significant that in this species aerial hyphae are very short and often become transformed into spore chains.

Translocation towards the apex is clearly of great ecological significance because it allows the hyphal apex to elongate through regions, whether these be in the gas, liquid or solid phase, that provide no or inadequate nutrients, depending upon nutrients absorbed in those more distant portions of the mycelium which are in contact with the required nutrients. It also permits hyphae to grow through gas phases toward a new substrate, guided by concentration gradients of a volatile attractant (Koske, 1982). Translocation may also occur in the other direction, from younger to older hyphae and this, as we shall see, has important ecological consequences in some fungi of complex morphology and in mycorrhizas (sect. 5.2).

Growth Pattern of Mycelial Colonies

In the absence of limiting factors, the growth rate of unicellular organisms is exponential in form. Thus $dx/dt = \mu x$ and, on integration, $ln x - ln\ x_0 = \mu(t - t_0)$ and $\mu = ln\ 2/t_d$, where x is the mass of culture at time t, t_d is the doubling time, and μ is the specific growth rate. Intuitively, there seems no reason why young filamentous colonies, in which all portions are equally active, should not function similarly, yet observation shows that the rate of radial extension of the colony is linear. This apparent contradiction has given rise to much research to elucidate the developing structure of filamentous colonies and the control mechanisms involved (Plomley, 1959; Pirt, 1966, 1967; Trinci, 1969, 1971, 1978, 1979; Caldwell and Trinci, 1973a,b; Bull and Trinci, 1977; Prosser and Trinci, 1979; Righelato, 1979, Hutchinson et al., 1980). Key elements of current understanding are as follows.

The active vegetative mycelium generates components needed for cell wall synthesis at a constant rate per unit volume (length) of hypha, the rate varying however with species and imposed conditions of growth. The components become aggregated in vesicles that travel to the tips of hyphae where they fuse with the plasma membrane, liberate their contents into the wall and so increase the surface area of the hypha, producing elongation. The rate of elongation will thus be related to the length of hypha contributing metabolites to the apex. This is borne out by the exponential rate of growth of germ tubes, until they reach a length of about 0.2 mm.

With increasing length, the rate of arrival of vesicles at the apex may exceed the rate of their incorporation into the wall. Apical branching then occurs, providing two apices instead of one to utilize the wall-generating vesicles. Alternatively, and more commonly, the concentration of vesicles may reach a critical level at some point further back along the hypha from the apex, and a lateral branch will be initiated there. Such lateral branches often form immediately proximal to a septum, where the forward translocation of vesicles is likely to be impeded. It follows that growth of a filamentous colony may be envisaged as the repeated production of "growth units" consisting of an apex and a length of hypha contributing wall-generating vesicles (and no doubt other metabolites) to that apical region. The growth unit in a young colony may be measured as the ratio of total length of hyphae to number of apices and it has been found to be constant for a given species under a given set of environmental conditions. Growing in this way, exponential growth can be maintained in *Mucor hiemalis* until the total length of hyphae in the colony is about 10 mm.

The rate of hyphal elongation in many fungal species is remarkably insensitive to nutrient concentration. In the light of the foregoing, this is easily understood because reduced nutrient supply will reduce branching before reducing the rate of extension of established apices. Nutrient concentration therefore affects colony density more than colony extension rate. The ecological significance of this has been lucidly expressed by Trinci (1969):

> The hyphal density control mechanism is presumably of significant selective advantage as it enables moulds to spread across substrates containing low concentrations of nutrients at near maximum radial growth rates; when a substrate containing a high concentration of nutrients is encountered, the colony grows more densely, produces aerial hyphae and may sporulate. Unicellular micro-organisms do not possess a specific mechanism to control the density of growth within the colony and typically form small colonies.

The continued growth of a filamentous colony implies progressive differentiation. At its simplest, this is merely a change from an actively growing periphery to a central zone of reduced activity or even senescence. The differences between undifferentiated and differentiated mycelia and the features of mature colonies have been extensively discussed by Bull and Trinci (1977) and Trinci (1979). In differentiated colonies, it has been shown that $r = K_r t + r_0 = \mu wt + r_0$, where r and r_0 are the radii at time t and zero, respectively, K_r is the radial growth constant, μ is the specific growth rate and w is the width of the peripheral growth zone (often of the order of 1 mm).

Such a model of colony growth implies that changes in culture conditions that alter hyphal density or the width of the peripheral growth zone will lead to nonlinear relationships between radial growth rate and specific growth rate. Temperature and nutrient concentration provide examples of variables for which the rates for radial growth and specific growth are and are not, respectively, approximately linearly related.

Striking visual evidence of the integration existing within fungal colonies is provided by the patterns of growth exhibited by those species with growth rhythms (which are rarely circadian). Most evidence has been gained from mutants of ascomycete species that show morphological rhythms when grown on nutrient agars in Petri dishes. In most, the wild type hyphae branch monopodially and growth is uniform. In the mutant, however, the branching system is largely sympodial so that dense, highly branched mycelia are formed (Chevaugeon and van Nguyen, 1969; Lysek, 1978; Kubicek and Lysek, 1982). In some cases, rhythms arise in the mutants because at the normal oxygen partial pressures existing on the agar surface, branching is sympodial and hyphae extend at a declining rate but with increasing numbers of branches. Hyphae buried in the agar, exposed to lower oxygen partial pressures, grow at a constant rate and thus overtake those at the surface. Those hyphae then reaching the surface from below, and just ahead of the now slow-growing surface hyphae, lead to a repetition of the pattern. Synchronization of the developing pattern throughout the colony is effected by both tangential and longitudinal movement of metabolites. This morphological rhythm has been shown to be associated with underlying changes in carbohydrate metabolism, respiration, energy turnover and synthetic activities.

The colony of a filamentous fungus can thus be envisaged as a mass of protoplasm lying within a system of tubes. The degree of connection between one portion of protoplasm and another will usuaully be high, reduced by the presence of septa and dead hyphae but enhanced by extensive anastomosis. In the most usual form of anastomosis, hyphal side branches fuse at their apices, thereby permitting transfer of metabolites, organelles, and nuclei. The interconnected protoplasm of a single colony may thus come to have a diverse genotype (see below).

Complex Structures

Physiological integration within a fungal colony becomes most obvious in those species that form complex reproductive or vegetative structures. The formation of the large sexual reproductive structures of mushrooms and toadstools obviously involves the simultaneous mobilization and transport of large quantities of metabolites and water from the vegetative mycelium. The construction of the basidiocarp also involves considerable differentiation and coordination of function amongst the component hyphae (Smith, 1966).

Sclerotia are fungal resting-bodies composed of dense aggregates of somatic hyphae. They are usually rounded or elongated structures, often with a well-defined differentiation into a compact, pigmented rind, a pseudoparenchymatous cortex and a central medulla. Varying in size from ca. 100 μm to over 25 cm in diameter, sclerotia are important survival structures. Sclerotial morphogenesis and physiology have been much studied (Chet, 1975; Chet and Henis, 1975; Willetts, 1978).

For the present purpose, however, the formation of complex vegetative growth structures is of more significance. These are commonly divided into mycelial strands and rhizomorphs; the latter are characterized by the coordinated apical growth of the hyphal aggregate which in the most highly evolved forms takes on the character of a pseudomeristem. The morphology and physiology of these structures has been much studied and the literature is now extensive. Much of it can be traced through Butler (1966), Motta (1969), Garrett (1970), Smith and Griffin (1971), Rishbeth (1978), Watkinson (1979), Clarke et al. (1980), Brownlee and Jennings (1981, 1982), Jennings and Watkinson (1982), and Thompson and Rayner (1982).

Ecologically, the significance of strands and rhizomorphs lies in the ability they confer for the transport of nutrients, water or oxygen from sources to remote sinks. The origin of a strand or rhizomorph is usually extensively developed mycelium lying within a moist, well-aerated nutrient substrate. Characteristic substrates are roots, leaf litter, or housing timbers. Absorbed nutrients are then translocated, or in some species oxygen diffuses or water passes, along the rhizomorph to the apices, which are thereby able to extend through sites which in themselves could not support growth. New foci of colonization occur if the apex comes into contact with a suitable substrate. Strands and rhizomorphs often extend for a length of 1 m and exceptionally to nearly 10 m, their growth sometimes depending entirely upon nutrients absorbed at the origin but occasionally absorption by the strand itself may be significant (Morrison, 1978).

Attachment and Penetration

In most situations, nutrient concentrations are higher within or adjacent to solids than in the surrounding aqueous phase. Adsorption to surfaces is therefore often critical to bacterial growth and survival, whether this be physical adsorption to clay micelles in soil or site-specific attachment to cells in the animal mucosa (Berkeley et al., 1980; Bitton and Marshall, 1980). Much less is known about the detailed factors involved in the attachment of fungal spores to surfaces, and in particular it is uncertain whether group-specific binding is involved in fungal adhesion to plants (Young and Kauss, 1982). Some fungal spores become impacted onto surfaces, others become entangled in leaf or nasal hairs, for instance, but in almost all cases except yeasts continued growth is dependent on the development of hyphae on and within the solid substratum. Thus, in the predominently aquatic environments of oceans and lakes, filamentous fungi grow within the solid substrates available, whether they be wharf timbers, algae or fish, and only their spores become free of the substrate (Jones, 1976; Kohlmeyer and Kohlmeyer, 1979). Yeasts are the only pelagic fungi (Fell, 1976), often being associated with algal blooms and thus approaching the life-style of some marine bacteria.

The possibility of unidirectional growth conferred by the hypha is important in the penetration of solids by fungi. Early studies were mainly investigations of how fungal germ-tubes penetrated the cuticle and epidermal walls of leaves. These showed that fungi produced from the germ-tube modified cells or clusters of cells, termed appressoria, which became firmly attached to the leaf surface and which gave rise to infection pegs that pierced the cuticle and epidermal walls (Emmett and Parbery, 1975). The adhesive material attaching the appressorium to the leaf has been shown to be a hemicellulose in at least one species (Lapp and Skoropad, 1978). Much of the earlier work indicated that penetration was largely mechanical through pressure exerted at the tip of the infection peg (so that even gold foil could be penetrated), but recent electron microscope studies make it most likely that penetration occurs through a combination of enzymatic weakening and mechanical pressure. Once within the leaf, penetration from cell to cell usually occurs without the production of any obvious appressorium and is thought to occur primarily by enzymatic means (Aist, 1976). English (1963, 1965, 1968) has also shown the importance of the hyphal form in permitting the saprophytic growth of fungi on keratinized substrates. Boring hyphae of nonkeratinolytic fungi appear to penetrate the substrates entirely by mechanical pressure, but enzymatic attack predominates with the more complex perforating organs of keratinolytic fungi. In the hard, dense keratin of nails and hair, and in other substrates with a laminated structure such as cellophane or bark, growth by fungi frequently takes the form of frondlike hyphae that erode the substrate along planes of structural weakness.

Regardless of whether it is by mechanical or enzymatic means, or both, the penetration of solids as diverse as leaf cutin, wood and keratin by well-defined narrow tubes is to be attributed to the possession of hyphae and this mechanism is not available to unicellular organisms.

The exertion of mechanical pressure by hyphae perhaps occurs in its most spectacular form when fruiting bodies (hyphal aggregates) of some higher fungi burst through soil or even such hard surfaces as tennis courts or paved streets, as recorded by Ramsbottom (1953).

PHYSIOLOGICAL ASPECTS

General

Some of the contrasting physiological features of bacteria and fungi are so well-accepted that they pass almost unnoticed. For example, the temperature set for incubators in most bacteriological laboratories is 37°C, contrasting with that of 25°C in mycological laboratories. Further, standard bacteriological media tend to have much lower carbon to nitrogen ratios than those used for fungi. The ratio in nutrient broth is 10:1, whereas that of potato dextrose agar is 180:1. Such accepted generalities, however, are sometimes derived from experimentation with a relatively narrow range of organisms.

The low carbon:nitrogen ratio favored by many bacteria is associated with their preference for utilizing amino acids and other nitrogen-containing compounds as substrates, whereas fungi generally prefer carbohydrates. Some fungi have developed to an extraordinary degree the ability to grow on substrates with extreme carbon:nitrogen ratios, such as wood with a ratio of 1600:1. This ability is associated with the internal translocation of cytoplasmic constituents or their autolytic products from older to younger parts of a colony, to an extent where growth can persist for appreciable periods through reutilizing various fractions of the existing mycelium as the sole nitrogen source. Nitrogen is also preferentially allocated to those metabolic systems that are highly efficient in utilizing the substrate (Merrill and Cowling, 1966; Levi et al., 1968; Levi and Cowling, 1969).

In recent years, there has been much research aimed at characterizing the norms and extremes of microbial growth and a number of related topics will now be considered.

Maximal Activity Rates

In much of the literature, the growth rates of filamentous fungi are expressed as rates of linear extension on solid (agar) substrates. Less frequently, rates of dry weight increase are given. Neither measure permits easy comparison with bacteria, particularly as only the marginal areas of conventional fungal colonies contribute to growth. In relatively recent years, however, following upon the analysis of colony growth described above, it has been realized that the earliest stages of fungal growth on solid media and long-term growth in chemostats can be analyzed to permit direct comparison with bacteria (Righelato, 1975, 1979; Solomons, 1975; Bull and Trinci, 1977; Forage and Righelato, 1979). On the assumptions made above μ (the specific growth rate) becomes a useful parameter for comparison. Values of μ_{max} (the maximal specific growth rate) for many bacteria lie around 0.8 hr^{-1} and may even approach unity; but some species of genera less frequently used in laboratory studies (*Arthrobacter, Spirillum*) have lower values of about 0.35 hr^{-1}. Values of μ_{max} for most fungi investigated lie within the range 0.1 to 0.3 hr^{-1}, but values of 0.61 (*Geotrichum candidum*) and 0.81 hr^{-1} (*Achlya bisexualis*) have been reported. As with bacteria, μ_{max} depends on temperature, that for *Aspergillus nidulans* increasing from 0.148 hr^{-1} at 25°C to 0.360 hr^{-1} at 37°C.

Reports of respiration rates are scattered widely through the literature. They indicate maximal oxygen uptake rates of 700 $\mu l\ mg^{-1}\ hr^{-1}$ (on a dry wt. basis) for many bacteria, with 2000 $\mu l\ mg^{-1}\ hr^{-1}$ being attained by some species. Rates for fungi are generally at least an order of magnitude lower. The difference between fungi and bacteria in terms of maximal specific growth rates and respiration rates is therefore as might be expected from the large dimensions of fungi.

Copiotrophs and Oligotrophs

Poindexter (1981a,b) has coined the term "copiotroph" to refer to microbes that "grow and multiply in the presence of an abundance of nutrients, and their natural distribution implies that nutrient abundance favours their survival and is an important factor in their competitiveness." Such organisms are to be contrasted with oligotrophs "whose survival in nature depends on their ability to multiply in habitats of low nutrient flux" and that characteristically occur only in such habitats. Because of their ease of isolation and growth, most bacteria that have been intensively studied are copiotrophs and there are relatively few precise data concerning oligotrophs.

Hirsch (1979) and Poindexter (1981a) have attempted to define the characteristics and consequences of the oligotrophic state, although it should be noted that some of their views have been challenged by Koch (1979). Amongst Hirsch's list of oligotrophic features are the following:

1. High surface/volume ratio.
2. Metabolic energy used preferentially for nutrient uptake, with low endogenous metabolic rate.
3. Constantly capable of uptake.
4. Large proportion of catabolic enzymes would be inducible whereas carriers would be constitutive.
5. Uptake systems of high affinity, with possibility for uptake and use of several carbon and energy sources simultaneously.
6. Uptake used preferentially for reserve accumulation so that net biosynthesis would occur only when uptake system and metabolic pools were saturated—hence relatively low maximal growth rates.

Hirsch (1979), Jannasch (1979), and Poindexter (1981a) have discussed the oligotrophic attributes of a number of bacteria, including *Arthrobacter* spp. and *Caulobacter crescentus*, and Matin and Veldkamp (1978) have studied experimentally the contrasting attributes of a *Spirillum* sp. and a *Pseudomonas* sp., selected for and against, respectively, in a chemostat with low lactate flux. Those studies have revealed that the supposed oligotrophic organisms do indeed exhibit such anticipated characteristics as low μ_{max} (0.35 hr^{-1}), very low μ_{min} (c. 0.01 hr^{-1}), high affinity for substrate ($K_s < 25$ μM, where K_s is the substrate concentration permitting half-maximal growth rate), high surface area/volume ratio and low energy of maintenance [less than 0.02 g g^{-1} hr^{-1} (dry wt.)]. In comparison, for copiotrophic bacteria such as *Escherichia coli* and *Pseudomonas* sp. μ_{max} is high (greater than 0.8 hr^{-1}), $K_s > 50$ μM, surface area/volume ratio is lower and the energy of maintenance exceeds 0.06 g g^{-1} hr^{-1}.

There are no comparable physiological studies of fungi directed at contrasting copiotrophs and oligotrophs. None the less, relevant data can be found in Righelato et al. (1968), Pirt (1973), Streensland (1973), Fiddiy and Trinci (1975), Righelato (1975), Solomons (1975), Bull and Bushell (1976), Mason and Righelato (1976), Trinci and Thurston (1976), and Bull and Trinci (1977). These reveal that the ratio μ_{min}/μ_{max} for *Aspergillus nidulans* and *Penicillium chrysogenum* is about 0.1, a higher ratio than in the oligotrophic bacteria, but comparable to that of copiotrophic bacteria. Maintenance energy requirements (glucose) for fungi in the same genera lie between 0.018 and 0.03 g g^{-1} hr^{-1} (0.04 for *Mucor hiemalis*), considerably lower than that reported for copiotrophic bacteria, but comparable to *Spirillum*.

The affinity for energy-yielding substrates reported varies widely, K_s

being about 500–600 μM for *Aspergillus nidulans* and *Saccharomyces cerevisiae* (Bull and Bushell, 1976), but less than 25 μM for *Fusarium aquaeductuum*, and *Geotrichum candidum* (Streensland, 1973; Fiddy and Trinci, 1975). *Neurospora crassa* has two glucose transport systems with K_s of 8 mM and 10 μM, the high-affinity system being repressed at high glucose concentration (Scarborough, 1970a,b; Schulte and Scarborough, 1975). The K_s values for the high-affinity systems in fungi are therefore similar to those reported for oligotrophic bacteria. It should not be assumed that high and low K_s values under different conditions necessarily imply the existence of two distinct systems because the change in affinity for phosphate with change in pH shown by an alga is associated only with the protonation of the carrier (Tanner, 1974; Bull and Trinci, 1977).

Temperature

The generally higher optimal temperatures found for bacterial, compared to fungal, growth are in accord with the greater development of thermophily in bacteria. Growth at high temperatures has received much attention (Crisan, 1973; Brock, 1978; Tansey and Brock, 1978; Amelunxen and Murdock, 1978; and several articles in both Heinrich, 1976 and Shilo, 1979), but its physiological basis is still uncertain. The term "thermophile" has been defined in many ways, and commonly adopted definitions in connection with prokaryotes and fungi differ widely (Cooney and Emerson, 1964; Farrell and Campbell, 1969).

Whereas some chemoheterotrophic and chemoautotrophic bacteria can grow at temperatures higher than 90°C and the ability to grow at 70–73°C is spread throughout all the prokaryotic groups, fungi fail to grow at temperatures higher than 62°C. Thermophilic prokaryotes are therefore characteristically found in high-temperature aquatic environments, such as hot springs and hot water systems, whereas thermophilic fungi are most often found, along with bacteria (especially actinomycetes), in moist particulate ecosystems, such as composts, stored products and wood chip piles. In these latter systems, spontaneous heating occurs because of microbial thermogenesis arising from the utilization of readily available soluble substrates (Lacey, 1980), with prokaryotes dominating at the peak temperatures (Fergus, 1964; Greaves, 1975). Many thermophilic and thermotolerant fungi can be isolated from a great variety of soils but the significance of this is still a matter for debate (Tansey and Jack, 1976, 1977; Tansey and Brock, 1978). Most soils are rarely simultaneously sufficiently damp and hot from insolation alone to permit growth of thermophilic organisms, but the occurrences of these conditions may be sufficient to permit growth and reproduction adequate to maintain soil populations. It is certainly difficult to believe that they consist entirely of spores derived from other ecosystems.

As with thermophile, the term "psychrophile" has been variously defined (Baross and Morita, 1978), although most definitions incorporate the concept of growth at 0°C. The ability to grow at 5°C is a widespread attribute and a number of microbes can grow at 0°C and some even at −5°C (Cochrane, 1958; Deverall, 1968; Griffin, 1972; Baross and Morita, 1978). Again, prokaryotes are most important in aquatic environments, whereas fungi are most significant as snow moulds of crops (Bruehl et al., 1966) and as agents of deterioration of food products in cold storage. The mechanisms that permit microbial activity at low temperatures have been reviewed by Inniss and Ingraham (1978), but knowledge is insufficient to make any comparisons of bacterial and fungal physiologies in this regard.

Hydrogen Ion Concentration

Standard bacteriological media are near-neutral in reaction, whereas many fungal media are acidic (pH 5–6.5). This might be supposed to reflect general pH optima, but this is not precisely so. Many fungi have a very broad pH optimum, whereas the tolerance of many bacteria to departures from near-neutrality is less. Thus, the isolation and continued cultivation of fungi free from bacterial contamination is aided if the medium is of a hydrogen ion concentration that reduces the effectiveness of bacterial antagonism. Similarly bacterial populations characteristically decline as soil acidity increases from neutrality, whereas the converse is true for fungi, largely because of changes in bacterial antagonism (Alexander, 1961). Nevertheless, exceptions abound and specialized species or strains of both bacteria and fungi are able to grow at extremes of pH 1 and pH 11 (Langworthy, 1978).

Thiobacillus spp. are perhaps the best known microbes active at high hydrogen ion concentrations. Through the oxidation of sulphur or its reduced compounds to produce metabolically useful energy, the microbes generate sulphuric acid and are able to grow at pH 1-2. They are characteristic of hot acid springs, mine residues, and some acidic, reduced muds, and are obligate extreme acidophiles. In contrast, fungi tolerant of extremely low pH values also grow on neutral media. Two tolerant fungal strains have been isolated from, and grew on, media containing 2.5 N sulphuric acid (Starkey and Waksman, 1943).

Extreme alkaline environments are usually associated with arid soils, or lakes in arid areas. Few fungi appear to grow naturally in environments with pH values exceeding 9. Langworthy (1978) has reviewed the little that is known about the physiological adaptations related to growth at extreme pH values.

Water Stress

Microbes are frequently subject to stress because of some factor in the environment directly associated with the water regime. An obvious example is the stress caused by the presence of high concentrations of solutes in the extracellular solution, whether for instance this be the ocean or the solution in the nectaries of a plant. Microbes also experience a fundamentally similar stress when they are enveloped in an unsaturated porous matrix, such as soil. In all such cases, the key concept is that of water moving along gradients of potential energy. Water will be available to a microbe only if it can establish the appropriate potential energy gradient between the cell and its environment, and this will usually involve the organism in the expenditure of chemical energy.

In the Systeme International d'Unites (SI), energy is measured in joules and in the case of potential energy of water as joules per unit quantity of water. The unit quantity of water may be the mole, kilogram or cubic metre. As, on the volumetric basis, Jm^{-3} is the same as the pressure unit pascal (Pa), much of the relevant modern literature expresses water potential in kilopascals or megapascals (kPa, MPa) and these units will be used here. The bar, where 1 bar = 10^5 kPa, is also still often used as a pressure unit, but its use as an SI unit is restricted to meterology and it is therefore invalid in the present context. I shall also simplify the system by assuming that the water potential to which an organism is exposed has only two components. The first, solute potential, refers to the reduction in potential produced by the presence of solutes. The second, matric potential, refers to the reduction in potential produced by the water being held within a porous system (matrix). There, the forces involved are associated with the interfaces between liquid, solid and gas. In most cases, environmental water potentials are less than that of pure, free water and hence the energy values are negative.

The water potential of the cell itself also has matric and solute components but in addition has a third, turgor potential. This arises, for instance, as a positive balancing wall pressure of 1 MPa for an organism with an internal solute potential of -1.2 MPa when immersed in a broth of solute potential -0.2 MPa (Luard and Griffin, 1981).

Another major system for expressing energy values uses the concept of water activity. For both the underlying theory and the methods for converting data between the systems the articles by Griffin (1981b) and Papendick and Campbell (1981) should be consulted. Current knowledge of the relationships between water and microbial growth and stress has been reviewed (Brown, 1976, 1978; Griffin, 1978, 1981b; Kushner, 1978; Harris, 1981; a number of articles in Shilo, 1979), and only a few salient aspects will be noted here. Of prime importance is that the water potential of cells must be less than that of their environment if they are to gain water from that environment. In most habitats, most of this reduction in cellular water potential is obtained through the presence of intracellular solutes. As high concentrations are often required, these solutes must be compatible with continued metabolic activity and especially enzyme functioning (Brown, 1978, 1979).

Living cells, regardless of water stress, contain basal metabolic solutes and their concentration sets the highest (least negative) solute potential attainable by the cell. Some microbes also accumulate reserves, such as polyols, and these may also contribute significantly to cellular solute potentials. When the cytoplasmic solutes of a microbe are those present solely because of their function as basal metabolites or reserves, its adaptability to water stress is very limited. Harris (1981) has suggested that such fresh-water bacteria as *Spirillum* may be of this type. If, however, the microbe can produce, or change the concentration of, internal solutes in response to changes in environmental water potential, then its ability to withstand and adapt to water stress is greatly increased. Most microbes are probably able to produce controlled adjustments to their cytoplasmic solute potential in this way, and a number of compatible solutes used in osmoregulation have been identified (Brown, 1978; Griffin, 1981b; Luard and Griffin, 1981; Luard, 1982a,b,c). Potassium glutamate and proline are characteristic of many bacteria whereas polyols, especially glycerol, are characteristic of fungi. Proline, however, is predominant in the few lower fungi (Mastigomycotina, Zygomycotina) investigated. The compatible solutes of actinomycetes are unknown.

Variation in ability to grow at reduced external solute potentials is great, with limiting potentials varying from −1.5 to −69 MPa. Many organisms show maximal growth at solute potentials of −0.1 to −1 MPa; others are obligately xerophilic and are unable to grow at high solute potentials. Griffin (1981b) has suggested that microbes might be divided into five groups on the basis of their reported growth response to reduced solute potentials (Table I).

Whereas it might be anticipated that the response of unicellular and filamentous organisms to reduced solute potential in an essentially aqueous environment would be fundamentally similar, this is not true for reductions in matric potential within porous solid systems. Within such a matrix, e.g. soil, pores of decreasing radii will be progessively emptied of solution, and filled with the gas phase, as matric potential declines. Hyphae will still be able to bridge air-filled pores and thus continue to spread from one particle of substrate to another, but bacteria and other unicellular organisms will progressively be restricted in their mobility, becoming virtually imprisoned within remaining water-filled pores and lenses. The ability of a bacterial species to grow at a given matric potential per se may then be of no benefit if its cells are unable to contact new substrates when those originally available to the cell have been utilized (Griffin, 1981a).

Some filamentous fungi produce mobile zoospores and these, like bacteria, are dependent upon the presence either of surface water or of water-filled pores of the appropriate dimensions within soil for their appreciable movement. The infection of roots by such fungi is therefore extraordinarily sensitive to changes in the soil water regime and matric potentials exceeding −0.1 kPa are generally required for unimpeded zoospore migration (Griffin, 1981a; Benjamin and Newhook, 1982).

Within systems such as soil, therefore, changes in matric potential will be a potent selective factor in determining activity, both through direct effects as discussed earlier and through changes in ability to spread (Griffin, 1978, 1981a,b). Taking into account the direct and indirect effects of solute and matric potentials, it becomes possible to produce a classification of xeric en-

TABLE I

Response of Microorganisms to Solute Potential[a]

Group	Solute potential (MPa)		Organisms
	Optimum	Minimum	
(1) Extremely sensitive	−0.1	−2	Wood-decay basidiomycetes
			Some soil basidiomycetes
			Some gram-negative bacteria
(2) Sensitive	−1	−5	Most phycomycete and coprophilous fungi
			Many other fungi
			Most gram-negative rods
			Many soil actinomycetes
(3) Moderately sensitive	−1	−10 to −15	Many fungi
			Some yeasts
			Many soil actinomycetes
			Many gram-negative bacteria
(4) Xerotolerant	−5	−20 to −50	Many yeasts
			Fungi *(Aspergillus, Eurotium, Penicillium, Eremascus, Wallemia)*
			Some saline lake bacteria
(5) Xerophilic	−4	−40 to −69	Fungi (*Monascus bisporus*), *Chrysosporium fastidium*, some isolates of *Aspergillus restrictus* and *Saccharomyces rouxii*,
			Actinospora halophila (halophilic bacteria)

[a]Adapted from Griffin (1981b).

TABLE II

A Classification of Xeric Environments and Their Characteristic Microfloras[a]

Group	A		B	C	
	a	b		a	b
Major constituent	Liquid		Porous solid	Porous solid	
Solute potential	Low		High	Low	
Matric potential	High		Low	Low	Variable
Nutrient concentration	Low	High	Low	Low	High
Examples	Salt lake	Syrups and preservative brines	Soil of low water content	Saline soil of low water content	Stored food
Characteristic microorganisms	Bacteria and algae (*Halobacterium, Halococcus, Dunalliela*)	Yeasts (*Debaryomyces hansenii, Saccharomyces rouxii*)	Filamentous fungi (*Aspergillus, Penicillium*)	Filamentous fungi (*Aspergillus, Penicillium*)	Filamentous fungi (*Aspergillus, Penicillium, Chrysosporium, Eremascus, Monascus, Wallemia*)

[a]From Griffin (1981b).

vironments (with lower water potentials) and to indicate their characteristic microfloras (Table II). In accord with the above considerations, unicellular microbes are seen to be most important in systems where the liquid phase predominates, filamentous fungi where the solid phase predominates.

The water relations of fungi and bacteria have been particularly well-studied in two contexts and appropriate reviews should be consulted for more detailed information. I refer here to biodegradation (Ayerst, 1968; Duckworth, 1975) and plant pathology (Griffin, 1977; Kozlowski, 1978; Duniway, 1979; Cook and Duniway, 1981).

Genetics

Knowledge of bacterial genetics has leapt forward in recent years, but here I shall merely note that the vegetative bacterial cell possesses a single circular chromosome and that variation occurs through mutation, conjugation, transformation, and transduction. Fungi possess typical eukaryotic nuclei dividing by mitosis, and the nuclei of many species undergo meiosis in connection with a haploid-diploid life cycle (Fincham et al., 1979). In most groups, the vegetative hyphae are basically haploid although the latest evidence is that those of the Mastigomycotina are diploid. Members of the Basidiomycotina, however, have vegetative hyphae which are predominantly dikaryotic (Casselton, 1978). The dikaryotic state is haploid but pairs of nuclei of different mating types associate, generally undergo simultaneous mitosis, yet they do not fuse to form diploids until later in the life cycle. Dominant-recessive characteristics of genes may be fully or partly expressed in dikaryotic hyphae. The genetics of fungal mating systems is remarkably complex (Stamberg and Koltin, 1981).

Although the haploid, dikaryotic and diploid natures of the vegetative hyphae of various species are often assumed, they are in fact gross simplifications of commonly occurring real situations. Vegetative anastomosis between hyphae of different genotypes (see above) often leads to hyphae becoming heterokaryotic, that is containing nuclei of different genotypes. (The dikaryotic condition is but one special form of heterokaryotism). A fungal colony can thus come to contain a mixed population of nuclei, with the frequency of genotypes changing, at least sometimes, in response to environmental selection.

Further, a parasexual cycle exists in many fungi whereby nuclei fuse in the vegetative hyphae and the resultant diploid multiplies alongside the original haploids (Caten, 1981). Such diploids sometimes undergo mitotic crossing-over and subsequent haploidization, yielding haploids with genotypes different from either of the originals. In the vegetative hyphae of the basidiomycete *Phellinus noxius,* Bolland (1980) has demonstrated extreme concurrent heteroploidy, from haploid to octoploid, although diploidy was the commonest state. A simplistic view of fungal cytogenetics would have led to the expectation that this species would have been haploid dikaryotic; no doubt further surprises are in store. Cytoplasmic inheritance has also been demonstrated in fungi (Turner, 1978).

The diversity of cytogenetical systems in fungi no doubt has implications for their ecology, but these have been relatively little investigated beyond the specialized areas of plant pathology (Webster, 1974; Day, 1978; Ellingboe, 1981) and industrial mycology (Johnston, 1975). It is of interest to note that existence of the "gene-for-gene" systems has been demonstrated in many plant host:fungal pathogen systems, but that this has not occurred for bacterial plant pathogens (Sidhu, 1975). The gene-for-gene hypothesis concerning the specificity of host:pathogen interactions supposes that for each resistance gene in the host there is a specific, related virulence gene in the pathogen. The apparent absence of such a relationship involving bacteria may be associated with the intercellular habitat of plant pathogenic bacteria, which must to some degree militate against the development of specific, cell-based resistance mechanisms.

SOME IMPORTANT MICROBIAL ECOSYSTEMS

Soil

Original and review articles on soil microbiology are legion and here attention will be focused on only a few aspects involving the comparison of fungal and bacterial roles as aerobic chemoheterotrophs. Fuller treatments have been given by Alexander (1961), Gray and Parkinson (1968), Gray and Williams (1971a), Griffin (1972), and Richards (1974). Further, many aspects of the roles of fungal communities in various ecosystems have been considered by a number of authors in Wicklow and Carroll (1981).

Soil is an unusually complex environment and the description and quantitation of its physical, chemical and biological components present immense problems that have yet been only partly solved. Direct microscopic examination of the soil, or of sections prepared from it, is fraught with difficulties. If enumeration in culture is relied upon, all media and cultural conditions are more or less selective. Quantitation of fungi raises particular problems in deciding on the meaningful unit. Spores represent the product of past rather than current activities, and the intensity of spore production varies greatly with species. Measurement of hyphal lengths, or of number and size of colonies, is usually extremely difficult if not impossible. Further, the differentiation between active, senescent, dead, and dormant cells or hyphae is a problem with all microbes.

Despite these problems, it has been established that the numbers of bacterial cells in soil greatly exceed those of fungal propagules, where propagule here implies any unit that gives rise to a colony. The difference is usually at least of the order of one hundredfold. The actual biomass of fungi, however, exceeds that of bacteria, again often by greater than a hundredfold. Because of the higher maximal rates of growth and respiration of many bacteria, it is easy to assume that this group is more important that fungi in the general decomposition of soil organic matter. This may be true in soils at very high matric potentials, when solute diffusion will occur rapidly (Papendick and Campbell, 1981) and bacterial movement, whether active or passive, is possible (see above). As matric potentials fall, however, it may well be that fungal activities predominate over bacterial, and this is supported by the limited experimental data available (Dommergues, 1962; Anderson and Domsch, 1973; Wilson and Griffin, 1975; Faegri et al., 1977). Certainly the total rate of decomposition is extremely sensitive to water potential (Sommers et al., 1981).

Whereas a fairly clear understanding has emerged of the effects of the water regime on microbial activity in soil (see above), the same can scarcely be said concerning the influence of soil nutrient substrates. In early studies, Winogradsky divided soil bacteria into two groups, autochthonous and zymogenous. The former included those which maintained relatively uniform levels of population and activity regardless of the changing availability of ephemeral substrates, whereas the latter included those species whose populations fluctuated greatly. Previously (Griffin, 1972) I noted that "all soil fungi for which we have much information are zymogenous," and I went on to support Clark's (1967) contention that "the autochthonous bacteria are those for which the environmental and nutritional conditions needed to evoke a zymogenous response are still unknown." I do not think this position can now be sustained in the light of the work on oligotrophic bacteria described by Poindexter (1981a,b) (see above). There is now considerable evidence, although most of it is still circumstantial, that species of *Arthrobacter* are autochthonous, oligotrophic soil bacteria. Whether there are comparable fungi in soil remains to be demonstrated.

Although exact quantitative definition is difficult, it has long been apparent that in most soils most of the time the amount of utilizable carbon provides a severe limitation on heterotrophic microbial activity (Gray and Williams, 1971b; Lockwood and Filonow, 1981). The immediate consequence is that the rate of cell division must be very low, and many studies indicate less than one hundred, and sometimes less than one generation in each year. The further consequence is that most of the microflora must be inactive for most of the time and that when it is active, specific growth rates will probably be far below maximal. The value derived for *Arthrobacter* sp. in soil is 0.029 hr^{-1} (μ_{max} = 0.37 hr^{-1}). No data exist that permit comparisons of the growth patterns and metabolic activities of known species of bacteria and fungi within soil sites of graded nutrient substrate availability.

As the availability of substrates declines, there will be a progressive transition by the microflora to the nongrowing state (Dawes, 1976; Gray, 1976; Trinci and Thurston, 1976). Persistence in the growing state as nutrient availability declines is associated with a number of interrelated factors, including low minimal specific growth rate (μ_{min}), high substrate affinity (low K_s) and low maintenance energy requirement. These factors have been discussed above in association with oligotrophs (see above), but any worthwhile body of comparative data for bacteria and fungi is lacking.

Persistence in the nongrowing state may be in two forms, as vegetative cells and as dormant spores or other specialized resistant cells. Many bacteria survive in soil and elsewhere as apparently unspecialized vegetative cells, and some fungi also survive as hyphae. Fungi in the great majority of instances, however, respond to decreased nutrient availability by sporulating (Smith, 1978), and the spore forms the resting structure. Some bacteria, of course, also form specialized dormant spores.

The length of survival of both vegetative cells and the spores of fungi and actinomycetes depends upon the amount of reserve energy-yielding metabolites stored within the cells and on the rate at which they are utilized through endogenous respiration (Sussman, 1968). Obviously survival rates will vary with the organism, but they are also greatly influenced by the conditions in which the cell was formed and in which it must survive. Some bacteria do not accumulate reserve materials and on starvation are forced to utilize RNA or cell wall components. Such organisms die rapidly. In other bacteria, the free amino acid pool acts as an endogenous substrate, but many accumulate glycogen (polyglucans). In fungi the main reserve is carbohydrates, especially polyglucans and polyols or lipids (Blumenthal, 1976). Longevity will be affected by changes in endogenous respiration induced by temperature and by other environmental factors to which the organism is exposed during the nongrowing period.

Bacterial endospores are highly adapted cells conferring exceptional resistance to heat and survival over very long periods, perhaps associated with the lack of any demonstrable metabolic activity. They have received great attention, as evidenced by the series of volumes entitled "Spores" published by the American Society for Microbiology. Ellar (1978) has discussed the relationship between structures specific to endospores and their function. Among these are certain membrane structures and the spore cortex, while chemical features are high contents of calcium ions and dipicolinic acid and low water content.

Detailed knowledge of the micromorphology and physiology of fungal spores is of relatively recent origin and less in quantity than that for bacterial endospores (Weber and Hess, 1976). Most fungal spores have been shown to have complex multilayered walls but many other attributes are still uncertain. The water content of spores is reported to vary with species from 6–88% of fresh weight, indicating the likelihood that functions vary equally widely. Indeed, longevity is known to differ greatly with species and spore type, perhaps in association with short-term dispersal in distance, or long-term survival where first formed.

Fungal spores in soil exhibit two types of dormancy, constitutional and exogenous (Sussman, 1966). Constitutional dormancy is exhibited by oospores, zygospores, and some ascospores, in which development is prevented by some innate characteristics such as the need for some nuclear, membrane, or physiological change of state sometimes associated with exposure to temperatures of 50–60°C (Anderson, 1978). Exogenous dormancy is exceedingly common and is a condition imposed by an unfavorable chemical or physical environment. In this case, the spore germinates immediately on transferral to a favorable environment. Low temperature is an obvious causal factor in exogenous dormancy but is clearly incapable of explaining the dormancy of most fungal propagules in soil. The reasons for this phenomenon, often referred to as fungistasis or mycostasis, have been fiercely debated, but it now appears that nutrient insufficiency is the major cause, aided by a generalized antibiosis (Lockwood, 1977; Lockwood and Filonow, 1981). Bacterial-fungal relations are involved because nutrient deprivation is in part caused by bacteria adjacent to the spore; their metabolic activities provide an immediate sink for nutrients exuding from the spore and thus increase the concentration gradient.

Lichens

Lichens are symbiotic associations between fungi (usually Ascomycotina but rarely Basidiomycotina) and either cyanobacteria (prokaryotes, although often referred to as blue-green algae) or green algae. Despite their obvious presence in most terrestrial environments, the experimental study of lichens is a relatively recent development, and much of existing knowledge can be traced through the publications of Ahmadjian and Hale (1973), Smith (1975, 1979), Brown et al. (1976), and Seaward (1977).

The photosynthetic partner in the lichen forms less than 10% of the dry mass and is more or less localized in a layer just under the lichen surface. Photosynthetic rates are high, but net primary productivity of the lichen as a whole is exceptionally low and the radial growth rate is usually of the order of 1 mm $year^{-1}$. The reasons for this unusual situation are slowly being clarified.

Although the photosynthetic partner releases negligible carbohydrate into the culture medium when grown in pure culture, up to 80% of the net carbon fixed within the lichen is transferred from the photosynthetic component to the fungus. This transfer occurs as glucose in the case of cyanobacteria and as a polyol in the case of green algae (Richardson, 1973; Hill, 1976). Within the fungus, conversion of the carbohydrate to one of the characteristic fungal polyols (Lewis and Smith, 1967), predominantly mannitol, occurs rapidly. Up to 10% of the dry weight of some lichens consists of polyols and it has been suggested that these compounds assist in physiological buffering whereby the lichen is protected against the extremes of environment in which they occur.

Smith (1975, 1979), in particular, has argued that lichens have evolved to exploit environments, such as rock, bark and bare soil surfaces, that are both persistent and characterized by continual severe cycles of drying and wetting. Lichens both lose and regain water very rapidly in response to environmental change, and are able to resume physiological activity within minutes of rewetting. The basis of this tolerance is still not fully understood, but is likely to involve the high concentrations of polyols present. These compounds are compatible solutes that will greatly reduce the water potential of the lichen (Griffin, 1981b) and may help to maintain the hydration, and hence structural integrity, of biopolymers through water substitution (Schobert, 1977; Schobert and Tschesche, 1978). Smith (1979) has further argued that these physiological attributes, and not nutrient deprivation, lead to the slow growth rate characteristic of lichens. Experimental augmentation of nutrient supply does not greatly increase growth rate but, on the contrary, often results in the breakup of the symbiotic association. Stability of the association is also enhanced by cycles of drying and wetting (Smith, 1975).

Lichens occur in polar areas, hot arid lands, and in the central Asian deserts, which experience extremes of both hot and cold (Kappen, 1973). Clearly, terrestrial temperatures as such do not restrict the presence of lichens, although they may affect activity. Some lichens are able to photosynthesize slowly at freezing temperatures, but even a desert lichen suffered irreversible damage when exposed in a water-saturated condition at 38°C. In the dry state, however, 65°C was survived.

In all lichens, the fungus is dependent on its associate for supply of carbohydrates, but only if the associate is a cyanobacterium is there a comparable transfer of fixed nitrogen (Millbank and Kershaw, 1973; Millbank, 1976). It is likely, however, even if the associate is a nitrogen-fixing prokaryote, that lichens obtain most of their nitrogen from the environment in an already combined form, and this must be entirely so if the associate is a green alga. The obtaining of combined nitrogen compounds is unlikely to be a problem because lichens are extremely efficient in absorbing many chemicals present in the environment in low concentration. This ability makes lichens extremely sensitive indicators of environmental pollution (Ferry et al., 1973; Gilbert, 1973, and chapters in Brown et al., 1976).

Associations between Microbes and Higher Organisms

Bacteria and fungi are involved in a wide range of interactions amongst themselves and with animals and plants; comprehensive classifications of interactions involving fungi have been given by Cooke (1977) and Culver (1981).

A considerable number of bacterial species are consistent associates on and within the bodies of animals, for examples, in gastrointestinal tract and in the rumen of herbivores (Schlegel and Jannasch, 1981). The main existence of these organisms is in these sites and most are ill-adapted to a saprophytic existence apart from their animal hosts. This is true whether the bacterium forms a normal part of the microbiota of the healthy animal or is a pathogen (Isenberg and Balows, 1981). The situation with fungi is quite contrary. Few fungi, with the dramatic exceptions of the Laboulbeniomycetes (Benjamin, 1973) and some chytridlike fungi in the sheep rumen (Emerson and Natvig, 1981), are constant and characteristic components of the microbiota associated with animals, and it is significant that the most common fungal associate of man, *Candida albicans,* is largely yeast-like in its morphology. Some other yeasts are also often isolated from various parts of the mucous membrane. Most fungi parasitizing man and animals, however, have their major reservoir in a saprophytic existence, often in soil or in organic deposits of plant or animal origin (Austwick, 1972). Unlike bacteria, relatively few fungal pathogens are therefore transmitted directly from animal to animal, so that epidemics caused by fungi are unusual and at no time approach the scale of those caused by bacteria and viruses. Apart from the members of the Laboulbeniomycetes, which are obligate parasites of insects, the nearest approximation is perhaps in those anthropogenic members of the dermatophyte fungi that have no saprophytic existence independent of their human host. The essentially opportunistic nature of most fungal infections of animals has the corollary that many fungal species prove to be unexpectedly able to colonize tissues to which they have fortuitously gained entry, often through a wound or by inhalation (Conant et al., 1954). Examples are infection of man by *Basidiobolus ranarum,* more usually associated with the faeces of frogs and lizards, and *Cercospora apii,* a pathogen of celery and tomatoes.

The maximum temperature for growth of many fungi is about 37°C and this was at one time thought to be a major reason for the relative unimportance of fungal diseases of animals. A significant number of fungi can, however, grow at 35–40°C and the surface tissues, particularly of the extremities, of animals are often somewhat below 37°C. It therefore seems likely that the establishment of infections by many fungi fails because of active host resistance.

Progressive infection by fungi is often aided by host debility or certain forms of chemotherapy. Thus, candidiasis is often associated with the administration of oral antibiotics, and the spread of phycomycete hyphae within the walls and lumina of blood vessels is associated with both diabetes and cortisone therapy.

Adhesion to specific sites on cell surfaces is important in the survival of many bacteria on and within the animal mucosa, but the same appears not to be so for fungi (Berkeley et al., 1980; Bitton and Marshall, 1980). After attaining initial adventitious entry, subsequent growth can take a number of forms. Ramifying hyphal growth within the cutaneous tissues occurs in dermatomycoses, circumscribed hyphal growth within phagocytes in aspergillosis and maduromycosis, and growth in the yeast-phase within phagocytes in many systemic mycoses. It is noteworthy that the granular aggregations characteristic of maduromycosis are formed by both fungi and actinomycetes, the filamentous morphology of each responding similarly to host defense mechanisms.

Toxins are involved in disease causation by bacteria and fungi in both plants and animals. Toxic action in plants is, however, generally far more circumscribed and takes a less overt and dramatic form than in animals (Patil, 1974; Rudolph, 1976; Scheffer, 1976). This may reflect the absence of both an active circulatory system and an internal sensitive mucous membrane in plants. Toxins produced by fungi while growing saprophytically on food stuff of plant origin are often a cause of severe metabolic disturbance in animals that eat them (Mirocha and Christensen, 1974).

In sharp contrast, diseases of plants are far more frequently caused by fungi than bacteria (Horsfall and Dimond, 1959; Agrios, 1969; Perombelon and Kelman, 1980; Starr, 1981) and the involvement of some prokaryotes (mycoplasmas, spiroplasmas, and rickettsialike organisms) is still being elucidated (Schneider, 1970; Maramorosch, 1976; Nienhaus and Sikora, 1979). The prime reason for this difference is the general inability of bacteria to breach the external envelope of the plant; infection is entirely dependent upon wounds or natural openings such as stomata and lenticels (Goodman, 1976). Fungi, however, are able to penetrate without difficulty the cuticle and cell walls of plants by means of appressoria, infection pegs, and hyphae (see above). Again, in contrast with bacteria infecting animals, most plant pathogenic bacteria have a saprophytic existence independent of the host, many surviving for long periods in soil (Leben, 1981). Many plant pathogenic fungi, however, are either obligate parasites or facultative saprophytes, the latter implying that the species is able to grow as a saprophyte, e.g., in culture, but is unlikely to be able to compete with the microflora in a sapropytic existence.

Plant pathogenic bacteria grow intercellularly or even essentially outside the plant, for instance, between bud scales. Such an occurrence has close parallels in animal pathology, because of the extracellular location of most bacterial pathogens of animals. Many species of fungi, however, grow within the living cells of their plant host and their pathogenicity, in distinction from their parasitism, becomes only slowly manifest. Such a biotrophic life-style is incompatible with toxin production. Amongst plant pathogenic fungi, tissue-decomposing enzymes and toxins are most commonly produced in abundance by those necrotrophs which grow in the host cells only after host cell damage and tissue disintegration are well-advanced (Bateman and Basham, 1976).

Crown gall disease of plants has uniquely bacterial aspects (Merlo, 1978). *Agrobacterium radiobacter* var. *tumefaciens* is able to enter roots through wounds and subsequently a plasmid is transferred to the host cells. The plasmid DNA is reproduced and expresses itself through unregulated host cell division and gall formation. Only the variety *tumefaciens* of the species contains the tumor-inducing plasmid.

The success of fungi as epidemic plant pathogens (Horsfall and Cowling, 1978; Scott and Bainbridge, 1978) is also associated with their efficient means of dispersal especially in spore liberation. The dispersal of pathogenic bacteria is frequently in water, whether this be in drinking supplies, e.g., *Vibrio cholerae,* or air-borne droplets, e.g., *Pseudomonas tabaci.* Insect transmission is also often important, as with *Pasteurella pestis* and *Erwinia amylovora.* Whilst both splash and insect dispersal are important for fungi (Ingold, 1953), many have developed special structures which assist in the efficient liberation of their spores, particularly into moving airstreams (Ingold, 1965). This is of great significance in plant disease epidemiology (Meredith, 1970). All aerial surfaces are surrounded by a microscopically thin still-air layer and between this layer and the general turbulent air is a layer of laminar air flow, about 1 mm thick (Gregory, 1961). Air dispersal depends on spores becoming incorporated into the turbulent air flow, and splash-initiated liberation is a simple way of achieving this. The spore-bearing structures of many fungi, however, themselves grow through the still-air layer so that the spores are formed in a position ideal for dispersal. In other species, particularly of the Ascomycotina, spores are shot violently into the moving air.

A fundamental difference between fungal and bacterial diseases of plants is derived from the colonial morphology of the former. Fungal hyphae after ramifying through infected tissue can then translocate nutrients to sporulating structures to initiate spread of the species. Alternatively, the hyphae within a host can act as absorbing structures from which nutrients are translocated to specialized hyphae, strands or rhizomorphs, which act as nutrient sinks (see above). Such strands can then spread over the host surface a metre or more ahead of penetration or through the soil to a new host. These aggregations of hyphae are also important because of the high inoculum potential which they are able to bring to bear on small areas of host surface, leading to a readier overwhelming of host defenses. The ecological significance of the various aspects of this ectotrophic growth habit has been discussed by Garrett (1970).

The ability of fungal mycelium to translocate nutrients is also fundamental to the nature of mycorrhiza (Harley, 1972; Mosse et al., 1981). Mycorrhizas are a mutualistic association of roots and fungi in which the young root becomes invaded, either inter- or intracellularly, by a variety of fungi. In most mycorrhizas, advantage accrues to the host because of the efficient uptake of phosphorus from soil by the fungus and its subsequent translocation to the root. Other essential nutrients and water may be similarly transported from soil to root via the mycelium, which is often aggregated into strands. In mycorrhizas, the host may also be protected in part from infection by pathogens (Schenck, 1981). In turn, the fungus gains advantage through an assured supply of carbohydrates resulting from the photosynthesis of the host. Balanced mycorrhizas therefore tend to persist best on sites where soil nutrient levels are low and where photosynthesis rates are high. Fertilization or undue shading often lead to the disappearance of stable mycorrhizal associations, but many other factors affect the efficiency of the symbiosis (Bowen, 1978).

Mutualistic associations between bacteria and higher plants are of quite a different nature, determined in large part by the lack of translocation in prokaryotes. The best known associations are those between *Rhizobium* spp. and leguminous plants and between *Frankia* spp. (Actinomycetales) and a variety of perennial plants, notably species of *Alnus* and *Casuarina*. In these associations, the prokaryote penetrates roots hairs and eventually becomes entirely established within specialized root-nodules. There, nitrogen fixation occurs by the action of prokaryotic enzyme systems, to the advantage of the host, whilst the prokaryote receives energy-yielding substrates from the host (Sprent, 1979).

CONCLUSIONS

Fungi are exclusively chemoheterotrophic and most depend upon aerobic respiration for growth. Their ecological role is therefore far more restricted than that of bacteria. Considering the respective roles of fungi and bacteria as aerobic chemoheterotrophs, the similarities reflect the microscopic size of both, the differences mainly their contrasting morphologies. The development of hyphae, with the corollary of integrated, differentiated colonies in fungi, is of major ecological significance. Bacteria are characteristic of essentially aquatic environments, and especially of nutrient liquids and interfaces between solids and liquids. Fungi, however, by virtue of their hyphae, are able to penetrate solids and to bridge gaps that in themselves could not support growth. Their greatest development is therefore within solid matrices. The profound importance of hyphae is well illustrated by the fact that some filamentous bacteria, e.g., *Streptomyces* spp. come to have much the same ecological role as fungi whereas unicellular fungi, especially yeasts, have a role comparable to that of many bacteria.

REFERENCES

Agrios, G. N., 1969, *Plant Pathology,* Academic Press, New York.

Ahmadjian, V., and Hale, M. E. (eds.), 1973, *The Lichens,* Academic Press, New York.

Aist, J. R., 1976, Cytology of penetration and infection—fungi, in: *Physiological Plant Pathology* (R. Heitefuss and P. H. Williams, eds.), Springer Verlag, Berlin, pp. 197–221.

Alexander, M., 1961, *Introduction to Soil Microbiology,* John Wiley, New York.

Amelunxen, R. E., and Murdock, A. L., 1978, Microbial life at high temperatures: mechanisms and molecular aspects, in: *Microbial Life in Extreme Environments* (D. J. Kushner, ed.), Academic Press, New York, pp. 217–278.

Anderson, J. G., 1978, Temperature-induced fungal development, in: *The Filamentous Fungi,* Volume 3 (J. E. Smith and D. R. Berry, eds.), Edward Arnold, London, pp. 358–375.

Anderson, J. P. E., and Domsch, K. H., 1973, Quantification of bacterial and fungal contributions to soil respiration, *Arch. Mikrobiol.* **93:**113–127.

Austwick, P. K. C., 1972, The pathogenicity of fungi, *Symp. Soc. Gen. Microbiol.* **22:**251–268.

Ayerst, G., 1968, Prevention of biodeterioration by control of environmental conditions, in: *Biodeterioration of Materials* (A. H. Walters and J. J. Elphick, eds.), Elsevier, Amsterdam, pp. 223–241.

Baross, J. A., and Morita, R. Y., 1978, Life at low temperatures: ecological aspects, in: *Microbial Life in Extreme Environments* (D. J. Kushner, ed.), Academic Press, New York, pp. 9–71.

Bateman, D. F., and Basham, H. G., 1976, Degradation of plant cell walls and membranes by microbial enzymes, in: *Physiological Plant Pathology* (R. Heitefuss and P. H. Williams, eds.), Springer Verlag, Berlin, pp. 316–355.

Benjamin, M., and Newhook, F. J., 1982, Effect of glass microbeads on *Phytopthora* zoospore mobility, *Trans. Br. Mycol. Soc.* **78:**43–46.

Benjamin, R. K., 1973, Laboulbeniomycetes, in: *The Fungi: An Advanced Treatise,* Volume 4A (G. C. Ainsworth, F. K. Sparrow and A. S. Sussman, eds.), Academic Press, New York, pp. 223–246.

Berkeley, R. C. W., Lynch, J. M., Melling, J., Rutter, P. R., and Vincent, B., (eds), 1980, *Microbial Adhesion to Surfaces,* Ellis Horwood, Chichester.

Bitton, G., and Marshall, K. C. (eds.), 1980, *Adsorption of Microorganisms to Surfaces,* John Wiley, New York.

Blumenthal, H. J., 1976, Reserve carbohydrates in fungi, in: *The Filamentous Fungi,* Volume 2 (J. E. Smith and D. R. Berry, eds.), Edward Arnold, London, pp. 292–307.

Bolland, L., 1980, Variation in *Phellinus noxius* (Corner) G. H. Cunn., Ph.D. thesis, Australian National University, Canberra.

Bowen, G. D., 1978, Dysfunction and shortfalls in symbiotic responses, in: *Plant Disease: An Advanced Treatise,* Volume 3 (J. G. Horsfall and E. B. Cowling, eds.), Academic Press, New York, pp. 231–256.

Brock, T. D., 1978, *Thermophilic Microorganisms and Life at High Temperatures,* Springer Verlag, New York.

Brown, A. D., 1976, Microbial water stress, *Bacteriol. Rev.* **40:**803–846.

Brown, A. D., 1978, Compatible solutes and extreme water stress in eukaryotic microorganisms, *Adv. Microbial Physiol.* **17:**181–242.

Brown, A. D., 1979, Physiological problems of water stress, in: *Strategies of Microbial Life in Extreme Environments* (M. Shilo, ed.), Verlag Chemie, Weinheim, pp. 65–81.

Brown, D. H., Hawksworth, D. L., and Bailey, R. H. (eds.), 1976, *Lichenology: Progress and Problems,* Academic Press, London.

Brownlee, C., and Jennings, D. H., 1981, The content of soluble carbohydrates and their translocation in the mycelium of *Serpula lacrimans, Trans. Br. Mycol. Soc.* **77:**615–620.

Brownlee, C., and Jennings, D. H., 1982, Long distance translocation in *Serpula lacrimans:* velocity estimates and the continuous monitoring of induced perturbations, *Trans. Br. Mycol. Soc.* **79:**143–148.

Bruehl, G. W., Sprague, R., Fisher, W. R., Nagamitsu, M., Nelson, W. L., and Vogel, O. A., 1966, Snow Molds of Winter Wheat in Washington, Washington Agricultural Experimental Station Bulletin 677.

Bull, A. T., and Bushell, M. E., 1976, Environmental control of fungal growth, in: *The Filamentous Fungi,* Volume 2 (J. E. Smith and D. R. Berry, eds.), Edward Arnold, London, pp. 1–31.

Bull, A. T., and Trinci, A. P. J., 1977, The physiological and metabolic control of fungal growth, *Adv. Microbial Physiol.* **15:**1–84.

Burnett, J. H., 1968, *Fundamentals of Mycology,* Edward Arnold, London.

Butler, G. M., 1966, Vegetative structures, in: *The Fungi: An Advanced Treatise,* Volume 2 (G. C. Ainsworth and A. S. Sussman, eds.), Academic Press, New York, pp. 283–337.

Caldwell, I. Y., and Trinci, A. P. J., 1973a, The growth unit of the mould *Geotrichum candidum, Arch. Mikrobiol.* **88:**1–10.

Caldwell, I. Y., and Trinci, A. P. J., 1973b, Kinetic aspects of growth of *Geotrichum candidum* on various carbon sources, *Trans. Br. Mycol. Soc.* **61:**411–416.

Carlile, M. J., 1980, From prokaryote to eukaryote: gains and losses, *Symp. Soc. Gen. Microbiol.* **30:**1–40.

Casselton, L. A., 1978, Dikaryon formation in higher basidiomycetes, in: *The Filamentous Fungi,* Volume 3 (J. E. Smith and D. R. Berry, eds.), Edward Arnold, London, pp. 275–297.

Caten, C. E., 1981, Parasexual processes in fungi, in: *The Fungal Nucleus* (K. Gull and S. G. Oliver, eds.), Cambridge University Press, Cambridge, pp. 191–214.

Chet, I., 1975, Ultrastructural basis of sclerotial survival in soil, *Microbial Ecol.* **2:**194–200.

Chet, I., and Henis, Y., 1975, Sclerotial morphogenesis in fungi, *Annu. Rev. Phytopathol.* **13:**169–192.

Chevaugeon, J., and Nguyen van, H., 1969, Internal determinism of hyphal growth rhythms, *Trans. Br. Mycol. Soc.* **53:**1–14.

Clark, F. E., 1967, Bacteria in soil, in: *Soil Biology* (A. Burges and F. Raw, eds.), Academic Press, London, pp. 15–49.

Clarke, R. W., Jennings, D. H., and Coggins, C. R., 1980, Growth of *Serpula lacrimans* in relation to water potential of substrate, *Trans. Br. Mycol. Soc.* **75:**271–280.
Cochrane, V. W., 1958, *Physiology of Fungi,* John Wiley, New York.
Conant, N. F., Smith, D. T., Baker, R. D., Callaway, J. L., and Martin, D. S., 1954, *Manual of Clinical Mycology,* 2nd ed., Saunders, Philadelphia.
Cook, R. J., and Duniway, J. M., 1981, Water relations in the life cycle of soil-borne plant pathogens, in: *Water Potential Relations in Soil Microbiology* (J. F. Parr, W. R. Gardner and L. F. Elliott, eds.), *Soil Sci. Soc. Am.* Special Publication 9, pp. 119–139.
Cooke, R. C., 1977, *The Biology of Symbiotic Fungi,* John Wiley, London.
Cooney, D. G., and Emerson, R., 1964, *Thermophilic Fungi,* W. H. Freeman, San Francisco.
Crisan, E. V., 1973, Current concepts of thermophilism and the thermophilic fungi, *Mycologia* **65:**1171–1198.
Culver, D. C., 1981, Introduction to the theory of species interactions, in: *The Fungal Community* (D. T. Wicklow and G. C. Carroll, eds.), Marcel Dekker, New York, pp. 281–294.
Dawes, E. A., 1976, Endogenous metabolism and the survival of starved prokaryotes, *Symp. Soc. Gen. Microbiol.* **21:**19–54.
Day, P. R., 1978, The genetic base of epidemics, in: *Plant Disease: An Advanced Treatise,* Volume 2 (J. G. Horsfall and E. B. Cowling, eds.), Academic Press, New York, pp. 263–285.
Deverall, B. J., 1968, Psychrophiles, in: *The Fungi: An Advanced Treatise,* Volume 3 (G. C. Ainsworth and A. S. Sussman, eds.), Academic Press, New York, pp. 129–135.
Dommergues, Y., 1962, Contribution a l'etude de la dynamique microbienne des sols en zone semi-aride et en zone tropicale seche, *Ann. Agron., Paris,* **13:**265–324, 391–468.
Duckworth, R. B. (ed.), 1975, *Water Relations of Foods,* Academic Press, London.
Duniway, J. M., 1979, Water relations of water molds, *Annu. Rev. Phytopathol.* **17:**431–460.
Ellar, D. J., 1978, Spore specific structures and their function, *Symp. Soc. Gen. Microbiol.* **28:**295–325.
Ellingboe, A. H., 1981, Changing concepts in host-pathogen genetics, *Annu. Rev. Phytopathol.* **19:**125–143.
Emerson, R., and Natvig, D. O., 1981, Adaptation of fungi to stagnant waters, in: *The Fungal Community* (D. T. Wicklow and G. C. Carroll, eds.), Marcel Dekker, New York, pp. 109–128.
Emmett, R. W., and Parbery, D. G., 1975, Appressoria, *Annu. Rev. Phytopathol.* **13:**147–167.
English, M. P., 1963, Saprophytic growth of keratinophilic fungi on keratin, *Sabouraudia* **2:**115–130.
English, M. P., 1965, The saprophytic growth of non-keratinophilic fungi on keratinized substrata, and a comparison with keratinophilic fungi, *Trans. Br. Mycol. Soc.* **48:**219–236.
English, M. P., 1968, The developmental morphology of the perforating organs and eroding mycelium of dermatophytes, *Sabouraudia* **6:**218–227.
Eze, J. M. O., 1975, Translocation of phosphate in mould mycelia, *New Phytol.* **75:**579–581.
Faegri, A., Torsvik, V. L., and Goksoyr, J., 1977, Bacterial and fungal activities in soil: separation of bacteria and fungi by a rapid fractionated centrifugation technique, *Soil Biol. Biochem.* **9:**105–112.
Farrell, J., and Campbell, L. L., 1969, Thermophilic bacteria and bacteriophages, *Adv. Microbial Physiol.* **3:**83–109.
Fell, J. W., 1976, Yeasts in oceanic regions, in: *Recent Advances in Aquatic Mycology* (E. B. G. Jones, ed.), Elek Science, London, pp. 93–124.
Fergus, C. L., 1964, Thermophilic and thermotolerant molds and actinomycetes of mushroom compost during peak-heating, *Mycologia* **56:**267–284.
Ferry, B. W., Baddeley, M. S., and Hawksworth, D. L. (eds.), 1973, *Air Pollution and Lichens,* University of London, Athlone Press, London.
Fiddy, C., and Trinci, A. P. J., 1975, Kinetics and morphology of glucose-limited cultures of moulds grown in a chemostat and on solid media, *Arch. Mikrobiol.* **103:**191–197.
Fincham, J. R. S., Day, P. R., and Radford, A., 1979, *Fungal Genetics,* 4th ed., Blackwell, Oxford.
Forage, A. J., and Righelato, R. C., 1979, Biomass from carbohydrates, in: *Economic Microbiology,* Volume 4 (A. H. Rose, ed.), Academic Press, London, pp. 289–313.
Garrett, S. D., 1970, *Pathogenic Root-Infecting Fungi,* Cambridge University Press, Cambridge.

Giesey, R. M., and Day, P. R., 1965, The septal pores of *Coprinus lagopus* in relation to nuclear migration, *Am. J. Bot.* **52**:287–293.
Gilbert, O. L., 1973, Lichens and air pollution, in: *The Lichens* (V. Admadjian and M. E. Hale, eds.), Academic Press, New York, pp. 443–472.
Goodman, R. N., 1976, Physiological and cytological aspects of the bacterial infection process, in: *Physiological Plant Pathology* (R. Heitefuss and P. H. Williams, eds.), Springer Verlag, Berlin, pp. 172–196.
Gray, T. R. G., 1976, Survival of vegetative microbes in soil, *Symp. Soc. Gen. Microbiol.* **26**:327–364.
Gray, T. R. G., and Parkinson, D., 1968, *The Ecology of Soil Bacteria,* Liverpool University Press, Liverpool.
Gray, T. R. G., and Williams, S. T., 1971a, *Soil Microorganisms,* Oliver and Boyd, Edinburgh.
Gray, T. R. G., and Williams, S. T., 1971b, Microbial productivity in soil, *Symp. Soc. Gen. Microbiol.* **21**:255–286.
Greaves, H., 1975, Microbial aspects of wood chip storage in tropical environments, *Aust. J. Biol. Sci.* **28**:315–322.
Gregory, P. H., 1961, *The Microbiology of the Atmosphere,* Leonard Hill, London.
Griffin, D. M., 1972, *Ecology of Soil Fungi,* Chapman and Hall, London.
Griffin, D. M., 1977, Water potential and wood-decay fungi, *Annu. Rev. Phytopathol.* **15**:319–329.
Griffin, D. M., 1978, Effect of soil moisture on survival and spread of pathogens, in: *Water Deficits and Plant Growth,* Volume 5 (T. T. Kozlowski, ed.), Academic Press, New York, pp. 175–197.
Griffin, D. M., 1981a, Water potential as a selective factor in the microbial ecology of soils, in: *Water Potential Relations in Soil Microbiology* (J. F. Parr, W. R. Gardner, and L. F. Elliott, eds.), *Soil Sci. Soc. Am.* Special Publication 9, pp. 141–151
Griffin, D. M., 1981b, Water and microbial stress, *Adv. Microbial Ecol.* **5**:91–136.
Gull, K., 1978, Form and function of septa in filamentous fungi, in: *The Filamentous Fungi,* Volume 3 (J. E. Smith and D. R. Berry, eds.), Edward Arnold, London, pp. 78–93.
Harley, J. L., 1972, *The Biology of Mycorrhiza,* 3rd ed., Leonard Hill, London.
Harris, R. F., 1981, Effect of water potential on microbial growth and activity, in: *Water Potential Relations in Soil Microbiology* (J. F. Parr, W. R. Gardner, and L. F. Elliott, eds.), *Soil Sci. Soc. Am.* Special Publication 9, pp. 23–95.
Heinrich, M. R. (ed), 1976, *Extreme Environments—Mechanisms of Microbial Adaptations,* Academic Press, New York.
Hill, D. J., 1976, The physiology of lichen symbiosis, in: *Lichenology: Progress and Problems* (D. H. Brown, D. L. Hawksworth and R. H. Bailey, eds.), Academic Press, London, pp. 457–496.
Hirsch, P., 1979, Life under conditions of low nutrient concentrations, in: *Strategies of Microbial Life in Extreme Environments* (M. Shilo ed.) Verlag Chemie, Weinheim, pp. 357–372.
Horsfall, J. G., and Cowling, E. B. (eds.), 1978, *Plant Disease: An Advanced Treatise,* Volume 2, Academic Press, New York.
Horsfall, J. G., and Dimond, A. E. (eds.), 1959, *Plant Pathology,* Volumes 1–3, Academic Press, New York.
Howard, A. J., 1978, Translocation in fungi, *Trans. Br. Mycol. Soc.* **70**:265–269.
Hutchinson, S. A., Sharma, P., Clarke, K. R., and MacDonald, I., 1980, Control of hyphal orientation in colonies of *Mucor hiemalis, Trans. Br. Mycol. Soc.* **75**:177–191.
Ingold, C. T., 1953, *Dispersal in Fungi,* Clarendon Press, Oxford.
Ingold, C. T., 1965, *Spore Liberation,* Clarendon Press, Oxford.
Inniss, W. E., and Ingraham, J. L., 1978, Microbial life at low temperatures: mechanisms and molecular aspects, in: *Microbial Life in Extreme Environments* (D. J. Kushner, ed.), Academic Press, New York, pp. 73–104.
Isenberg, H. D., and Balows, A., 1981, Bacterial pathogenicity in man and animals, in: *The Prokaryotes,* Volume 1 (M. P. Starr, H. Stolp, H. G. Truper, A. Balows and H. G. Sehlegel, eds.), Springer Verlag, Berlin pp. 83–122.
Jannasch, H. W., 1979, Microbial ecology of aquatic low nutrient habitats, in: *Strategies of Microbial Life in Extreme Environments* (M. Shilo, ed.), Verlag Chemie, Weinheim, pp. 243–280.

Jennings, D. H., 1976, Transport and translocation in filamentous fungi, in: *The Filamentous Fungi,* Volume 2 (J. E. Smith and D. R. Berry, eds.), Edward Arnold, London, pp. 32–64.
Jennings, D. H., Thornton, J. D., and Galpin, M. F. J., 1974, Translocation in fungi, *Symp. Soc. Exp. Biol.* **28:**139–156.
Jennings, L., and Watkinson, S. C., 1982, Structure and development of mycelial strands in *Serpula lacrimans, Trans. Br. Mycol. Soc.* **78:**465–474.
Johnston, J. R., 1975, Strain improvement and strain stability in filamentous fungi, in *The Filamentous Fungi,* Volume 1 (J. E. Smith and D. R. Berry, eds.), Edward Arnold, London, pp. 59–78.
Jones, E. B. G. (ed.), 1976, *Recent Advances in Aquatic Mycology,* Elek Science, London.
Kalakoutskii, L. V., and Agre, N. S., 1976, Comparative aspects of development and differentiation in actinomycetes, *Bact. Rev.* **40:**469–524.
Kappen, L., 1973, Response to extreme environments, in: *The Lichens* (V. Ahmadjian and M. E. Hale, eds.), Academic Press, New York, pp. 311–380.
Koch, A. L., 1979, Microbial growth in low concentrations of nutrients, in: *Strategies of Microbial Life in Extreme Environments* (M. Shilo, ed.), Verlag Chemie, Weinheim, pp. 261–279.
Kohlmeyer, J., and Kohlmeyer, E., 1979, *Marine Mycology–the Higher Fungi,* Academic Press, New York.
Koske, R. E., 1982, Evidence for a volatile attractant from plant roots affecting germ tubes of a VA mycorrhizal fungus, *Trans. Br. Mycol. Soc.* **79:**305–310.
Kozlowski, T. T. (ed.), 1978, *Water Deficits and Plant Growth,* Volume 5, Academic Press, New York.
Kubicek, R., and Lysek, G., 1982, Morphogenesis of growth bands in the clock-mutant *zonata,* of *Podospora anserina, Trans. Br. Mycol. Soc.* **79:**167–170.
Kushner, D. J., 1978, Life in high salt and solute concentrations: halophilic bacteria, in: *Microbial Life in Extreme Environments* (D. J. Kushner, ed.), Academic Press, New York, pp. 318–368.
Kutzner, H. J., 1981, The family Streptomycetaceae, in: *The Prokaryotes,* Volume 2 (M. P. Starr, H. Stolp, H. G. Truper, A. Balows, and H. G. Schlegel, eds.), Springer Verlag, Berlin, pp. 2028–2090.
Lacey, J., 1980, Colonization of damp organic substrates and spontaneous heating, *Soc. Appl. Bacteriol.* Tech. Ser. **15:**54–70.
Langworthy, T. A., 1978, Microbial life at extreme pH values, in: *Microbial Life in Extreme Environments* (D. J. Kushner, ed.), Academic Press, New York, pp. 279–315.
Lapp, M. S., and Skoropad, W. P., 1978, Nature of adhesive material of *Colletotrichum graminicola* appressoria, *Trans. Br. Mycol. Soc.* **70:**221–223.
Leben, C., 1981, How plant-pathogenic bacteria survive, *Plant Disease* **65:**633–637.
Levi, M. P., and Cowling, E. B., 1969, Role of nitrogen in wood deterioration. VII. Physiological adaptation of wood-destroying and other fungi to substrates deficient in nitrogen, *Phytopathology* **59:**460–468.
Levi, M. P., Merrill, W., and Cowling, E. B., 1968, Role of nitrogen in wood deterioration. VI. Mycelial fractions and model nitrogen compounds as substrates for growth of *Polyporus versicolor* and other wood-destroying and wood-inhabiting fungi, *Phytopathology* **58:**626–634.
Lewis, D. H., and Smith, D. C., 1967, Sugar alcohols (polyols) in fungi and green plants. I. Distribution, physiology and metabolism, *New Phytol.* **66:**143–184.
Lockwood, J. L., 1977, Fungistasis in soils, *Biol. Rev. Cambridge Philos. Soc.* **52:**1–43.
Lockwood, J. L., and Filonow, A. B., 1981, Response of fungi to nutrient-limiting conditions and to inhibitory substances in natural habitats, *Adv. Microbial Ecol.* **5:**1–61.
Luard, E. J., 1982a, Accumulation of intracellular solutes by two filamentous fungi in response to growth at low steady state osmotic potentials, *J. Gen. Microbiol.* **128:**2563–2574.
Luard, E. J., 1982b, Effect of osmotic shock on some intracellular solutes in two filamentous fungi, *J. Gen. Microbiol.* **128:**2575–2582.
Luard, E. J., 1982c, Growth and accumulation of solutes by *Phytopthora cinnamomi* and other lower fungi in response to changes in external osmotic potential, *J. Gen. Microbiol.* **128:**2583–2590.

Luard, E. J., and Griffin, D. M., 1981, Effect of water potential on fungal growth and turgor, *Trans. Br. Mycol. Soc.* **76:**33–40.
Lucas, R. L., 1977, The movement of nutrients through fungal mycelium, *Trans. Br. Mycol. Soc.* **69:**1–9.
Lysek, G., 1978, Circadian rhythms, in: *The Filamentous Fungi,* Volume 3 (J. E. Smith and D. R. Berry, eds.), Edward Arnold, London, pp. 376–388.
Maramorosch, K., 1976, Plant mycoplasma diseases, in: *Physiological Plant Pathology* (R. Heitefuss and P. H. Williams, eds.), Springer Verlag, Berlin, pp. 150–171.
Mason, H. R. S., and Righelato, R. C., 1976, Energetics of fungal growth: the effect of growth-limiting substrates on respiration of *Penicillium chrysogenum, J. Appl. Chem. Biotechnol.* **26:**145–152.
Matin, A., and Veldkamp, H., 1978, Physiological basis of the selective advantage of a *Spirillum* sp. in a carbon-limited environment, *J. Gen. Microbiol.* **105:**187–197.
Meredith, D. S., 1970, Significance of spore release and dispersal mechanisms in plant disease epidemiology, *Annu. Rev. Phytophathol.* **11:**313–342.
Merlo, D. J., 1978, Crown-gall—a unique disease, in: *Plant Disease: an Advanced Treatise,* Volume 3 (J. G. Horsfall and E. B. Cowling, eds.), Academic Press, New York, pp. 201–213.
Merrill, W., and Cowling, E. B., 1966, Role of nitrogen in wood deterioration: amount and distribution of nitrogen in fungi, *Phytopathology* **56:**1083–1090.
Millbank, J. W., 1976, Aspects of nitrogen metabolism in lichens, in: *Lichenology: Progress and Problems* (D. H. Brown, D. L. Hawksworth, and R. H. Bailey, eds.), Academic Press, London, pp. 441–455.
Millbank, J. W., and Kershaw, K. A., 1973, Nitrogen metabolism, in: *The Lichens* (V. Ahmadjian and M. E. Hale, eds.), Academic Press, New York, pp. 289–307.
Mirocha, C. J., and Christensen, C. M., 1974, Fungus metabolites toxic to animals, *Annu. Rev. Phytopathol.* **12:**303–329.
Morrison, D. J., 1978, Effects of soil organic matter on rhizomorph growth by *Armillaria mellea, Trans. Br. Mycol. Soc.* **78:**201–207.
Mosse, B., Stribley, D. P., and LeTacon, F., 1981, Ecology of mycorrhizae and mycorrhizal fungi, *Adv. Microbial Ecol.* **5:**137–210.
Motta, J. J., 1969, Cytology and morphogenesis in the rhizomorph of Armillaria mellea, *Am. J. Bot.* **56:**610–619.
Nesbitt, H. J., Malajczuk, N., and Glenn, A. R., 1981, Translocation and exudation of metabolites in *Phytothora cinnamomi, Trans. Br. Mycol. Soc.* **76:**503–505.
Nienhaus, F., and Sikora, R. A., 1979, Mycoplasmas, spiroplasmas and rickettsia-like organisms as plant pathogens, *Annu. Rev. Phytopathol.* **17:**37–58.
Papendick, R. I., and Campbell, G. S., 1981, Theory and measurement of water potential, in: *Water Potential Relations in Soil Microbiology* (J. F. Parr, W. R. Gardner, and L. F. Elliott, eds.), *Soil Sci. Soc. Am.* Special Publication 9, pp. 1–22.
Patil, S. S., 1974, Toxins produced by phytopathogenic bacteria, *Annu. Rev. Phytopathol.* **12:**259–279.
Perombelon, M. C. M., and Kelman, A., 1980, Ecology of the soft rot Erwinias, *Annu. Rev. Phytopathol.* **18:**361–387.
Pirt, S. J., 1966, A theory of the mode of growth of fungi in the form of pellets in submerged culture, *Proc. Roy. Soc. Series B* **166:**369–373.
Pirt, S. J., 1967, A kinetic study of the mode of growth of surface colonies of bacteria and fungi, *J. Gen. Microbiol.* **47:**181–197.
Pirt, S. J., 1973, Estimation of substrate affinities (K_s values) of filamentous fungi from colony growth rates, *J. Gen. Microbiol.* **75:**245–247.
Plomley, N. J. B., 1959, Formation of the colony in the fungus *Chaetomium, Aust. J. Biol. Sci.* **72:**53–64.
Poindexter, J. S., 1981a, Oligotrophy: fast and famine existence, *Adv. Microbial Ecol.* **5:**63–89.
Poindexter, J. S., 1981b, The caulobacters: ubiquitous unusual bacteria, *Microbiol. Rev.* **45:**123–179.
Prosser, J. I., and Trinci, A. P. J., 1979, A model for hyphal growth and branching, *J. Gen. Microbiol.* **111:**153–164.
Ramsbottom, J., 1953, *Mushrooms and Toadstools,* Collins, London.

Read, D. J., and Stribley, D. P., 1975, Diffusion and translocation in some fungal culture systems, *Trans. Br. Mycol. Soc.* **64:**381–388.

Richards, B. N., 1974, *Introduction to the Soil Ecosystem,* Longman, Harlow.

Richardson, D. H. S., 1973, Photosynthesis and carbohydrate movement, in: *The Lichens* (V. Ahmadjian and M. E. Hale, eds.), Academic Press, New York, pp. 249–288.

Righelato, R. C., 1975, Growth kinetics of mycelial fungi, in: *The Filamentous Fungi,* Volume 1 (J. E. Smith and D. R. Berry, eds.), Edward Arnold, London, pp. 79–103.

Righelato, R. C., 1979, The kinetics of mycelial growth, in: *Fungal Walls and Hyphal Growth* (J. H. Burnett and A. P. J. Trinci, eds.), Cambridge University Press, Cambridge, pp. 385–401.

Righelato, R. C., Trinci, A. P. J., Pirt, S. J., and Peat, A., 1968, The influence of maintenance energy and growth rate on the metabolic activity, morphology and conidiation of *Penicllium chrysogenum, J. Gen. Microbiol.* **50:**399–412.

Rishbeth, J., 1978, Effects of soil temperature and atmosphere on growth of *Armillaria* rhizomorphs, *Trans. Br. Mycol. Soc.* **70:**213–220.

Romano, A. H., 1966, Dimorphism, in: *The Fungi: An Advanced Treatise,* Volume 2 (G. C. Ainsworth and A. S. Sussman, eds.), Academic Press, New York, pp. 181–209.

Rudolph, K., 1976, Non-specific toxins, in: *Physiological Plant Pathology* (R. Heitefuss and P. H. Williams, eds.), Springer Verlag, Berlin, pp. 270–315.

Scarborough, G. A., 1970a, Sugar transport in *Nerospora crassa, J. Biol. Chem.* **245:**1694–1698.

Scarborough, G. A., 1970b, Sugar transport in *Neurospora crassa, J. Biol. Chem.* **245:**3985–3987.

Scheffer, R. P., 1976, Host-specific toxins in relation to pathogenesis and disease resistance, in: *Physiological Plant Pathology* (R. Heitefuss and P. H. Williams, eds.), Springer Verlag, Berlin, pp. 247–269.

Schenck, N. C., 1981, Can mycorrhizae control plant disease? *Plant Disease* **65:**230–234.

Schlegel, H. G., and Jannasch, H. W., 1981, Prokaryotes and their habitats, in: *The Prokaryotes,* Volume 1 (M. P. Starr, H. Stolp, H. G. Truper, A. Balows, and H. G. Schlegel, eds.), Springer Verlag, Berlin, pp. 43–82.

Schneider, H., 1970, Cytological and histological aberrations in woody plants following infections with viruses, mycoplasmas, rickettsias and flagellates, *Annu. Rev. Phytopathol.* **11:**119–146.

Schobert, B., 1977, Is there an osmotic regulatory mechanism in algae and higher plants? *J. Theor. Biol.* **68:**17–26.

Schobert, B., and Tschesche, H., 1978, Unusual solution properties of proline and its interaction with proteins, *Biochem. Biophys. Acta* **541:**270–277.

Schulte, T. H., and Scarborough, G. A., 1975, Characterization of the glucose transport systems in *Neurospora crassa* sl, *J. Bacteriol.* **122:**1076–1080.

Schutte, K. H., 1956, Translocation in the fungi, *New Phytol.* **55:**164–182.

Scott, P. R., and Bainbridge, A. (eds.), 1978, *Plant Disease Epidemiology,* Blackwell, Oxford.

Seaward, M. R. D., (ed), 1977, *Lichen Ecology,* Academic Press, London.

Shilo, M. (ed.), 1979, *Strategies of Microbial Life in Extreme Environments,* Verlag Chemie, Weinheim.

Sidhu, G. S., 1975, Gene-for-gene relationships in plant parasitic systems, *Sci. Prog. Oxford* **62:**467–485.

Smith, A. H., 1966, The hyphal structure of the basidiocarp, in: *The Fungi: An Advanced Treatise,* Volume 2 (G. C. Ainsworth and A. S. Sussman, eds.), Academic Press, New York, pp. 151–177.

Smith, A. M., and Griffin, D. M., 1971, Oxygen and the ecology of *Armillariella elegans* Heim, *Aust. J. Biol. Sci.* **24:**231–262.

Smith, D. C., 1975, Symbiosis and the biology of lichenized fungi, *Symp. Soc. Exp. Biol.* **29:**373–405.

Smith, D. C., 1979, Is a lichen a good model of biological interactions in nutrient-limited environments?, in: *Strategies of Microbial Life in Extreme Environments* (M. Shilo, ed.), Verlag Chemie, Weinheim, pp. 291–303.

Smith, J. E., 1978, Asexual sporulation in the filamentous fungi, in: *The Filamentous Fungi,* Volume 3 (J. E. Smith and D. R. Berry, eds.), Edward Arnold, London, pp. 214–239.

Solomons, G. L., 1975, Submerged culture production of mycelial biomass, in: *The Filamentous Fungi,* Volume 1 (J. E. Smith and D. R. Berry, eds.), Edward Arnold, London, pp. 249–264.

Sommers, L. E., Gilmour, C. M., Wildung, R. E., and Beck, S. M., 1981, The effect of water potential on decomposition processes in soils, in: *Water Potential Relations in Soil Microbiology* (J. F. Parr, W. R. Gardner and L. F. Elliott, eds.), *Soil Sci. Soc. Am.* Special Publication 9, pp. 97–117.

Sprent, J. I., 1979, *The Biology of Nitrogen-Fixing Organisms,* McGraw Hill, London.

Stamberg, J., and Koltin., Y., 1981, The genetics of mating systems: fungal strategies for survival, in: *The Funal Community* (D. T. Wicklow and G. C. Carroll, eds.), Marcel Dekker, New York, pp. 157–170.

Starkey, R. L., and Waksman, S. A., 1943, Fungi tolerant to extreme acidity and high concentrations of copper sulfate, *J. Bacteriol.* **45:**509–519.

Starr, M. P., 1981, Prokaryotes as plant pathogens, in: *The Prokaryotes,* Volume 1 (M. P. Starr, H. Stolp, H. G. Truper, A. Balows, and H. G. Schlegel, eds.), Springer Verlag, Berlin, pp. 123–134.

Starr, M. P., and Schmidt, J. M., 1981, Prokaryote diversity, in: *The Prokaryotes,* Volume 1 (M. P. Starr, H. Stolp, H. G. Truper, A. Balows, and H. G. Schlegel, eds.), Springer Verlag, Berlin, pp. 3–42.

Steensland, H., 1973, Continuous culture of a sewage fungus *Fusarium aquaeductuum, Arch. Mikrobiol.* **93:**287–294.

Stewart, P. R., and Rogers, P. J., 1978, Fungal dimorphism: a particular expression of cell wall morphogenesis, in: *The Filamentous Fungi,* Volume 3 (J. E. Smith and D. R. Berry, eds.), Edward Arnold, London, pp. 164–196.

Sussman, A. S., 1966, Dormancy and spore germination, in: *The Fungi: An Advanced Treatise,* Volume 2 (G. C. Ainsworth and A. S. Sussman, eds.), Academic Press, New York, pp. 733–764.

Sussman, A. S., 1968, Longevity and survivability of fungi, in: *The Fungi: An Advanced Treatise,* Volume 3 (G. C. Ainsworth and A. S. Sussman, eds.), Academic Press, New York, pp. 447–486.

Tanner, W., 1974, Energy coupled sugar transport in *Chlorella, Trans. Biochem. Soc.* **2:**793–797.

Tansey, M. R., and Brock, T. D., 1978, Microbial life at high temperatures: ecological aspects, in: *Microbial Life in Extreme Environments* (D. J. Kushner, ed.), Academic Press, New York, pp. 159–216.

Tansey, M. R., and Jack, M. A., 1976, Thermophilic fungi in sun-heated soils, *Mycologia* **68:**1061–1075.

Tansey, M. R., and Jack, M. A., 1977, Growth of thermophilic and thermotolerant fungi in soil in situ and in vitro, *Mycologia* **69:**563–578.

Thompson, W., and Rayner, A. D. M., 1982, Structure and development of mycelial cord systems of *Phanerochaete laevis* in soil, *Trans. Br. Mycol. Soc.* **78:**193–200.

Thrower, L. B., and Thrower, S. L., 1968a, Movement of nutrients in fungi. I. The mycelium, *Aust. J. Bot.* **16:**71–80.

Thrower, L. B., and Thrower, S. L., 1968b, Movement of nutrients in fungi. II. The effect of reproductive structures, *Aust. J. Bot.* **16:**81–88.

Trinci, A. P. J., 1969, A kinetic study of the growth of *Aspergillus nidulans* and other fungi, *J. Gen. Microbiol.* **57:**11–24.

Trinci, A. P. J., 1971, Influence of the width of the peripheral growth zone on the radial growth rate of fungal colonies on solid media, *J. Gen. Microbiol.* **67:**325–344.

Trinci, A. P. J., 1978, The duplication cycle and vegetative development in moulds, in: *The Filamentous Fungi,* Volume 3 (J. E. Smith and D. R. Berry, eds.), Edward Arnold, London, pp. 132–163.

Trinci, A. P. J., 1979, The duplication cycle and branching in fungi, in: *Fungal Walls and Hyphal Growth* (J. H. Burnett and A. P. J. Trinci, eds.), Cambridge University Press, Cambridge, pp. 319–358.

Trinci, A. P. J., and Thurston, C. F., 1976, Transition to the non-growing state in eukaryotic micro-organisms, *Symp. Soc. Gen. Microbiol.* **26:**55–80.

Turner, G., 1978, Cytoplasmic inheritance and senescence, in: *The Filamentous Fungi,* Volume 3 (J. E. Smith and D. R. Berry, eds.), Edward Arnold, London, pp. 406–425.

Watkinson, S. C., 1979, Growth of rhizomorphs, mycelial strands, coremia and sclerotia, in: *Fungal Walls and Hyphal Growth* (J. H. Burnett and A. P. J. Trinci, eds.), Cambridge University Press, Cambridge, pp. 93–113.

Weber, D. J., and Hess, W. M. (eds.), 1976, *The Fungal Spore,* John Wiley, New York.

Webster, R. K., 1974, Recent advances in the genetics of plant pathogenic fungi, *Annu. Rev. Phytopathol.* **12:**331–353.

Wicklow, D. T., and Carroll, G. C. (eds.), 1981, *The Fungal Community,* Marcel Dekker, New York.

Wilcoxson, R. D., and Sudia, T. W., 1968, Translocation in fungi, *Bot. Rev.* **34:**32–50.

Willetts, H. J., 1978, Sclerotium formation, in: *The Filamentous Fungi,* Volume 3 (J. E. Smith and D. R. Berry, eds.), Edward Arnold, London, pp. 197–213.

Williams, S. T., Sharples, G. P., and Bradshaw, R. M., 1973, The fine structure of the Actinomycetales, in: *Actinomycetales: Characteristics and Practical Importance* (G. Sykes and F. A. Skinner, eds.), Academic Press, New York, pp. 113–130.

Wilsenach, R., and Kessel, M., 1965, Micropores in the cross-wall of *Geotrichum candidum, Nature,* London **207:**545–546.

Wilson, J. M., and Griffin, D. M., 1975, Water potential and the respiration of microorganisms in the soil, *Soil Biol. Biochem.* **7:**199–204.

Young, D. H., and Kauss, H., 1982, Agglutination of mycelial cell wall fragments and spores of *Colletotrichum lindemuthianum* by plant extracts and by various proteins, *Physiol. Plant Pathol.* **20:**285–297.

INDEX